Grown but not made

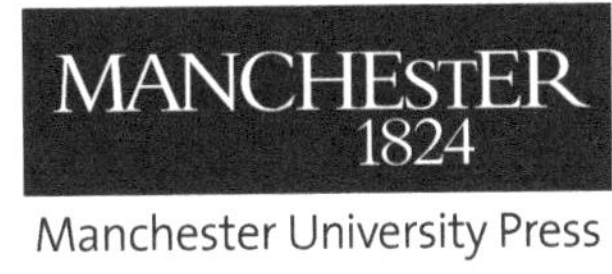

MANCHESTER
1824

Manchester University Press

Grown but not made

British Modernist sculpture
and the New Biology

Edward Juler

Manchester University Press

Published by Manchester University Press
Altrincham Street, Manchester M1 7JA
www.manchesteruniversitypress.co.uk

British Library Cataloguing-in-Publication Data
A catalogue record for this book is available from the British Library

Library of Congress Cataloging-in-Publication Data applied for

ISBN 978 1 5261 0653 7 paperback

First published by Manchester University Press in hardback 2015

This edition first published 2016

Typeset in Bodoni and Futura by R.J. Footring Ltd

To my mother

Contents

Plates

Figures

Acknowledgements

Over the years, this project has benefited – in ways both tangible and intangible – from the advice of colleagues and friends, all of whom have been kind enough to share with me their expertise and knowledge. Their collective generosity, intellectual resourcefulness and patience have proved invaluable and this book is, in many ways, a testimony to the many kindnesses, both professional and personal, that I have been shown.

In the first instance, I would like to thank David Lomas and Andrew Causey, both of whom have enthusiastically and unwearyingly supported my work long after they ceased to be my doctoral supervisors. Together they whetted my appetite for the subject of art's relationship to science and helped me to navigate the meandering, interdisciplinary channels through which my interest in British Modernist sculpture and the New Biology could be developed. I am obliged to Andrew, in particular, for having carefully read and insightfully commented upon an early draft of this book. The financial support of the AHRC during my doctoral studies helped to sow the seed for what would become the manuscript – even if, in style and subject, the current book has far outgrown its original conception as a PhD thesis.

I would also like to gratefully acknowledge the generosity of the Henry Moore Foundation. The award of a two-year postdoctoral research fellowship grant by the Foundation enabled me to develop my doctoral research into a viable, book-length project. Without this support the manuscript would have taken far longer to produce and would, undoubtedly, have taken a very different form.

Thanks are also owed to the staff at the Scottish National Gallery of Modern Art. In particular, I would like to express my gratitude to Richard Calvocoressi and Ann Simpson, who were instrumental, at an early stage, in agreeing to host me as a collaborative doctoral student in 2005. Kirstie Meehan and Kerry Watson both demonstrated patience and generosity in helping me find my way around the Scottish National Gallery of Modern Art's archives. Patrick Elliott's kindness in sharing with me his expertise in artistic Modernism has also been greatly appreciated.

At the University of Edinburgh, I am grateful to the Department of Art History for having hosted me as a postdoctoral fellow and for providing me, over several years, with the facilities, community and supportive environment in which to complete my research. I would also like to acknowledge the financial support of the University of Edinburgh's Moray Endowment Fund, which helped to meet some of the illustration costs.

The generous support of a publication grant from the Paul Mellon Centre for Studies in British Art has enabled me to reproduce many wonderful images which would otherwise have been omitted. I would especially like to thank Mary Smith for her helpful guidance during the application process.

Martin Hammer acted tirelessly as my mentor throughout my Henry Moore postdoctoral fellowship and I am indebted to him for providing a thoughtful reading of an early version of the manuscript. I also owe a huge debt of gratitude to Viccy Coltman and Elizabeth Cowling, both of whom have championed my role within the department and continually supported my postdoctoral research.

Of the many friends and colleagues who have assisted me in countless ways, I would especially like to thank Janet Black, Marion Endt, Matthew Jarron, Ludmilla Jordanova, Alice O'Connor, Andrew Patrizio, Carol Richardson and Christian Weikop.

Special thanks should also go to Tony Millatt of the Mersea Museum, Essex, who patiently responded to endless requests about Richard Bedford and kindly supplied me with photographs of Bedford's work. I am also grateful to the Bedford family, most especially John Bedford and Chris Galvin, for allowing me to reproduce images of Richard Bedford's work.

Among the organisations that have granted me permission to reproduce images, I would particularly like to express my thanks to the Victor Batte-Lay Foundation (www.vblfcollection.org.uk), which provided me with digital images of Richard Bedford's drawings. The assistance of Robin and Jan Matthews and Evelyne Bell at the Foundation proved invaluable to the completion of the book. For granting permission (as well as, in certain instances, technical assistance) to reproduce various images within the book I would also like to take the opportunity to thank: Sophie Bowness at the Hepworth Estate; Sarah Gretton and the F. E. McWilliam Estate; Sonia Wiffen and the Leon Underwood Estate; Ann and Jürgen Wilde and Simone Förster at the Karl Blossfeldt Archive; and the Henry Moore Foundation. Special mention should go to Isla Robertson, whose resourcefulness in the art of picture research has been nothing short of an inspiration.

I owe a huge debt of gratitude to my wife, Arielle, whose unwavering love and support have sustained me throughout the long years it has taken to turn an idea into a book. Lastly, my parents have abidingly provided me with heartfelt encouragement throughout my time as a postdoctoral researcher and it is to the memory of my mother – who did so much to kindle in me an interest in art and culture – that this book is dedicated.

Every reasonable attempt has been made to obtain permissions for those images believed to be in copyright. If any proper acknowledgement has not been made, copyright holders are invited to contact the publisher.

Introduction

In our age the discovery and study of single-celled organisms has been followed by a search after the units, the source, the primitive form of expression; and no artist can live by himself, or live altogether in, or by, the impressions once vivid in the eyes of a dead generation.[1]
Geoffrey Grigson, *Henry Moore*, 1943

In the decade following the Great Depression and leading up to the Second World War, British society experienced a period of extraordinary cultural innovation, in which the impact of interwar regeneration and discovery uprooted old certainties, leading to an avowedly modern refashioning of British sculpture and biology. The emergence of a legitimate avant-garde as a response to the deadening influence of neo-classicism on contemporary art would lead to the birth of a new movement in sculpture – a powerful expression of artistic modernity that would, perhaps surprisingly, find an intellectual counterpart in the revisionary philosophy of modern biology. In his monograph *Henry Moore* (1943), the critic Geoffrey Grigson identified a common vision that drew together both Modernist sculpture and biology, censuring those who saw contemporary art as symptomatic of a 'distorted vision and [...] disordered mind', by claiming that, while not everyone may 'be familiar with the cells and organs and elements of life [...]. Biology must be acknowledged'.[2] A poet by vocation, Grigson founded the literary review *New Verse* in 1933 and would later acquire celebrity in art history for his introduction – in 1935 – of the term 'biomorphism' into the lexicon of art criticism. Borrowed from anthropology, the appellation did not describe a resemblance to natural form *per se* but rather emphasised that the organic qualities of contemporary art had, as a point of departure, the smoothly contoured shapes of the natural world.[3] Yet while the designation 'biomorphic' could, to a considerable extent, refer to natural form in the widest possible sense – encompassing objects as diverse as river-worn pebbles, nuggets of bone and the shapes of animals – it nonetheless relied upon the findings of modern biology to fully articulate the range of meanings to which it was subject. Thus was Grigson alive to the correspondence between the sinuous profile of a sculpture, such as Henry Moore's *Figure* of 1931 (Figure 0.1), and the globular, asymmetrical appearance of a single-celled organism, noting that the visual knowledge of the sculptor had been permanently enriched by the visions of

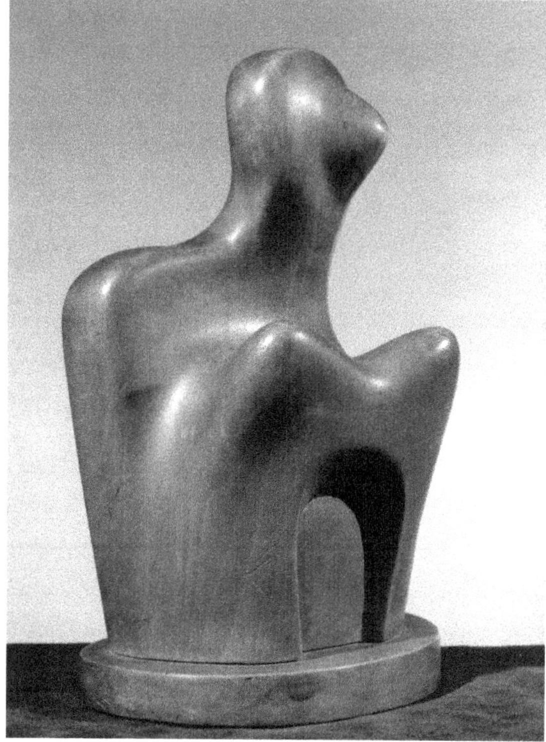

Figure 0.1

Henry Moore, *Figure*, 1931. Tate, London

modern biology: 'When I look at [Moore's] carvings, I sometimes have to reflect that so much of our visual experience of the anatomical detail and microscopical forms of life comes to us, not direct, but through the biologist'.[4]

The distinctively *modern* implications of biomorphist imagery demonstrate the extent to which biological references in Modernist art criticism of the 1930s moved far beyond any orthodox emphasis upon what might be otherwise described as the traditional aspects of art's relationship to nature.[5] Characterised by a fluid vocabulary of smooth-edged curves, flowing outlines and gentle protuberances, Modernist sculpture therefore lent itself particularly well to biologistic interpretations by critics, as it seemed to echo – albeit abstractly – the vital, burgeoning forms newly captured by scientific photography or theorised by experimental biology.[6]

The startling breadth of discovery enjoyed by biology in the years following the turn of the century encouraged many Modernists to view the subject as the paradigmatic science of the epoch and adopt a 'biocentric' attitude which privileged the life sciences and emphasised the centrality of nature in culture.[7] In artistic terms, a growing disenchantment with the fruits of industrialisation and mechanisation – what one reviewer of the time described as humankind's psychological 'inability to deal with the machine' – had served to heighten hostility towards those forms of artistic Modernism, specifically geometric abstraction, that appeared to cravenly mimic mechanical imagery and therefore ignore the biological needs of humankind.[8] By contrast, the incorporation of biological principles into art was seen as a positive development by those Modernists who sympathised with a biocentric viewpoint, seeing this biologistic tendency as an antidote to the alienating, machinist aesthetic of contemporary abstraction – a development which ostensibly healed the rift between humankind and nature that the industrial age was understood to have precipitated.[9] Perhaps nowhere was this predilection for biological leitmotifs more evident in modern art than in the folds, swells and pleats of Modernist sculpture, which seemed to represent natural forces and generative energies abstractly conceived.

Biocentrism, the New Biology and British Modernist sculpture

Bernard Reynolds noted in a 1937 defence of Henry Moore in the BBC's weekly *The Listener* that, unlike those Modernists who sought to parallel the products of the machine age through geometry, a sculptor such as Moore – who was sensitively attuned to the deeper rhythms of nature – could produce 'biomorphic' artworks that were characteristically 'vital' and life-affirming:

> Most contemporary sculptors kill Nature, reduce it to non-sensitive geometric forms (they call it decorative), and translate it into stone. Henry Moore puts life into stone; his forms are 'biomorphic'. In his opinion, if a piece of sculpture defies the vital laws of natural growth and construction, it is bad [...]. However, he does not copy Nature, he creates within its life-principles.[10]

Reynolds' usage of the term 'biomorphic' to connote the way in which Moore's sculptures, by virtue of their turgescency and undulant curvilinearity, metaphorically alluded to natural forms without directly copying their outward appearance, possessed the unmistakable watermark of Grigson's critical thinking on biomorphism.[11] Certainly, given that Grigson had, just two years earlier, explicitly labelled Moore as the *only* 'biomorphist producing viable work' in England, it was perhaps understandable that Reynolds should have chosen this particular designation to signpost the biophilic character of Moore's oeuvre.[12] Nevertheless, that he elected to emphasise the *biological* vitality of Moore's art exemplifies the biologistic disposition of Modernist sculptural discourse and the scientific temper of a biocentric aesthetic philosophy that was channelled by (but not restricted to) the stylistic thematics of biomorphism.[13]

Needless to say, the biologistic readings of Reynolds and his fellow Modernist critics reflected profounder neo-romantic, nature-centric and biophilic trends in European art and design. While this book will underscore the importance of positioning this biologistic sculptural discourse within a British cultural context, it is nonetheless important to situate such an approach within the broader framework of the emergence of 'biocentrism' as an intellectual phenomenon in early twentieth-century European Modernist culture. Artistically located at the convergence of various movements in European Modernism (including Art Nouveau, Expressionism, Primitivism, Surrealism and Constructivism), biocentrism, in general terms, rejected anthropomorphism and machinist culture, privileged the life sciences as an epistemological model and highlighted the idea of existential fluidity in nature.[14] Within this philosophical setting, those artworks which partook of a biomorphic appearance were interpreted by critics as dependent upon biological law and therefore emblematic of humankind's inseparability from the totality of nature.[15] Thus Modernists – such as Wassily Kandinsky and Henry Moore – who incorporated zoological and embryological motifs into their compositions were self-consciously aligning themselves with a biocentric attitude that laid emphasis on notions of regeneration, procreativity and the common origins of life.[16]

In its broadest sense, biocentrism represented a particular neo-romantic worldview – prevalent at the turn of the century – which philosophically refuted what many saw as the extreme positivism and materialism of nineteenth-century science. The speedy decline of faith in materialism and mechanism as meaningful epistemological systems during the *fin de siècle* was matched by an upsurge of interest in romantic approaches to nature, many of which favoured instinctive, idealistic, holistic and/or metaphysical attitudes towards nature and which prized the experiential unity of life.[17] The First World War, in this respect, acted as a catalytic event in the development of European biocentrism, functioning as a lightning rod for anti-mechanistic sentiment and encouraging the appearance of a neo-romantic nature philosophy. Just as widespread intellectual disillusionment with materialism fanned the philosophical flames of biocentrism in the years leading up to the First World War, so too anxiety at the devastation caused by mechanised conflict alongside a more general dissatisfaction with the socio-environmental degradation triggered by postwar industrialisation led to the strengthening of biocentric approaches during the 1920s and 1930s.[18] In Britain – as in the rest of Europe – the postwar impression of existential crisis would lead to a profound questioning of mechanistic values and the advent of holistic and biocentric philosophies, providing the spark that would ultimately lead to the establishment of a *biologistic* conception of Modernist sculpture.

Significantly, it was the emergence of the 'New Biology' as a powerful force within reformist factions in interwar science that was to have the greatest impact upon the language of British Modernist sculpture, furnishing it with a biologistic vocabulary that privileged holistic concepts, metamorphic imagery and epigenetic themes. A complex phenomenon in the history of science, the New Biology represented an array of neo-idealistic scientific perspectives – including such movements as neo-Lamarckism and neo-vitalism – which variously sought to question the legitimacy of the predominant mechanistic and positivistic scientific attitudes of the period.[19] The catalyst for this shift in outlook was the increasing prominence given to an integrative approach to physiology in the interwar years, which promised to supersede the reductive shortcomings of mechanistic theory. Rather than examining the parts of a biological system in isolation, as mechanistic biologists favoured, proponents of an integrative or holistic approach argued that the study of the properties of the whole was essential to any meaningful understanding of biological function. Instead of simply focusing attention on low-level physico-chemical interaction at a cellular level, the new generation of biologists attended to the higher-level interchanges that typically took place in the tissues or organs, concluding that new characteristics emerged from the parts of a bio-system operating in a state of dynamic interaction.[20]

While I have here considerably simplified a complicated development in the history of science that will be unravelled more completely as the book progresses, at this point it is worth simply remarking upon the centrality of the New Biology to biological theory in the 1930s and the degree to which it helped biology establish itself as the exemplary science of the period. As Oliver Botar has demonstrated, while the New Biology was not anti-materialistic *per se*, its roots in *fin de siècle* pantheism, mysticism and neo-vitalism ensured that it was a central component of the biocentric attitude which aimed to restore 'life' to the heart of contemporary experience.[21] Indeed, the perception that biology, through its role as the paradigmatic science, could act as a model for disciplines as diverse as aesthetics and politics – which similarly sought to respond to the anti-mechanistic sympathies of the age – meant that biologistic frames of thought acquired currency as a legitimate interpretative methodology for artistic Modernism.

Conspicuously, the emergence of a biologistic sculptural discourse occurred alongside the appearance of a new type of Modernist sculpture in Britain. Influenced to varying extents by the styles of synthetic Cubism, Constructivism and Surrealism, the manifestation of this movement represented – with respect to the conservative artistic sensibilities that had held sway in the country throughout the 1920s – something of a belated British response to the aesthetic challenges posed by pre- and postwar continental Modernism.[22] Certainly, in the text 'Going Modern and Being British', the artist and critic Paul Nash maintained that it was possible to achieve both ends, insisting that 'going modern' necessitated some familiarisation with European Modernist trends in art.[23] Demonstrating an industrious conjunction of Modernist practice, Primitivism and monumental classicism, Pablo Picasso and Jacques Lipchitz's form experiments of the 1920s offered a set of powerful precedents, in this respect, for the ambitious brand of sculptural Modernism that developed in Britain in the late 1920s and early 1930s.[24]

With its bold fusion of the classicism of Aristide Maillol, with the thick-set volumetry of primitive art, Moore's *Reclining Woman* of 1930 (Figure 0.2) can be seen to embody the quintessence of the new sculpture, evincing a partiality towards a vocabulary of heavy-set form that found its aesthetic counterpart in

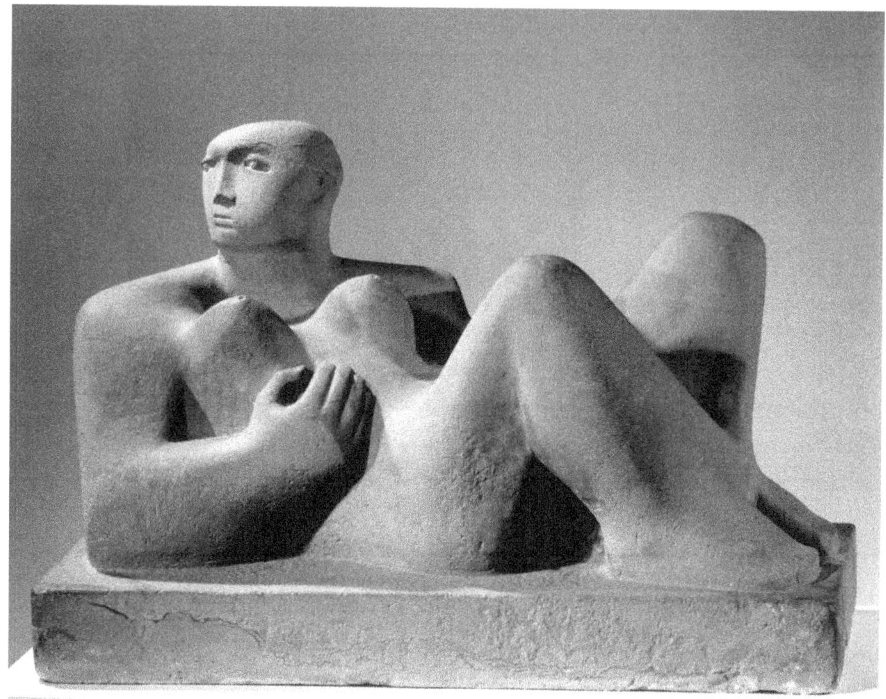

Figure 0.2

Henry Moore, *Reclining Woman*, 1930. National Gallery of Canada, Ottawa

the Mesoamerican sculptures of the British Museum and the paintings of bathers that Picasso created at this time. Such artworks, by virtue of their conceptual audacity, paved the way for the radically abstract sculptures produced in the middle years of the 1930s, such as Hepworth's sexually suggestive *Two Forms* (1933) (Figure 0.3), which critics characteristically appreciated as symptomatic of either a solipsistic type of nihilism or evidence of an artistic interest in contemporary design.[25] So profoundly did the objects of the new sculpture differ from the Neo-Cubist and neo-classical styles that preceded them that Modernist critics agreed that only a wholesale change in terminology could effectively detail the aesthetic transformations wrought by the modern movement in art. In 1935, the chief apologist of artistic Modernism in Britain – the critic Herbert Read – remarked in the first edition of *Axis* magazine on the bewildering proliferation of terms used to describe modern art:

> In the criticism of modern art we have reached a stage at which the everyday vocabulary of criticism is proving inadequate and therefore confusing. Developments of the last twenty years have given rise to various new types of art, which, although they may have their parallels in past epochs, have never existed as self-conscious entities.[26]

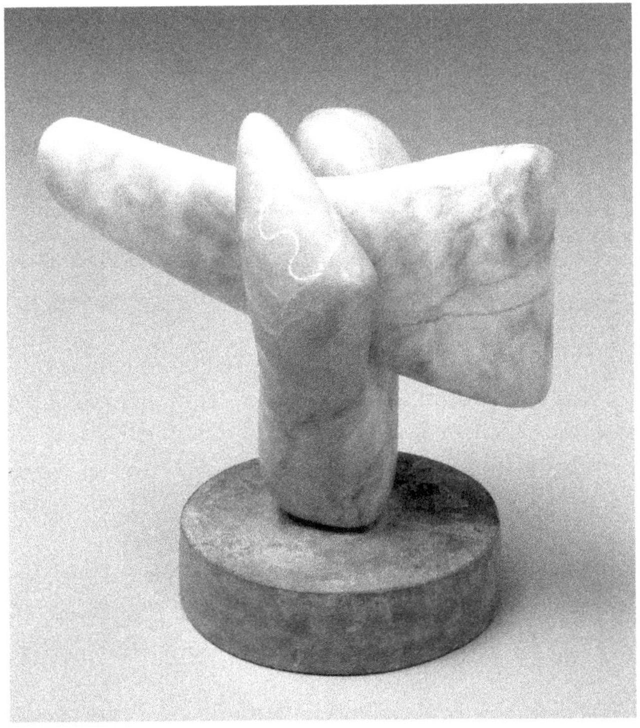

Figure 0.3

Barbara Hepworth, *Two Forms*, 1933. Tate, London

The editor of *Axis*, Myfanwy Evans, similarly despaired of being 'lost amongst individuals who protest at clarification'.[27] However, despite the proliferation of 'isms' in modern British art in the 1930s, both critics and practitioners generally recognised no implicit connection between ideological persuasion and style, applying biologistic frames of interpretation to naturalistic, biomorphic and geometric artworks alike. And while biomorphic abstraction was perhaps particularly susceptible to biocentric readings, due to its unique ability to abstractly evoke the forms of life, artworks of a more geometric temperament (such as the Neo-Constructivist bas-reliefs of Ben Nicholson) were equally subject to biologistic interpretations.[28] Henry Moore – who flitted between the varying styles of Primitivism, Surrealism and abstraction throughout the 1930s – was continually subject to biologistic analyses, even as Naum Gabo, a dyed-in-the-wool Constructivist, whose stark experiments in geometric design would seem to have singled him out as an unrepentant adherent to a machine-age aesthetic, emphasised the nature-centric character of his work and its biologistic capacities. Just as biomorphic abstraction was deemed to shun superficial appearances while simultaneously representing the quintessence of natural form, so Gabo saw the hard-edged orthogonality and ascetic parabolic curves of Constructivism as indicative of the artist's aptitude for discerning the underlying structures of the natural world. His notion that 'the constructivist has renounced the representation of natural forms' while concomitantly aiming 'not to reproduce Nature but to create and enrich it' verbalised a belief that Constructivism possessed a deep understanding of bio-systems, creating within their limits, *even as* it eschewed artistic naturalism.[29] The certainty that the contemporary artist was somehow privy to biological principles was linked to the conviction – here expressed by the film-maker John Grierson – that 'biology [was] getting into [the] blood' of interwar society through such agents as cinema and the popular press and was therefore reshaping the aesthetic of Modernist sculpture.[30] Consciousness of the efficiency of modern media in transmitting biological ideas to a wider audience was contemporaneously bolstered by the birth of 'decorative micrography' – a genre of art photography which delighted in artful images of microscopic forms, rendered pictorially but in a way that was wilfully respectful of scientific fact.[31] Thus such trends in popular culture shaped the impression that the ubiquity of biological imagery in contemporary society had profoundly impacted upon the psyche of the artist, subconsciously equipping him or her with a biologistic knowledge of life which percolated through into the very fabric of the artworks themselves. As Grierson noted in 1930, while reviewing an exhibition of sculpture by Hepworth, John Skeaping and Moore: 'It is not that they are anything of scientists, but they do seem to appreciate the principles on which living things stand up from the earth and get around it'.[32]

Far from being marginal to Modernist aesthetics, it is my central premise that the New Biology *fundamentally* influenced critical responses to the new sculpture, furnishing critics and artists with a complex biologistic vocabulary through which to elucidate the stylistic originality of sculptural Modernism. To be sure, the New Biology was seen by artistic Modernists to have particular significance for the arts, not least because it had become increasingly engrossed by questions of form.[33] For Modernist sculpture, especially, biological concepts appeared prescient, as – like a sculptor – the modern biologist was concerned with accounting for the production of form in hitherto undifferentiated matter, explaining how the spatial relations of the organism affected its developmental trajectory and how the flux of time impressed itself upon the form of living flesh.[34] Certainly, the very *three dimensionality* of the sculptural arts – in which forms are purposefully arranged so as to heighten the viewer's appreciation of volume – necessitates not only the consideration of *space* but also of *time*, a notion best illustrated by a passage from Gotthold Lessing's eighteenth-century aesthetic thesis, *Laocoön*:

> All bodies, however, exist not only in space, but also in time. They continue, and at any moment of their continuance, may assume a different appearance and stand in different relations. Every one of these momentary appearances and groupings was the result of a preceding, may become the cause of a following, and is therefore the centre of a present action.[35]

Just as the fossilised state of an organism was understood to be the manifestation of petrified time and its embryological state was seen to embody functional time (either side of which could be hypothesised a sequence of forms stretching endlessly forwards or backwards in time)[36] so – in the words of Rosalind Krauss – Modernist sculpture existed at 'the juncture between stillness and motion, time arrested and time passing', the instant at which the sculpture was visually apprehended hence representing a single moment in a countless series of possible moments, all of which collectively symbolised the perceptual continuity of experience.[37]

Conceived at a time when Henri Bergson's philosophy of an *élan vital* still possessed popular currency in art and science, the dynamic and fluxional epigenetic forces presupposed by the New Biology acted as an inspiration to those still in thrall to the creed of neo-vitalism and provided a powerful interpretative methodology to those who sought to explain the extraordinary spatio-temporal qualities of Modernist sculpture.[38] Even as the new sculpture was busy incorporating aspects of synthetic Cubism, Primitivism and Surrealism into its stylistic repertoire, the New Biology was effecting a synthesis between the hitherto separate areas of embryology, cytology, genetics and biochemistry, producing the basis for a unified theory of living systems within a pioneering holistic disciplinary framework.[39] Thus it is my intention to tell how these developments came to be linked in the minds of British Modernist sculptors and their critics, who recognised in the New Biology a ready set of concepts and metaphors through which to explicate the biologistic aesthetic of the new sculpture.

Theories and methods

The centrality of biology to the Modernist project has, in recent years, been subject to a range of scholarly studies which have located biologistic currents in a wide array of artistic practices, ranging from historical analyses of 'biocentric Constructivism' in the early twentieth century to theoretical explorations that seek to identify a tangible 'Bio-Art' strand in the polychromatic tapestry of contemporary art.[40] Naturally, the area of contemporary scholarship that is most pertinent to my current study is that which has focused upon biocentrism as a historical phenomenon. Seen as part of a blossoming environmental consciousness that swept through Europe in the early twentieth century, biocentrism (or, more generally, 'nature-centrism') has been identified in several studies as a series of overlapping, though not necessarily analogous, discourses which shared a group of attitudes and rhetorical formulae pertaining to nature, biology and epistemology. While these discourses diverged from one another in important ways, at the same time they mutually valued a number of key biocentric principles, including faith in the primacy of life processes, in biology as the prototypical science of the age, a deeply held anti-anthropomorphism and an implicit or categorical environmentalism.[41]

As a concept which helps to demarcate the interconnectedness of biological, nature-centric and cultural discourses, biocentrism provides a useful historical construct in which to situate British biologistic attitudes, as it allows a considerable variety of beliefs about biology and nature – present within artistic Modernism – to be accommodated within its conceptual frame. Insomuch as biocentrism recognises an anti-mechanistic, biophilic tendency in Modernist philosophy and aesthetics, it is a useful key term in helping to identify and categorise biologistic attitudes in British sculptural discourse, not least because of the philosophical overlap between European Modernism and its British variant.[42]

Biomorphism is, similarly, a critical term that is important to the subject of my enquiry – not least because of its rootedness in an English intellectual tradition.[43] Characterised 'by reference to an imagery of irregularly curvilinear shapes that are closed or tending towards closure', biomorphism's stylistic relatedness to the objects of the natural world (and biocentrism more generally) makes it a logical bedfellow for a study that seeks to understand the interrelatedness of the New Biology and sculptural Modernism in Britain.[44] While art historical studies of biomorphism have stressed its origins within the diverse influences of Victorian anthropology, psychoanalysis, museology and philosophy – disciplines and epistemologies which are beyond the thematic remit of this book – the part that the New Biology played in forming the biomorphic hypotheses of Grigson and others necessitates that biomorphism is considered as an integral component of the biologistic discourse that emerged around British Modernist sculpture.[45]

That said, the broad array of influences to which biomorphism was subject has inadvertently resulted – in terms of the discussion of the New Biology – in a degree of scientific generalisation: technical terms such as organicism, mechanism and neo-vitalism can be discussed in the literature imprecisely, meaning that the subtleties of interwar scientific debate are occasionally lost.[46] By therefore concentrating upon the epistemological relationship between the New Biology and Modernist sculpture, I intend to flesh out wider the scientific context in which biomorphism operated, identifying it as one factor among many that helped to shape the biologistic parameters of sculptural discourse in interwar Britain. Certainly, as a stylistic category, 'biomorphism' has epistemological limitations when considered as a byword for biologistic intention in artistic Modernism. Though receptive to the imagery of the New Biology, biomorphism's expressive form-vocabulary of sweeping curves and whorls – which evocatively recall the shapes of unicellular organisms, embryos and buds – was by no means the only style through which biocentric attitudes found articulation. As Oliver Botar and Isabel Wünsche have proven, there was 'no *necessary* connection between ideological background and style' and many Modernist artists – Naum Gabo and Ben Nicholson among them – did not feel compelled to work in a biomorphic style *even as* they espoused biologistic worldviews.[47] Indeed, as Jennifer Mundy has argued, biomorphism's iconographic kinship to the forms of nature means that certain biologistic concepts – such as organicism – which 'embrace a wide range of possible visual expressions [are often] incompatible with any description of biomorphic art'.[48] For example – as I will demonstrate in Chapter 3 – the idea that certain Modernist artworks could appear 'organic' – that is referring to *internal* organisational principles rather than *external* form appearance – meant that it was possible for contemporary critics and sculptors to think of a composition as adhering to biologistic laws without it necessarily appearing to be *biomorphic*.[49] As such, this book will chart the deeper currents of biologistic meaning flowing between Modernist sculpture and the New Biology that biomorphism, as a particular category in the history of art, would otherwise overlook.

By examining the *biologistic* parameters of Modernist sculpture, I am identifying interwar sculptural discourse with a type of *biologism*, by which I mean a particular kind of aesthetic philosophy which employed knowledge structures consciously drawn from biology to elucidate the principal features of contemporary art. The *Oxford English Dictionary* defines 'biologism' as the 'interpretation of human life and behaviour from a (purely) biological point of view' and thus a biological offshoot of the wider philosophical movement of *scientism*, which assumes that the methods of study appropriate to science are epistemologically applicable to other areas of human knowledge.[50] Biologism as a philosophical construct has many variants, although all forms of biologistic discourse take biological knowledge as their epistemological starting point.[51] More broadly, a biologistic attitude intersects with a biocentric worldview as – like biocentrism – it assumes 'the growing ascendency of biologically based accounts of human life' and so affirms biology's epistemic status as the paradigmatic science of the epoch.[52]

In addition, I classify the psychological character of Modernist biologism as *biophilic* in nature. Defined as 'a love of life' or 'love or empathy with the natural world',[53] 'biophilia' refers to 'the psychological tendency in humans to be attracted to all that is alive and vital'.[54] Conceptually linked to the nature-centric, neo-romantic component of biocentrism,[55] 'biophilia' will here be used to indicate a specific Modernist empathy towards *biological* systems rather than a weaker philosophical identification with 'nature' as a general concept.

In sum, this book aims to move beyond current understandings of biology's relationship to artistic Modernism – as they are presently defined by the scholarly parameters of biocentrism and biomorphism – by exploring the latent biologism and biophilia of Modernist sculpture in Britain from within the epistemological framework of the New Biology.

British sculpture and Modernism

As Charles Harrison has observed, there is always a danger, when pursued uncritically, that a study of British Modernism will concern itself with 'achievements that were largely marginal [within] a climate of persistent resistance and retrenchment in the face of more uncompromising European developments'.[56]

However, to ignore the historical and cultural specificity of artistic Modernism is to perpetuate a discourse that has enabled Modernism to achieve epistemological hegemony – arguably to the detriment of our understanding of the ways in which the individuals and localities that collectively contributed to the Modernist project envisaged Modernism itself. Christopher Wilk has emphasised that artistic Modernism was 'not only international in its outlook and practice, but [also] "extraterritorial"' – a proto-imperialist discourse perpetuated by the global dissemination of Modernist ideas by nomadic artists, designers, architects and thinkers in the 1920s and 1930s.[57] Correspondingly, Modernism's discursive supremacy within art history is somewhat problematic when considering biologistic discourse through the conceptual prism of biomorphism. Despite having its origins in late-nineteenth-century British culture (first emerging as a term in the writings of the anthropologist Alfred Court Haddon),[58] biomorphism is commonly discussed by scholars as an intercontinental stylistic phenomenon that – as an offshoot of artistic Modernism – largely ignored boundaries, be they cultural or political.[59] Thus although biomorphism clearly informed the stylings of the international avant-garde, to position it (however understandably) within a multinational milieu inevitably means downgrading the importance of the 'local' conditions under which the New Biology came to inform sculptural Modernism in Britain.

Late interwar Britain makes a particularly compelling case study through which to understand the relationship between Modernist sculpture and the New Biology. In the first instance, the 1930s is a significant decade when attempting to understand the reception of artistic Modernism in Britain as a whole. While Modernism had briefly flourished in Britain through the machinations of the prewar Vorticist movement, it had been rapidly supplanted in the interwar period by a French-inspired variety of Post-Impressionism. It was only during the 1930s that artistic Modernism came to be truly consolidated through the tardy arrival of Constructivism, synthetic Cubism and Surrealism, the 'dispersal of German culture throughout Europe under Nazi pressure',[60] and the establishment of home-grown avant-garde periodicals and journals – such as *Unit One* (1934) and *Axis* (1935–37) – which served as mouthpieces for Modernist activity.[61] Jane Beckett has spoken of the unparalleled velocity of this change, recognising it as the product of a 'complex interaction of political and economic forces' which impacted upon 'housing, education, industry and social issues' as well as 'the social organization of art production', through new configurations within the art market and novel critical approaches.[62] Certainly, the pervasive sense of existential crisis which emerged in response to these social and economic transformations provoked a wide range of reactions from Modernists, stretching from a self-conscious regression into the art of the past to a neo-romantic engagement with nature and the New Biology.[63]

In terms of the avant-garde, Surrealism – which had first appeared as a coherent force in 1924 following the publication of André Breton's *Manifesto of*

Surrealism – would finally emerge as a legitimate, if somewhat only tentatively supported, movement within the British art scene during the mid-1930s, with the 1935 publication of David Gascoyne's *First Manifesto of English Surrealism* and the 1936 International Surrealist Exhibition at the New Burlington Galleries in London.[64] Historically the antithesis to Surrealism's deeply held anti-rationalist aesthetic, Constructivism would also gain a firm foothold in the British avant-garde as the influx of ex-Bauhaus faculty members fleeing Nazi persecution – such as Naum Gabo, Walter Gropius and László Moholy-Nagy – acted as a powerful stimulus to the arts, introducing new trends in European Modernist abstraction and leading to the production of exhibitions, such as Nicolette Grey's 'Abstract & Concrete' show, and periodicals, such as *Circle*, which provided forums through which Constructivist values could be disseminated.[65] In particular, the commitment that Constructivist artists displayed towards an ideology of abstract art and industrial design that was rooted in scientific theory played a key role in creating the conditions through which British sculptural Modernism and the New Biology could be seen to productively interact.[66] Both Surrealism and Constructivism dominated understandings of contemporary art in Britain during the mid-1930s yet – despite their traditionally fraught relationship – many Modernists moved freely between both ideologies, seeking, furthermore, to intertwine international Modernism with home-grown conceptions of neo-romanticism and handicraft.[67]

Caught between the opposing ideological crosswinds of Surrealism and Constructivism, biomorphism – as we will see in Chapter 4 – emerged explicitly as an attempt to synthesise the diverse stylistic trends to which contemporary art was subject in the 1930s.[68] Conceived as a type of 'impure' abstraction, Grigson defined biomorphism as embodying those abstractions that 'exist between Mondrian and Dalí', unambiguously locating biomorphism at the stylistic junction halfway between the severe orthogonality of geometric abstraction and the psychologically inflected hyperrealism of Surrealism.[69] Importantly – from the standpoint of this thesis – his theorisation of biomorphic abstraction was published in 1935 (in the Modernist magazine *Axis*) under the title 'Comment on England', and formed part of a larger discussion of the special cultural conditions under which contemporary British art operated. Although Grigson sought to apply European aesthetic criteria to his argument, his essay left little doubt that his coinage of 'biomorphism' was 'a recognition of the context in which modern artists necessarily had to operate in England, and a warning against two extremes which [he] believed to be sterile and derivative'.[70] Indeed, he specifically singled out 'Moore and Wyndham Lewis as the only English artists of maturity in control of enough imaginative power to settle themselves actively between the new Pre-Raphaelites of *Minotaure* and the unconscious nihilists of extreme geometric abstraction'.[71] And – as Alan Powers has demonstrated – the emphasis upon the re-enchantment of nature and national tradition which typified Grigson's neo-vitalist conception of biomorphic abstraction was an integral component of the wider neo-romantic revival that took place in Britain in the 1930s and which, in turn, was a symptom of the biocentric neo-romanticism which arose in Modernist circles in early-twentieth-century Europe.[72]

While the New Biology was a phenomenon which spanned Europe and North America and was not, in any meaningful sense, intellectually limited to Britain, many of its principal advocates and theorists – including such luminaries as the philosopher Alfred North Whitehead, the evolutionary biologist Julian Huxley, the physiologist John Scott Haldane and the experimental embryologist Joseph Needham – were British and lived and worked in England during the 1920s and 1930s. Many of these figures contributed, in some way or other, to the burgeoning market in self-educational literature that developed in Britain during the interwar years and which saw scientific writers – such as James Jeans and Lancelot Hogben – publish 'popular' books on scientific topics which went on to sell in the tens of thousands. Certainly, publishers were keen to capitalise upon the growing market of individuals who were willing to part with modest sums of money in order to educate themselves about a range of scientific topics. Many of these works – as in the case of H.G. Wells, Julian Huxley and G.P. Wells' encyclopaedic *The Science of Life* (1929) – tackled the findings of the New Biology.[73] Moreover, the attention paid by British newspapers to the marvels of the 'Biological Revolution', which, as science pundits emphasised, augured the possibility of test-tube babies and lengthened life spans for humankind, created a climate of interest in the New Biology within Modernist circles that helped cement the impression that biology was the exemplary science of the epoch.[74]

This book traces a broadly thematic trajectory, the overall aim of which will be to demonstrate the palpability of biologistic ideas in Modernist sculptural discourse in Britain between the years 1930 and 1939. Each chapter will discuss a different biological topic and examine its potency in relation to the work of Moore, Nicholson, Gabo, Hepworth *et al.* in the tumultuous years sandwiched between the Great Depression and the start of the Second World War.

Chapter 1 adopts the most theoretical tone in the book, as it attempts to theorise an epistemological frame in which to understand the complex inter-relationship that existed between Modernist sculpture and the New Biology in the interwar years. Here, I sketch out a brief overview of the ways in which art–science relations have been hypothesised by historians and examine the critical legacy of C.P. Snow's, now infamous, 'Two Cultures' paradigm. This chapter lays down some of the historical groundwork for the chapters that follow by analysing the popular science industry in the 1920s and 1930s and exploring the ways in which the New Biology was publicly presented by proselytising scientists and the popular press. To contextualise the narrative arc of the book, this chapter will study how the biologistic character of interwar sculptural discourse was part of a much wider Modernist response to modern science. I will here investigate the wider avant-garde response to the new science and argue that science was perceived by many Modernists both positively and negatively – as both an inspiration and a disenchantment to the modern artist – and resulted in the emergence of a distinctive *scientistic* discourse within contemporary aesthetics which paved the way for the biologism of the 1930s.

Chapter 2 will explore how 'metamorphosis' (and attendant ideas about evolution, embryology, growth and development) influenced Modernist sculpture in the 1930s. I begin with an assessment of the 'new evolutionary synthesis' and how this scientific development problematised notions of evolutionary change through positing competing neo-vitalist and materialist conceptions of embryological growth. I will focus especially on the work of Hans Driesch, whose concept of the embryo as a self-adjusting, epigenetic field did much to strengthen the philosophical fortunes of neo-vitalism within the Modernist community. I will then address the role that a neo-vitalist understanding of metamorphosis played in the development of the biologistic art theories of Henri Focillon and Stephen Haden-Guest. This neo-vitalist theorisation of art history will be seen to have had a significant impact on Modernist sculpture, not least in the readiness with which critics and artists were willing to ascribe 'vital' principles to objects that appeared to be derived in some way from evolutionary or developmental forces. Indeed, this chapter will focus on the impact of embryology on Modernist sculpture (most particularly in relation to Surrealist theories of the object) by exploring the iconography of 'the egg shape' which preoccupied sculptors like Hepworth and Moore throughout the 1930s. I will conclude by gauging how neo-vitalist philosophy and experimental embryology conspired to produce a kinaesthetic understanding of Modernist sculpture, whereby the asymmetry of the sculptural object was seen to activate a new relationship between viewer and sculpture in which spatio-temporal extension was viewed – in objecthood terms – as symptomatic of vital, developmental forces.

In Chapter 3, I will discuss the subject of organicism in relation to Modernist sculpture. Essentially, a philosophical idea that tackles the ontological problem of the compositional relationship between the *parts* of an object and the *whole*, organicism will thus be seen to re-emerge in the New Biology as a means of comprehending the physiological constitution of living things. I will here argue that at the same time as organicist sensibilities were developing in the New Biology, Modernist sculptors (such as Moore and Hans Arp) began experimenting with a wholly new sculptural idiom in which the archetype of the sculptural object was disbanded, to be replaced by a collection of objects arranged seemingly haphazardly upon a plinth. I will contend that organicism provided Modernist sculptural discourse with a powerful holistic and biologistic philosophy through which to explain the rationale behind 'multipart' sculpture, one which epitomised the anti-reductionist, neo-romantic and organicist tendencies to which British culture was contemporaneously subject.

Chapter 4 will look at the relationship between morphology and sculptural Modernism. Advances in biochemistry and experimental embryology, obtained by the New Biology in the interwar period, led to a scientific reappraisal of the importance of morphology to biological study. I will contend that this resurgence of interest in biological 'form' was mirrored by the curiosity Modernist theoreticians exhibited towards the morphology of art. Spellbound by the recurrence of certain types of 'form constant' in both art and nature (most especially the Golden Section ratio), art critics – such as Herbert Read and Matila Ghyka – drew upon romantic nature philosophy, Goethean idealist morphology and the New Biology to explain the ubiquity of the spiral form as a compositional standard in artworks and the natural world. I will show that Modernist sculptors, keen to align themselves with the neo-romantic sensibilities of neo-vitalism, recognised spirals as symbolic of life's quintessence and represented spiriform shapes in their art as ciphers of neo-vitalist intent. The morphological revelation that certain forms predominated in the natural world will subsequently be examined from the standpoint of monism and how the perceived unity of nature – promulgated, most especially, by crystallography – fed into a Modernist debate over the synchrony of modern art. Despite the apparent divergence between Constructivist abstraction and Surrealist hyperrealism – which appeared to represent inorganic and organic nature, respectively – critics hypothesised their essential unity: a monistic vision of art which drew inspiration from the monistic suppositions of the New Biology. I will conclude here with a discussion of the highly politicised form–function debate – which was driven partly by the political imperatives of Constructivist theory – which took place in artistic Modernism and how morphology provided Modernist critics with a scientific means through which to demonstrate the social utility of modern sculpture. The 'biotechnical' hypotheses of the science writer and design theorist Raoul Francé will, in particular, be seen to have provided Modernist sculpture with a functionalist, biologistic base.

In Chapter 5 I will focus on the impact that new scientific visualising technologies (primarily photomicrography) had on modern sculpture in the 1930s. Beginning with a lengthy discussion of the popularisation of photomicrographic and microcinematographic imagery in the interwar years, I will assess the artistic influence of these close-up images of nature initially from the standpoint of the critical reception of Karl Blossfeldt's macrophotographs by the Modernist community – chiefly in relation to the 'bioromantic' and 'associationist' theories of the art critics Paul Nash and Reginald Wilenski. I will here also take the opportunity to suggest that, far from simply embodying positivist values, the troubling distortions of scale permitted by microscopy appealed to contemporary artists due to their *de*-sublimatory potential. At this point I will indicate a far closer relationship between the philosophy of the Georges Bataille and British Modernist sculptural theory than has hitherto been suggested. I will then explore the analogical relationship Modernist critics and artists – such as Geoffrey Grigson and John Piper – staged between contemporary sculpture and biological imagery. Aside from providing biomorphic abstraction with its distinctively modern stylistic connotations, I will also suggest that microscopy furnished Modernism with an authoritative, scientistic terminology which rhetorically conflated perspicacity of vision with misogynistic empowerment.

A brief epilogue will conclude the book by investigating the continuation of a biologistic discourse in modern art in the years immediately following the Second World War. In this way I hope to demonstrate that the biologistic aesthetic formulated by Modernist critics and sculptors in the years 1930–39 had a tangible impact on the artistic philosophy of the subsequent generation of artists.

Notes

1 G. Grigson, *Henry Moore* (Harmondsworth: Penguin Books, 1943), pp. 8–9.

2 *Ibid.*, p. 9.

3 J. Mundy, 'Comment on England', in C. Stephens (ed.), *Henry Moore* (London: Tate Publishing, 2010), p. 28.

4 Grigson, *Henry Moore*, pp. 8–9.

5 J. Mundy, *Biomorphism*, unpublished PhD thesis (London: University of London, 1986), pp. 111–12.

6 On the wider European avant-garde engagement with images of generation and procreativity, see G. Maldonado, *Le cercle et l'amibe: Le biomorphisme dans l'art des années 1930* (Paris: CTHS/INHA, 2006), pp. 82–4, 200–2.

7 See O. Botar & I. Wünsche, 'Introduction: Biocentrism as a Constituent Element of Modernism', in O. Botar & I. Wünsche (eds), *Biocentrism and Modernism* (Farnham: Ashgate, 2011), pp. 2, 5.

8 Anon., 'Circle Review', in *The Listener*, Vol. XVIII, No. 447, 4 August, 1937, p. 259.

9 O. Botar, *Prolegomena to the Study of Biomorphic Modernism: Biocentrism, László Moholy-Nagy's 'New Vision' and Ernő Kàllai's Bioromantik*, unpublished PhD thesis (Toronto, University of Toronto, 1998), p. 23.

10 B. Reynolds, 'Letters', in *The Listener*, Vol. XVIII, 15 September 1937, p. 575.

11 See also Mundy, 'Comment on England', p. 28.

12 G. Grigson, 'Comment on England', in *Axis*, No. 1, January 1935, p. 10.

13 For a brief history of the term 'biomorphism' and the various intellectual currents from which it originated, see J. Mundy, 'The Naming of Biomorphism', in O. Botar & I. Wünsche (eds), *Biocentrism and Modernism* (Farnham: Ashgate, 2011), pp. 61–73.

14 Botar, *Prolegomena to the Study of Biomorphic Modernism*, pp. 6–7.

15 *Ibid.*, p. 23.

16 On Kandinsky's usage of biological motifs in the 1930s, see V.E. Barnett, 'Kandinsky and Science: The Introduction of Biological Images in the Paris Period', in O. Botar & I. Wünsche (eds), *Biocentrism and Modernism* (Farnham: Ashgate, 2011), pp. 222–3. For a general discussion of this topic, see Maldonado, *Le cercle et l'amibe*, pp. 82–4.

17 O. Botar, 'Defining Biocentrism', in O. Botar & I. Wünsche (eds), *Biocentrism and Modernism* (Farnham: Ashgate, 2011), pp. 15–16.

18 Botar, *Prolegomena to the Study of Biomorphic Modernism*, p. 9.

19 Botar, 'Defining Biocentrism', p. 23.

20 G. Allen, *Life Science in the Twentieth Century* (Cambridge: Cambridge University Press, 1978), pp. 103–6.

21 Botar, *Prolegomena to the Study of Biomorphic Modernism*, p. 200.

22 See: F. Gore, 'Introduction', in S. Compton (ed.), *British Art in the 20th Century – The Modern Movement* (London: Royal Academy, 1987), pp. 10–11; A. Causey, 'Formalism and the Figurative Tradition in British Painting', in S. Compton (ed.), *British Art in the 20th Century – The Modern Movement* (London: Royal Academy, 1987), pp. 20–3.

23 P. Nash, 'Going Modern and Being British', in *Weekend Review*, 12 March 1932, quoted in C. Harrison, *English Art and Modernism, 1900–1939* (New Haven: Yale University Press, 1994), p. 254.

24 Harrison, *English Art and Modernism*, p. 225.

25 On the perceived nihilistic nature of abstract art, see: D.S. MacColl, 'Visual and Vocal Art', in *The Listener*, 9 May 1934, pp. 799–800; a formidable response to modern art, aspects of MacColl's criticism shall be discussed in later chapters. On the relationship between modernist sculpture and interior design see: A. Bertram, 'Artists Indoors', in *The Listener*, 1 November 1933, pp. 660–1; in this article Bertram linked abstract art to interior design by suggesting that the simplicity of abstract art allowed for a harmonious rapport with the pared-down aesthetics of Modernist decorative taste. Anne Wagner discusses the sexual nature of *Two Forms* in: A.M. Wagner, 'Miss Hepworth's Stone *Is* a Mother', in D. Thistlewood (ed.), *Barbara Hepworth Reconsidered* (Liverpool: Liverpool University Press, 1996), pp. 63–4.

26 H. Read, 'Our Terminology', in *Axis*, No. 1, January 1935, p. 6.

27 M. Evans, 'Order, Order!', in *Axis*, No. 6, summer 1936, p. 6.

28 For a fuller discussion of the relationship between biocentric interpretative models and style, see Botar & Wünsche, 'Introduction', pp. 3–4.

29 Naum Gabo, 'Constructive Art', in *The Listener*, Vol. XVI, No. 408, 4 November 1936, p. 846.

30 J. Grierson, 'The New Generation in Sculpture', in *Apollo*, Vol. XII, November 1930, p. 350.

31 A. Thomas, 'The Search for Pattern', in A. Thomas (ed.), *Beauty of Another Order: Photography in Science* (New Haven: Yale University Press, 1997), p. 106.

32 Grierson, 'The New Generation in Sculpture', p. 350.

33 On the importance of form to biology, see D. J. Haraway, *Crystals, Fabrics and Fields: Metaphors That Shape Embryos* (Berkeley: North Atlantic Books, 2004), pp. 38–45.

34 *Ibid.*, p. 39.

35 G. Lessing, *Laocoön*, E. Frothingham (trans.) (New York: Noonday, 1957), p. 91.

36 See G. Canguilhem, 'On the History of the Life-Sciences Since Darwin', in *Ideology and Rationality in the History of the Life Sciences*, A. Goldhammer (trans.) (Cambridge: MIT Press, 1988), p. 108.

37 R. Krauss, *Passages in Modern Sculpture* (Cambridge: MIT Press, 1981), p. 5. Krauss takes her methodological cue from the fact that 'the history of modernist sculpture coincides with the development of two bodies of thought, phenomenology and structural linguistics, in which meaning is understood to depend on the way that any form of being contains the latent experience of its opposite: simultaneity always containing an implicit experience of sequence' (pp. 4–5). While I harbour similar beliefs about the pertinence of biology to Modernist sculpture, I aim to demonstrate its importance to interwar interpretative strategies rather than using biological principles as methodological tools in their own right.

38 Although Bergson's faith in an immaterial life force had waned in importance due to developments in physics and biochemistry, its unambiguous rejection of determinism made it something of a *cause célèbre* among neo-vitalists and freethinkers, who questioned the rigid determinism presupposed by contemporary science. As George Woodcock has demonstrated, Bergson was a formative influence upon Read, helping him to productively move beyond 'a bleak rationalism'. On the scientific influence of Bergson, see P.J. Bowler, *Reconciling Science and Religion: The Debate in Early-Twentieth-Century Britain* (Chicago: University of Chicago Press, 2001), pp. 348–9. On Read and Bergson, see G. Woodcock, *Herbert Read: The Stream and the Source* (London: Faber & Faber, 1972), p. 195.

39 See Allen, *Life Science in the Twentieth Century*, pp. 113–14.

40 For an example of the diversity of art historical approaches to biologistic currents in modern and contemporary art, see Eduardo Kac (ed.), *Signs of Life: Bio Art and Beyond* (Cambridge: MIT Press, 2007).

41 See Botar & Wünsche, 'Introduction', p. 2. Oliver Botar has, in particular, been at the vanguard of scholarship seeking to define biocentric currents in Modernist culture and art practices. See Botar, *Prolegomena to the Study of Biomorphic Modernism*.

42 Biocentrism was undoubtedly a wider philosophical trend to which British sculptural Modernism was conceptually beholden, yet its very flexibility, even porousness, as a term means that it casts its art historical nets wider than the parameters of my study demand. Though intimately connected to the emergence of the New Biology, biocentrism was concerned with a multiplicity of concepts – including psychobiology, mysticism, anti-anthropomorphism, environmentalism and *Naturphilosophie* – which, though pertinent to biologistic thought, somewhat stray beyond the

parameters of my study, which is narrower in its emphasis upon the impact of the New Biology on British Modernist sculpture.

43 See Mundy, *Biomorphism*, p. 15.

44 *Ibid.*, pp. 13–14.

45 *Ibid.*, pp. 114–21.

46 Mundy, for instance, states that 'the organicist tradition in biology' was a 'minor [though] still forceful tradition' at the start of the twentieth century but overlooks the fact that mechanism and vitalism precisely found synthesis in the 1930s through the organicist paradigm. See Mundy, *Biomorphism*, p. 120. Maldonado, on the other hand, bases her discussion of the influence of biology almost entirely on the work of D'Arcy Wentworth Thompson – thus severely limiting the evidential scope of her analysis. Moreover, she confusingly links the evolutionary theories of Thompson and Bergson together, even though both thinkers had quite antithetical comprehensions of the mechanism of biological development. See Maldonado, *Le cercle et l'amibe*, pp. 195–7, 199–207.

47 Botar & Wünsche, 'Introduction', p. 4.

48 Mundy, *Biomorphism*, pp. 13–14.

49 Mundy agrees that while certain biomorphic artworks, in terms of their dynamic compositional arrangements, could perhaps be described as 'organic', by and large the term referred to biologistic principles that operated beyond the stylistic parameters of biomorphism. Indeed, she asserts that with 'equal justification [...] artworks which tend to the geometric, and the regular, may also be described as "organic" in reference to nature's constructive principles'. Mundy, *Biomorphism*, pp. 13–14.

50 *Oxford English Dictionary* online, 'biologism, *n.*', at www.oed.com/view/Entry/19225 (accessed 25 February 2013).

51 See A. Laitinen & G. Maude, 'Biologism, Politics and International Politics', in *Acta Sociologica*, Vol. XXIX, No. 2, 1986, p. 113.

52 See D. Skinner, 'Racialized Features: Biologism and the Changing Politics of Identity', in *Social Studies of Science*, Vol. XXXVI, No. 3, June 2006, p. 461.

53 *Oxford English Dictionary* online, 'biophilia, *n.*', at www.oed.com/view/Entry/251066 (accessed 25 February 2013).

54 J.P. Simaika & M.J. Samways, 'Biophilia as a Universal Ethic for Conserving Biodiversity', in *Conservation Biology*, Vol. XXIV, No. 3, June 2010, p. 903.

55 See Botar, *Prolegomena to the Study of Biomorphic Modernism*, p. 7.

56 Harrison, *English Art and Modernism*, p. viii. Although Harrison here is discussing *English* art (a term that is even more fraught with ethno-centric prejudices than *British* art), many of the points he raises on the epistemological difficulties of assessing the particular nature of English Modernism can be seen to apply to the study of British Modernism in general.

57 Christopher Wilk, 'Introduction: What Was Modernism?', in C. Wilk (ed.), *Modernism: Designing a New World* (London: V&A Publications, 2006), p. 15.

58 Mundy, *Biomorphism*, pp. 15, 85–8.

59 Maldonado, for example, speaks of tracking 'the ramifications of biomorphism in England, in Germany, in Northern and Eastern Europe, in Spain and in Italy, as well as the United States'. Maldonado, *Le cercle et l'amibe*, p. 14.

60 Causey, 'Formalism and the Figurative Tradition in British Painting', p. 22.

61 See Harrison, *English Art and Modernism*, pp. 231–93.

62 J. Beckett, 'Circle: The Theory and Patronage of Constructive Art in the Thirties', in J. Lewison (ed.), *Circle: Constructive Art in Britain 1934–40* (Cambridge: Kettle's Yard Gallery, 1982), p. 11.

63 David Peters Corbett, Ysanne Holt and Fiona Russell have written of Modernists attempting to 're-situate English art historically and geographically by redrawing the map of art history, re-orientating English art and re-mapping its history' during the early 1930s. D. Peters Corbett, Y. Holt & F. Russell, 'Introduction', in D. Peters Corbett, Y. Holt & F. Russell (eds), *Geographies of Englishness: Landscape and the National Past, 1880–1940* (New Haven: Yale University Press, 2002), p. xii.

64 For information on the establishment of Surrealism in Britain, see Michel Remy, *Surrealism in Britain* (Aldershot: Ashgate, 1999), esp. pp. 71–100.

65 Beckett, 'Circle', pp. 16–17.

66 *Ibid.*, pp. 17–18.
67 Peters Corbett, Holt & Russell, 'Introduction', pp. xv–xvii.
68 Mundy, *Biomorphism*, p. 80.
69 Grigson, 'Comment on England', p. 8.
70 A. Powers, 'The Reluctant Romantics: *Axis* Magazine 1935–37', in D. Peters Corbett, Y. Holt & F. Russell (eds), *The Geographies of Englishness: Landscape and the National Past, 1880–1940* (New Haven: Yale University Press, 2002), p. 253.
71 Grigson, 'Comment on England', p. 10.
72 Powers, 'The Reluctant Romantics', pp. 264–5.
73 P.J. Bowler, *Science for All: The Popularization of Science in Early Twentieth-Century Britain* (Chicago: University of Chicago Press, 2009), p. 10.
74 J. Huxley, 'Tissue Culture and Human Habits', in *The Listener*, Vol. IX, No. 231, 14 June 1933, p. 953.

1. Bridging the two cultures: relations between art and science in the 1930s

C.P. Snow's theory of 'The Two Cultures' – presented as a Rede Lecture in 1959 – has, since its inception, coloured responses to the theorisation of art and science, establishing what has been called an 'economy of the binary' in which both disciplines have been historically sustained through a set of mutually opposing terms. By degrees, science has been seen to be rational, systematic, cognizant and empirical, whereas art has been characterised as impulsive, haphazard, instinctive and creative – key terms that have served only to perpetuate the binary paradigm that science *discovers* while art *invents*.[1] Snow's assumption that the 'two cultures [of art and science] can't talk to each other' as they lack a common point of cultural interchange,[2] though expressed in the 1950s, actually had its roots in Victorian responses to modernity, in which science became indissolubly tied to technology, industrialisation and social mobility, while art came to be associated with traditionalism, the preservation of class hierarchies and time-honoured values.[3] Of course, even in the Enlightenment, maintaining such a binary divide was problematic, as, for instance, visual skills were essential to the knowledge of the eighteenth-century European naturalist, obligatory for both the collecting and classifying of material as well as the investigation of nature. As Andrew Graciano has confirmed: '[The] visualization of knowledge is a phenomenon common to both art and science from the Enlightenment to the present. Art and science are both fundamentally visual in terms of how they understand, collect, organise, create and disseminate knowledge'.[4] And, indeed, the visual output resulting from scientific experimentation has, historically, made it a particularly edifying bedfellow for the avant-garde. It has been well documented, for example, that the pointillist technique developed by the Neo-Impressionists resulted from a concerted engagement with the physics of chromatics as they were exemplified by physicists and experimental psychologists in the nineteenth century.[5] And the opposition traditionally harboured by the avant-garde towards institutionalised modes of practice meant that hostility towards the academy was often reflected by a pronounced interest in scientific and technological themes – ranging from X-rays to engineering – in ways that cut across the nominal binary divide of art versus science.[6]

Given that Snow formulated his binary model partly as a response to his experience of working as a writer *and* scientist in the 1930s, it is astonishing that his familiarity with both fields should have led to his 'Two Cultures' paradigm. Evinced by T.S. Eliot's belief that '[pure] literature is a chimera of sensation' and that a topical literary journal should include 'the results of contemporary work on history, archaeology, anthropology, even of the more technical sciences when those results are of such a nature to be valuable to the man of general culture',[7] the years between the wars witnessed a flourishing moment of interaction between the arts and sciences. Indeed, the sizable influx of political refugees during this period, such as Walter Gropius and László Moholy-Nagy, who had a background in the biocentric philosophies of design of the Bauhaus, would result in an allegiance to an ideology of Modernist abstraction that was deeply steeped in scientific and biologistic analogies.[8] Buoyed up by a resurgent popular science industry, this sensibility found expression in such radically collaborative ventures as the 1937 Constructivist project *Circle*, which brought together artists, architects and scientists. As early as 1919, Roger Fry penned an essay entitled 'Art and Science' for the *Athenaeum* in which he sought to demonstrate the methodological parallels between the two disciplines. Claiming that curiosity drove science to accumulate facts about the physical world and that an aesthetic sensibility led to the organisation of those facts into a satisfying scientific theory, Fry noted that '[both] these aspects – the particularising and the generalising – have their counterparts in art':

> Curiosity impels the artist to the consideration of every possible form in nature: under its stimulus he tends to accept each form in all its particularity as a given, unalterable fact. The other kind of intellectual activity impels the artist to attempt the reduction of all forms, as it were, to some common denominator which will make them comparable with one another. It impels him to discover some aesthetically intelligible principle in various forms, and even to envisage the possibility of some kind of abstract form in the aesthetic contemplation of which the mind would attain satisfaction.[9]

Recognising a parallel between this manner of aesthetic satisfaction and the intellectual pleasures obtained by the abstract fruits of scientific knowledge, Fry suggested that the emotional desire to perceive harmony in nature was 'so similar in art and in science that it is difficult not to suppose that they are psychologically the same', differing only in terms of medium and technique.[10] Faith in this type of epistemological parallelism would be profitably reappraised in the 1930s as Modernists sought to identify the points of intersection – methodological and theoretical – between science and artistic Modernism. While sketching out a biologistic theory of Modernist sculpture in his 1943 monograph on Henry Moore, Geoffrey Grigson explained how 'art or the forms of art' transform with the appearance of new kinds of knowledge about the natural world:

> In the eighteenth century, Stubbs painted an exquisite bunch of flowers held in a woman's hand up to one of his anatomically correct horses. Botanical classification and research in gardening helped to make flowers an empirical object under eighteenth-century eyes. An eighteenth-century physiologist investigated the way in which a sunflower follows the sun, so Blake and other writers used images about the sunflower. Humphrey Davy gave lectures on science: Coleridge went to them to 'increase his stock of metaphor'.[11]

Similarly, the distinctive references to natural form that characterised biomorphic Modernism was the result of twentieth-century biological imagery impressing itself upon the psyche of the artist. Nowhere was this phenomenon more apparent to Grigson than in the sculpture of Moore, whose interest 'in the rounded, solid shapes into which life builds itself' was precisely the result of the new age of biological discovery.[12] This chapter will provide the foundations for examining such biologistic concepts in greater detail, by studying the phenomenon of the popularisation of science in interwar Britain and how it led, in turn, to excited claims about the importance of science (especially biology) in the intellectual life of the citizen. It will then analyse the general climate of interest in science evinced by interwar Modernists as a means of theorising how the New Biology acquired significance for British sculptural Modernism.

Broadcasting nature's wonders: popular science in the years between the wars

In the introduction to her *Philosophy and the Physicists* of 1937, the philosopher Susan Stebbing professed that the book was written 'for that section of the reading public who buy in large quantities and, no doubt, devour with great earnestness the popular books written by scientists for their enlightenment'.[13] Indeed, sustained by fresh developments in physics, popular science had evolved rapidly as a field, with the publication of numerous articles, books and pamphlets – a point noted by the journal *Nature*, which observed in 1921 that almost a thousand papers had been already printed on the subject of the new physics.[14] Two books which symbolised the contemporary appetite for an accessible introduction to relativity theory were Arthur Eddington's *The Nature of the Physical World* (1928) and James Jeans' *The Mysterious Universe* (1930), both of which underwent repeated reprints and sold in the tens of thousands.[15]

Importantly, this phenomenon was not limited to expositions of modern physics, as the New Biology was similarly rendered into user-friendly and consumable formats for the general public. The publishing house Newnes was particularly active in this respect, seeking to rekindle the earlier triumph of H.G. Wells' secularist apologia *The Outline of History* (1920), by commissioning the veteran biologist J. Arthur Thomson to write a companion series on science. Eventually published as a fortnightly series, *The Outline of Science* (1921) conspicuously reflected Thomson's own interest in the life sciences, although his quasi-vitalist attitude to biology clashed somewhat confusedly with the more rationalist approach espoused by his collaborators, E.R. Lankester and Julian Huxley. Despite its conceptual inconsistencies, however, *The Outline of Science* was something of a hit on both sides of the Atlantic; and, on the basis of its success, Newnes issued a serialised version of Thomson's general survey of animal life, *New Natural History*, in 1926.[16]

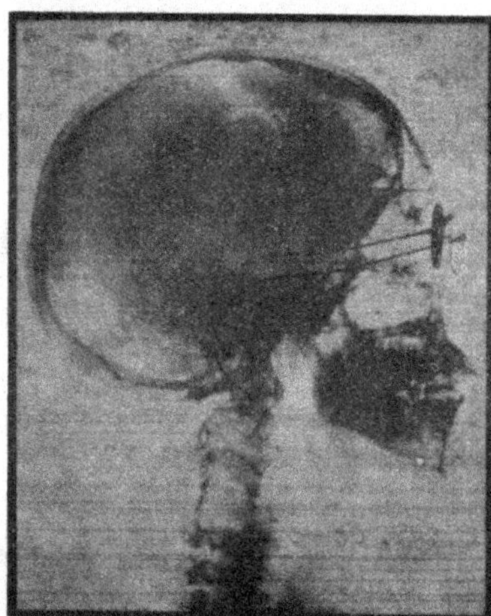
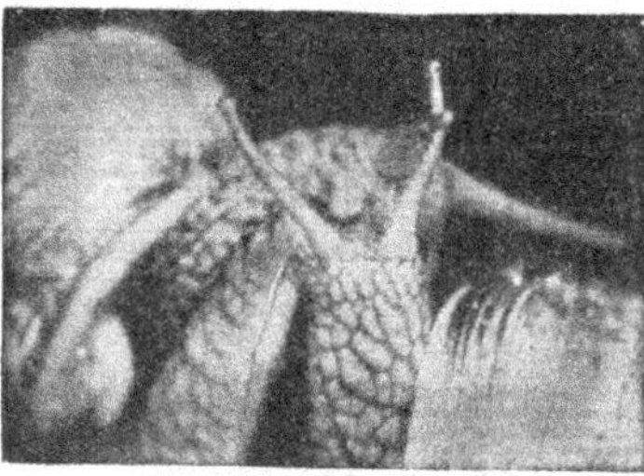

Figure 1.1

Advertisement for *The Science of Life* in *The Listener*, 18 November 1931, p. 886

For his part, Wells was disappointed by the fruits of Thomson's labours and felt compelled to broaden his own secularist way of thinking into a popular exposition of the New Biology. Working in partnership with his biologist son, G.P. Wells, Wells' pet project only really began to gather pace through the input of the evolutionary biologist Julian Huxley, who was offered a large enough advance by the publisher to make it worthwhile for him to relinquish academia in order to focus completely on the collaborative venture. Initially published in thirty-one parts between 1929 and 1930 (starting at the relatively cheap price of 1s 3d per part), the resulting *Science of Life* was generally well received by critics – despite Wells' self-professed puzzlement at its 'relative unsuccessfulness' – and actively endorsed by a public who continued to buy the text in both its serial and book format right through the following decade. In proselytising a staunchly mechanistic approach to the life sciences, Wells and his co-writers sought to steel the public against the ever-present threat of a resurgence of neo-vitalist doctrine that might otherwise derail the progress made by mechanistic science.[17]

Eye-grabbing scientific illustrations significantly assisted the interests of popularisation. Publicity for *The Science of Life*, for instance, stressed the publication's 'entrancing' nature by drawing attention to the '1015 Revealing New Science Illustrations' that generously accompanied the main text (Figure 1.1). Close-up photography was, of course, a key weapon in the armoury of those writers and their publishers who sought to educate their readers, in as engaging and entertaining a way as possible, about the underlying mechanisms of nature: a rhetorical strategy that played upon the dichotomy staged between objectivity and artistry that had been a component of scientific photography since its inception.[18] Not only had printing technology advanced sufficiently to allow for the comparatively inexpensive reproduction of close-up prints without any great loss of detail but recent technological developments (such as ultraviolet and, by 1938, electron microscopy) had greatly boosted the degree of magnification permitted by microscopes, literally bringing a hidden world into view and thus aiding those who sought to visually convey the biological marvels recently discovered by the scientific vanguard.[19] Certainly, buoyed by technological innovations and burgeoning public enthusiasm for the subject, the 1930s experienced a rapid expansion in the market for high-quality photo-albums that featured enlarged, two-tone reproductions of photomicrographic plates. Proof of this contemporary hunger for close-to images of nature was provided by the runaway success of publications such Karl Blossfeldt's *Art Forms in Nature* (which underwent no fewer than three reprints between 1928 and 1935), which analogised between the delicate structures of the nature and manmade art forms and, in so doing, contributed to the construction of an intercessional discourse between aesthetics and biology.[20]

The healthy sales figures enjoyed by popular science in the years 1920–39 stemmed, at least partly, from a genuine public excitement at the revolutionary discoveries of modern science, at 'the investigations into the fundamental units of matter, and by descriptions of vast interstellar distances'.[21] This interest was compounded by the tendency of critics to lump together the narratives of modern science and literature into one common discourse. In 1919, the science writer J.W.N. Sullivan, for instance, noted that popular-scientific books such as the editions published in the Home University Library would become 'serious rivals' to modern novels and poetry as they were 'more dramatic, [opened] up larger vistas, [and were] as well written, and [...] cheaper'.[22] Equally, in the course of undertaking a survey of modern poetry in *The Listener* in 1931, Henry Newbolt claimed that the ordinary reader was more intrigued by 'the wonders of science and its transitory universe' than by poetry, shunning metaphysics for the new physics.[23]

This co-option of territory previously occupied by literature into the remit of popular science was no doubt facilitated by the sensationalist manner in which popular-scientific texts were typically advertised. *The Science of Life*, for example, was billed as 'infinitely more thrilling than any short-lived play at the theatre or cinema' and 'stranger than any flights of fiction fancy' (see Figure 1.1). Through perusal of the text, readers were told 'You will realise your part as an actor in this gargantuan epic drama of Life' – thus granting to the reader of popular science an *active* role as opposed to the passive one adopted in fiction reading, and so, presumably, increasing the dramatic appeal of the publication.

The apparent relatedness of modern science to metaphysics assisted its popularisation. Of course, with the advent of the theory of evolution in the mid-nineteenth century, science and religion had been in conflict – and the emergence of relativity theory exacerbated this philosophical clash. Many theologians believed that relativity had substituted idealism for materialism and, as a result, had weakened the objectivity of conventional science; paradoxically, this appeared to construct a space in which the religious and the marvellous could exist. Part of this belief stemmed from the notion that as relativity questioned the basis of time and space, it was related, in some way, to spiritualism and mysticism.[24] J.W.N. Sullivan observed that scientific hypotheses like relativity had led to the creation of a new branch of literature, in which 'works [...] seem to result from a close collaboration between, say, a professor of physics and a Bond Street Crystal Gazer'.[25]

These metaphysical suppositions also found articulation in relation to the New Biology. The resurgence of vitalism in the interwar period encouraged those who wished to see life as the manifestation of divine will. Although neo-vitalism did not strictly endorse divine intervention in the creation of life, by maintaining that life was distinct and non-material, neo-vitalism explicitly rejected the claims of the materialists – who saw life as simply the result of intricate mechanical laws. Indeed, the neo-vitalist notion of an independent non-material life force opened the doors to theological speculation that the origin of life must ultimately have been supernatural[26] – an idea that was inherent to Henri Bergson's influential neo-vitalist thesis *Creative Evolution* (1907, first English edition 1911), which defined the continual evolution of living form as an expression of some active although undisclosed divine spirit.[27] Bergson's theory acted as a panacea to those Modernists who found the idea of a universe run solely according to mechanistic principles hard to stomach. Herbert Read spoke for many sympathetic to neo-vitalist dogma when he professed that Bergson had 'offered an interpretation of the universe that was neither mechanistic nor finalist [and which] provided a way out of a closed system of predestined fact'.[28]

On the one hand, these debates led to animated discussions in the popular press over the exact nature of living matter, and its ontological relation to theological concepts. Some thinkers, such as James Young Simpson, connected to the New Natural Theology movement, were keen to establish a means by which the findings of modern science could be used to justify faith in the divine. In 'Science and the Idea of God' (1933), for instance, Simpson discussed the effectiveness of science in proving the theory of evolution and in demonstrating the relatedness of inorganic substances to organic matter. Yet, at the same time, he argued that biology ultimately failed to explain the subtle interrelations between organisms and environment, and that only religious belief could communicate the full significance of life. On this basis, modern science required 'an enlarged, more refined and more exalted conception of God' as it stimulated a sense of awe and wonderment 'in the presence of forces [that the researcher] is investigating that is not far removed from worship in spirit and in truth'.[29]

The proximity of the new science to ontology meant that works on popular science typically introduced novel scientific themes through recourse to philosophical ideas. James Jeans' foreword to his hugely successful book *The Mysterious Universe* (1930), for example, dwelt at length on the philosophical implications of the new physics – quoting extensively from Plato's *Republic*, Book VII, in order to illustrate humankind's fundamental blindness to 'ultimate reality'.[30] 'There is a widespread conviction', wrote Jeans, 'that the new teachings of astronomy and physical science are destined to produce an immense change on our outlook on the universe as a whole, and on our views as to the significance of human life'. While the new science was warranted to ascertain the basic facts governing the nature of the universe, Jeans therefore concluded that any deliberation over the *meaning* of these findings should 'legitimately pass into the realms of philosophy'.[31] Likewise, Alfred North Whitehead's tome *Science and the Modern World* (1925), which launched an influential attack on scientific materialism, was reviewed in terms of its relevance to the whole of human knowledge. 'It embodies', observed Herbert Read, 'the material of a revolution of our whole concept of life or being, and seeks to reinterpret, not only the categories of science and philosophy, but even those of religion and art'.[32]

That modern science was perceived to have weighty philosophical implications meant that philosophers and cultural commentators saw its importance to society as being too great for it to stay critically unexamined. While Stebbing justified her trespass into the terrain of physics on the basis that scientific developments had 'important and far-reaching consequences for philosophy',[33] other writers were more concerned about the unchecked and rapid expansion of science and technology into all areas of human life. As Aldous Huxley (the brother of Julian) remarked in a 1932 BBC radio broadcast: 'Science is power as well as truth [...]. Science as an end in itself directly concerns only scientific workers and philosophers. As a means it concerns every member of a civilized community'.[34] Indeed – looked at through the prism of the economic woes of the 1930s – many thinkers blamed the *laissez-faire* outlook of modern science for the financial and political problems besetting interwar society. Huxley claimed that:

> The very arts and sciences which we have used to conquer Nature have turned on their creators and are now conquering us. The present economic disasters are of our own making: we have brought them on ourselves by allowing our mechanical and agricultural science to develop more rapidly than our economic science.[35]

Nevertheless, to reproach science in this way for its societal irresponsibilities was not the same as subscribing to a form of intellectual Luddism. Most scientific commentators believed that science, well employed, could aid in the restoration of a healthily functional society. Huxley, for one, quipped that '[t]he only cure for science is more science, not less. We are suffering from the effects of a little science badly applied. The remedy is a lot of science, well applied'.[36] Huxley's faith in scientific humanism found resonance in the work of John Burdon Sanderson Haldane (the son of J.S. Haldane), who, likewise, believed that modern science could act as corrective to the unsystematic character of interwar society. 'In our present society', he wrote, 'industry is not even organised so as to find a job for every able and willing citizen. But in a scientific society the attempt would be made to find the best possible jobs for everybody'.[37]

For many progressive scientists, biology was fundamental to the execution of this utopian agenda. In the context of recent embryological advances, Julian Huxley was emboldened to state that just as '[t]he last century witnessed the Industrial Revolution, brought about by the application of mechanical and chemical science to practical affairs', the twentieth century would observe 'the Biological Revolution – the application of biological science to practical affairs'.[38] While the scientific revolutions of earlier periods had transformed knowledge of inanimate matter, the chief benefit of the biological revolution was that it would have a direct effect on human beings, their habits and their nature. In this respect, Huxley was adamant that the biological revolution had already commenced: 'Applied biological and medical knowledge has already increased the average height and weight of the population. It has [also] lengthened the average life'.[39] '[A] wholesale change in our attitude towards love, sex and reproduction' was one conclusion that he drew from the advent of birth-control technologies, whereas discoveries in embryology would lead to 'the first beginning of the possibility of babies in bottles, motherhood without birth'.[40] The enormity of biology's potential impact on society led many scientific thinkers to encourage the public to consider how modern science might affect their everyday life. Huxley advised his readership that: 'It is well worth thinking about the quite remote effects of science in order to see the general trend of our evolution and think out a general philosophy to fit it'.[41] Additionally, writers on science believed that scientific progress was dependent 'not only on the progress of medical science, but on whether the average man or woman can be got to think biologically'.[42] Alongside sharing Huxley's belief in the revolutionary potential of the New Biology, J.B.S. Haldane maintained that the struggle against infectious disease and ill-health could be successful only if 'politicians thought in terms of biology as they now think in terms of economics', as only then would they 'realise that it is at least as important to keep out foreign diseases as foreign imports'.[43]

The sizable increase in the number of popular-scientific texts published in the interwar years can be directly linked to the contemporary belief that public ignorance of science was, in general, harmful to society. The capacity of science to transform society for better or for worse meant that, increasingly, leftist commentators believed that an awareness of natural science was fundamental to a humanistic education. For the philosophical grandee Bertrand Russell, the far-reaching changes that science wrought on society meant that 'the influence of science upon our thoughts, our hopes, and our habits is continually increasing and likely to increase for several centuries at least'. Outside the practical uses of scientific technique, he understood science to have a 'further consequence of which the implications are not yet fully evident, namely, that it makes possible, even necessary, new forms of human society'.[44] To inform individuals of the perils of unchecked scientific research, Lancelot Hogben intended to make science 'an essential part of the education of a citizen, because scientific discoveries affect the everyday lives of everyone'.[45] In this spirit, his primer *Science for the Citizen* was 'partly written for the large and growing number of intelligent adults who realise that the *Impact of Science on Society* is now the focus of genuinely constructive efforts. It is also written for the large and growing numbers of adolescents who realise that they will be the first victims of the new destructive powers of science misapplied'.[46] Part of a newly radicalised generation of socialist scientists, Hogben felt that only by educating people in a scientific way of thinking could they be made to grasp that this knowledge, effectively applied to the running of society, would bring about a social revolution.[47]

Though subscribing to a more humanistic and, hence, apolitical agenda, Huxley and the Wellses, like Hogben, marvelled at the ways in which biology had transformed everyday existence:

> A great and growing volume of fact about life as it goes on about us and within us becomes available for practical application. It reflects upon the conduct of our lives throughout; it throws new light upon our moral judgments; it suggests fresh methods of human cooperation; imposes novel conceptions of service, and opens new possibilities and freedoms to us.[48]

For the most part, this information was deemed remote to the average reader due to obscure technical terms and contentious material. Consequently, in 'the care of his health and the conduct of his life, the ordinary man [...] draws far less confidently upon the resources of science than he might do [and] is unavoidably ignorant of much that is established and reasonably suspicious of much that he hears'.[49] *The Science of Life* was subsequently described as an attempt at 'clearing up and simplifying [...] the science of life' to make it more intelligible to the general reader.

If, traditionally, a few specialist publishers and generalist periodicals had had the lion's share of the popular-science market, by the 1920s the development of the radio would forever transform the habits of knowledge consumption. From the very beginning, the BBC was granted a monopoly on the new technology and, at first consisting of just one station, broadcast a 'National Programme', which ranged from light music to highbrow lectures, appealing, in the process, to a surprising variety of listeners.[50] Effectively placed under the custodianship of the state, from early on the BBC adopted a paternalistic approach to its broadcasting, using the new medium as a means of both fostering a coherent sense of national identify as well as effecting a type of social control. This strategy was supported by the BBC's unveiling, in January 1929, of a weekly magazine, *The Listener*, the main aim of which was to publish the texts of its radio talks, with appropriate illustrations. Prior to 1929, science was presented on the BBC in a fairly dry manner, reminiscent of the didactic approach preferred by much self-educational literature of the time, in which an 'expert' would provide an account of the facts in as dispassionate a way as possible. Following the launch of *The Listener*, however, the organisation was able to revamp its science broadcasting by hiring the Cambridge-trained biologist Mary Adams as a talks editor. Favouring talks that highlighted the societal implications of science, she took it upon herself to build up a stable of regular broadcasters, consisting of scientists and science commentators (such as Gerald Heard). The importance of the role that the BBC played in popularising science was considerable, as, already by 1930, it was received by roughly 40 per cent of households, a portion of the market that rose to almost blanket coverage by the end of the decade.[51]

The opinion that science was just too important to the wellbeing of society to be ignored meant, of course, that the BBC increasingly took the issue of scientific popularisation seriously. In a talk of 1930, the political activist Margaret Cole bewailed that while 'on the future of science depends the question of whether the human race is to continue to exist on this planet [...] most of us are profoundly and shockingly ignorant of what science is, what it has done, is doing, and is trying to do'.[52] Claiming that specialisation obstructed general understandings of science, Cole petitioned scientists to 'popularize' their work, as 'compared with practitioners of other subjects, they do not do nearly as much as they might do for our enlightenment'. In the main, she believed that scientific books were ineffectual as beginners' guides, and that the philosophy of science needed clarification to demonstrate 'how far scientists think that science does or can go towards explaining the world as a whole, and how far and in what circumstances they believe that their conclusions are "true"'.[53]

Initially, however, the BBC's official position had been to focus on explaining the scientific method while avoiding any detailed exposition of scientific findings.[54]

For instance, an editorial of 1929, entitled 'Broadcasting and Popular Science', disparaged the public's 'thirst' for the conclusions of science in preference to its methods. 'It is not so easy', the writer explained, 'to collect a group of students to observe, demonstrate and discuss the process by which scientists reach their conclusions, as it is to find a public ready to devour textbooks in its armchair or to indulge its curiosity by taking in the "illustrated fortnightly parts" of a serial production'.[55] While commending efforts at 'popularizing' scientific conclusions, the editorial was resolute in stating that '[i]t ought to be the business of the scientist to go a step further and popularize the ways in which he acquires the information which he disseminates', as little progress had thus far been made in this field.[56] A series of forthcoming talks and articles on biology, however, aimed to rectify this inadequacy by taking 'listeners step by step through the actual process of experiment and observation'. Indeed, the non-mathematical sciences were typically favoured in this way as a more effective means of propagating the ideal of method.[57]

After rising to the relatively senior position of a talks editor, Adams was one of several prominent BBC contributors who regularly lamented the paucity of general knowledge that the public held about scientific theory. For instance, in a book review in 1930, she took the opportunity to query why the names of the founding fathers of science were 'unknown to our children leaving school' and she felt that 'too often the young student of science, for lack of better information, regards his science as the parthenogenetic progeny of laboratory apparatus'. 'Modern science', Adams concluded, 'has to a large extent lost its historical consciousness', yet given the controversies that still influenced scientific thought, an awareness of the history of science could offer valuable insights into the cultural construction of the subject as '[s]ocial and political theories hold up the mirror to scientific thought, and science, religion and reality are perpetually adjusting their boundaries'.[58] Such considerations regarding the character of scientific popularisation were naturally the subject of heated debate. The BBC's regular science correspondent, A.S. Russell, took the view that the history of science was irrelevant, as 'it [was] all wrong' and largely had no bearing on contemporary practice. Although he agreed that there were not many accessible texts for mainstream consumption, this was due to the process of specialisation and the inherent difficulty of the material. 'The books *are* too hard', he noted, 'but this is largely because those who might best popularize it are [...] engaged in their own research work'.[59] Moreover, to combat the complexity of the discipline, Russell advised that 'the reader must make a real effort, for even with the most lucid exposition of the subject it can never be very easy'. To be sure, unlike non-scientific subjects, science suffered from having 'the difficult bits removed', meaning that '[n]o popular book of science [could be] of any value unless it [was] written by men who are themselves good scientists'.[60]

Yet even if these views broadly coincided with the didactic overtones of *The Listener*, by the early 1930s such an elitist construal of popular science was increasingly being challenged. In the course of introducing a series of talks on science to be broadcast by the BBC, for instance, the science writer Gerald Heard rehearsed some of the criticisms hitherto levelled at the popularisation of science. 'Why cannot we be kept *au fait* with scientific advance', he wrote, 'as we are with the latest plays, pictures and politics?'[61] Part of the reason, Heard claimed, was that science failed to connect with the public because of the obscurity of the endeavour, even though 'the unintended consequences of pure research have now become too serious for us to remain indifferent'.[62] And in this sense, popularising lectures would represent a noticeable shift in direction for the BBC, as they recognised both the complexities of science and the need for the public to be kept up to date with the conclusions of the subject: 'These two considerations – the difficulty of the subject itself and the need for us to grasp its outline and implications have led the BBC to try these talks [as] a fortnightly review of current discovery'.[63] And as we shall see, the effective promotion of scientific ideas to the public through such agents as the popular-science industry and the BBC played no small part in garnering the interest of Modernists in the aesthetic implications of the New Biology.

A constructive partnership: art and science in the 1930s

Contemporary calls for a greater degree of interaction between society and the findings of science were echoed by Modernist critics, who demonstrated a tacit awareness of science's growing cultural influence as well as a newfound willingness to acknowledge the importance of scientific thinking to art. Taking as their *bête noir* the unsystematic ideologies that had ostensibly defined earlier forms of artistic Modernism, critics operating at the vanguard of artistic practice alleged, in the scientific spirit of the age, that objectivity and empirical criteria had newly become the watchwords of criticism. In an article on psychoanalysis and criticism that was published in 1925, at the very start of his career as a critic, Herbert Read confidently asserted that: 'We have become more empirical, and the general effect of the growth of science has been to discredit transcendental reasoning altogether'. Siding with those popularisers of science who similarly sought to dismantle the edifice of metaphysics and, in its place, erect the archly rationalist temple of modern science, Read argued that traditional criticism bore but little semblance to the everyday workings of reality and so 'to save it from becoming the province of emotional dictators [it needed to be related] to those systems of knowledge which have to a great extent replaced transcendental philosophy'.[64]

As a literary critic, Read was aware of the degree to which philosophy's obituary had already been written by those columnists sympathetic to the popular appeal of the new physics.[65] Indeed, he singled out physics, 'demanding as it does such impressive modifications of aspect and attitude', as providing 'the most general background for subsidiary efforts', although he felt that, ultimately, for 'the literary critic psychology [gained] an intimate importance because it [was] so directly concerned with the material origins of art'.[66] Though chiefly preoccupied with the utility of psychology to literary criticism (in terms of determining the origins of symbolism), Read's early forays into critical technique belied a profound dissatisfaction with what he saw as the empty rhetoric surrounding 'significant form' and thus foreshadowed his later attempts to forge a biologistic approach to art criticism. While Read's interest in modern art really began only after he started contributing articles on art to *The Listener* in 1929, throughout the 1920s he worked as a curator at the Victoria and Albert Museum, where his curiosity in discovering common links between divergent cultures and his love of categorisation piqued a latent interest in how modern science might better serve to clarify meaning in visual art.[67]

The subject of physics certainly provided a wellspring of metaphor through which modern artists and their critics, working in the highly charged creative atmosphere of the 1930s, could explain the innovatory stylistic characteristics of contemporary art. In the course of explaining to readers of *The Listener* the significance of Picasso's art, the Modernist painter Ben Nicholson openly drew analogies between the formal innovations of the Cubists and the findings of relativity physics, claiming that Cubism had 'an almost exact parallel in the world created by Einstein'.[68] Though the apparent connection between the fragmented picture-plane common to Cubist artworks and the hyper-geometries posited by relativity theory had, by 1933, been well rehearsed in Francophone art criticism, this symbolic pairing was comparatively new to English writing on art and significantly coincided with the somewhat belated emergence of artistic Modernism in Britain during the early 1930s.[69]

In other statements, most notably that which he produced a year later for the publication *Unit 1* (1934), Nicholson consistently demonstrated an interest in the new physics that overlapped markedly with the ontological concerns of the period. Famously, his *Unit 1* statement adopted the phraseology of the New Natural Theology movement by using modern physics to justify a type of quasi-religious mysticism, earnestly quoting Sir Arthur Eddington's nebulous conclusion that: 'If we are to know anything about [the nature of the universe] it must be through something like religious experience'.[70] Though Nicholson's keenness to link abstract art to metaphysics ostensibly ran counter to Read's patent distaste for transcendental reasoning, his readiness to exploit the findings of contemporary science to justify modern art similarly highlights the extent to which science had become established as a rhetorical tool through which various artistic positions could be articulated. Indeed, the neo-vitalist or pantheistic undertones of Nicholson's statement indicate the centrality of biocentric modes of thought to those British Modernists who comprised the Unit 1 group.[71]

Though psychology and physics, as potent scientific symbols of the modern age, were clearly seen as significant to contemporary art, it was ultimately the subject of biology that was to prove most instructive to art critics in their quest to decipher the logic of Modernist aesthetics. In a 1942 review of D'Arcy Wentworth Thompson's morphological magnum opus, *On Growth and Form* (1942), Read noted that, while he was unsure about the book's usefulness to philosophy, he nonetheless felt confident of its importance to the theory of art.[72] By the 1940s, Thompson's hefty tome on experimental morphology had, of course, long been an important sourcebook for avant-garde painters and sculptors working in Britain; yet Read's appraisal of the second edition was particularly significant in explaining the attractiveness of the volume to a non-specialist audience.[73]

Claiming that it 'would normally fall to a scientist – a physicist, biologist or mathematician – to review this book', Read argued that 'there are certain scientific classics, and this is one, which have such a wide significance that they must become, as it were, layman's property and be given general currency'. Following in the footsteps of those who campaigned for science's integration into the lexicon of everyday thought, Read recommended *On Growth and Form* on the strength of its humanistic sympathies and philosophical depth, factors which showed Thompson to be 'not a narrow specialist but a humanist, familiar with the speculations of philosophers and the visions of poets no less than with the multitudinous facts of natural science'. Thompson was thus 'of the same "blood and marrow" as Plato and Pythagoras, [having] something of the geniality of a Goethe or a Henri Fabre'. And just as J.W.N. Sullivan flew the flag for the intellectual thrill and literary sophistication provided by much popular science of the day, Read noted that *On Growth and Form* had 'an endless fascination merely as a collection of *faits divers*, of curiosities of nature', allowing it to be 'read for pleasure, like Buffon's *Histoire Naturelle*, long after its findings have become the unacknowledged commonplaces of science'.[74]

Although sympathetic to Read's polymathic leanings, Thompson, for his part, was somewhat bemused by the success of his work in contemporary art circles, confessing to a scant knowledge of 'modern men' and having 'got along without Picasso, easily enough'.[75] Yet Read was altogether convinced of the magnitude of art's debt to *On Growth and Form*, writing urgently to Thompson that '[such] references are inevitable – you have built the bridge between science and art. The danger lies in one's impulse to cross it too impetuously, without a proper understanding of the country on the other side'.[76] Of course, despite Thompson's original desire to construct a mathematically systematic discipline out of the wreckage of classical morphology, *On Growth and Form* itself was highly unusual in containing page after page of evocative, wordy descriptions of biological processes – a fact not lost on Read, who, adopting the attitude of interwar scientific popularisers, believed that, regardless of the difficulty of some of the mathematics involved, its literary merit alone made it 'a book which can be recommended to the general reader'.[77]

Read was especially struck by Thompson's ready use of the epithet 'beautiful', which he understood to deliberately bestow an element of aesthetic value on the forms of the physical world and, hence, show as 'never before [how] such a range of specific natural forms [...] possess the harmony and proportion we usually ascribe only to works of art'.[78] Thompson's eagerness to analogise between the forms of nature and those of engineering no doubt further encouraged Read in his assumption that *On Growth and Form* was something of a guidebook to how fundamental laws governed the production of form, irrespective of whether it be natural or artificial.[79] On this basis he was satisfied that 'efficient form in the organic world, in the inorganic world, and in the world of art, is shown to be determined by identical laws', although, as he was careful to note, that was 'not the whole story, either in biology or in art', as energy seemingly played a far more fickle role in the creation of form than had hitherto been recognised.[80]

In the same way that popular-science publications of the 1920s and 1930s often drew upon philosophical precedents so as to emphasise the ontological implications of some new theory or other in science, so Read's reception of *On Growth and Form* spotlighted how Thompson's desire to distinguish numerical harmony throughout the natural world ultimately had its roots in the philosophy of antiquity. In his autobiographical polemic *Annals of Innocence and Experience* (1940), Read spelt out this link to classical culture: '[Thompson's theory] was a question of exact measurements and demonstrable equations, and merely gave a contemporary scientific sanction to the intuitions of Pythagoras and Plato, who centuries ago had found in *number* the clue to both the nature of the universe and the definition of beauty'.[81] Though more abstract in purpose, Alfred North Whitehead's 'organic' critique of mechanism, *Science and the Modern World* of 1926, also incorporated a wide range of examples from philosophy and poetry in its primary argument and similarly earned Read's praise precisely because of its impressive conceptual and literary range.[82]

Modernist interest in the New Biology partly reflected the popular appeal of scientific photography, which contemporary commentators extolled as much for its aesthetic qualities as for its scientific merits. By the 1920s a variety of technical improvements in the qualities of lenses and photographic film, many of which originated in the German camera industry, had made the photography and appreciation of natural forms at varying levels of magnification both simpler and more wide-ranging.[83] In particular, the interwar years observed the birth of a form of art photography that was termed 'decorative micrography'. Epitomised by the shimmering photographic plates of mineral and botanical specimens made by Laure Albin Guillot, decorative micrography privileged the ornamental potential of microscopy, producing richly patterned and highly wrought photographs of magnified natural forms which critics wittily compared to the 'jewel bedecked fabric samples from the treasure chest of an eastern potentate'.[84] Whereas photomicrography had provoked aesthetic interest since 1858, when diatoms were first captured on film, technological advancements in the 1920s and 1930s – especially in the fields of X-ray crystallography, ultraviolet microscopy and electron microscopy – had brought a cornucopia of new photographic images to a public that was largely unaccustomed to nature made strange by extreme magnification or the penetrative vision of the X-ray.[85] Exhibitions such as the 1928 'Neue Wege der Photographie', held in Jena, and the 1926 'Deutsche Photographische Austellung', in Frankfurt, provided forums through which the aesthetic of the new scientific photography could be disseminated to interested audiences across Europe.[86] In Britain, the Zwemmer Galleries in London exhibited a selection of Karl Blossfeldt's close-up photographs of plant forms in 1932, at the same time as the documentary cinema movement thrilled cinema audiences with cutting-edge films of natural history.[87] Filmmakers like Percy Smith at British Instructional Films spearheaded a new age of natural history film production, bringing a variety of microcinematographic and time-lapse 'shorts' to classrooms and public cinemas throughout the British Isles.[88] Artists across Europe, including Modernist figureheads such as Fernand Léger and Henry Moore, responded enthusiastically to the imagery captured by scientific photography, populating their canvases with abstracted shapes which were loosely derived from the microscopic forms revealed by the New Biology.[89]

The integration of scientific themes and imagery into the discourse surrounding the reception of Modernist abstraction was, to an extent, a reply to the generally negative response that had greeted the emergence of a genuinely 'abstract' art, and which had seen established critics, such as the formidable Dugald Sutherland MacColl, round on its non-representational qualities as evidence of, at best, an inanely decorative bent or, at worst, an aesthetically nihilistic mind-set.[90] Certainly, it is true that, apart from the transitory, prewar Vorticist movement, artistic Modernism washed up on British shores at – by European standards at least – a relatively late date, only after a new generation of postwar critics and artists had had the opportunity to digest the formal innovations of Surrealism and Constructivism and mull over the critical implications of Marxist art theory. Throughout the early interwar years, the dominant artistic force in Britain had been the Bloomsbury Group, whose Post-Impressionistic experiments had been vigorously defended by the critic Roger Fry, whose own classical and Renaissance tenets were affronted by the onslaught of non-figurative and surreally distorted art stemming from the continent. After Fry's death in 1934, progressive critics – such as Read – reflected on the longevity of his cultural influence, decrying Fry's use of a formalist vocabulary as symptomatic of a solipsistic worldview that was grotesquely out of step with the realities of everyday life.[91]

If the introduction of scientific concepts into contemporary art criticism was partially an attempt by artistic Modernists to sidestep the interpretative shortcomings of Bloomsbury-era formalism, this endeavour was motivated by the conviction that science had somehow made the traditional role of the intelligentsia irrelevant. Commenting, in 1938, on the dwindling of intellectuals' 'priestly' status as a result of science's unstoppable advance, Bertrand Russell sardonically noted that:

> Science, in giving some real acquaintance with natural processes, has destroyed the belief in magic and therefore the respect for the intellectual [...]. The intellectuals,

finding their prestige slipping from them as a result of their own activities, become dissatisfied with the modern world. Those in whom the dissatisfaction is least take to Communism; those in whom it goes deeper shut themselves up in their ivory tower.[92]

Just as the new scientific photography seemed to justify non-figurative artworks by blurring the frontier between representation and abstraction, so Modernists, dissatisfied with the poverty of formalist jargon – which Herbert Read saw as an outmoded by-product of a division between ethics and beauty – and under pressure to validate their role as arbiters of taste, drew on the findings of science to legitimise the appearance of artistic Modernism.[93]

In the same year that the Unit 1 group produced its catalogue, *Unit 1* (1934), as a manifesto for the modern movement in English art, design and architecture, Hugh Gordon Porteus delivered a BBC radio talk entitled 'The Painter Speaks', in which he forwarded the idea that changes in scientific know-how had a penetrating impact upon the social function of the artist. Although public enthusiasm for modern art had grown steadily over the past few years – a fact reflected by the increasing number of exhibitions of contemporary work in London and by the spate of *Listener* articles on the subject by Paul Nash and Herbert Read[94] – hostility towards artistic Modernism was still ubiquitous and Porteus' lecture aimed to redress the balance by justifying the establishment of Unit 1 on the grounds of historical precedent and shifts in the cultural economy.[95] Claiming that the 'further ahead of his time an artist is, the more attention he is likely to deserve – and the less likely he is to get it', Porteus argued that 'probably the gap between the painter and the public has never been so great'.[96] Interestingly, Porteus saw the origins of this disparity in the unparalleled level of scientific and technological change that occurred at the latter part of the nineteenth century. For, at that moment, the 'scientist came along with new theories of light and new pigments for the painter to play with [while] the photographer robbed him of his traditional function – the camera could record appearances more accurately as well as more cheaply – and gave him reproductions of the art of the past that were at once an inspiration and a disenchantment'. Faced by these innovations, which obliged any artists worth their salt to urgently rethink their cultural role, any 'painter who is aware of all the changes in the modern world, and their implications, has practically no intelligent alternative but to go back to his studio and invent a sort of visual chamber music'.[97]

Of course, by the mid-1930s, the influence of photography on modern painting had become pretty well established. Yet what is striking is just how far science and technology had become the yardstick by which artistic Modernism was measured.[98] For example, in the first edition of the arts journal *Axis*, its editor, Myfanwy Evans, boldly contended that Cubism had come about precisely through 'the object [receding] further and further into the camera'. As machinery had thus rendered naturalism into a superfluous pursuit for the contemporary artist, avant-garde practitioners were necessarily drawn to increasingly non-figurative stylistic strategies: 'The object went because it was of no further use – but with it went also a restraint which had kept painting within the scope of its own tradition'. Quite clearly thinking about Surrealism and Constructivist abstraction, Evans insisted that the elimination of the object had been 'a great opportunity for painters not primarily interested in painting'.[99]

Also writing in *Axis* – which had become something of a nucleus around which vanguard artworks were keenly debated – the painter Jean Hélion toed the Modernist party line in averring that mechanical technology had undermined the integrity of the arts by bankrupting naturalism. Despite being French, Hélion was an important figure in the British arts scene, enjoying the friendship of many of its leading artists and providing Evans with the conceptual stimulus to set up *Axis* as an English-language counterpart to the hugely successful Paris-based journal *abstraction-création*.[100] Hélion's exposure to the biocentric flows of European Modernism meant that he customarily used the New Biology as a paradigm through which to hypothesise the principles behind Modernist abstraction and was an important figure in the development of a biologistic sculptural discourse in Britain.[101] 'When painting after the appearance of nature', he wrote in a keynote essay of 1935, 'respecting its order, its obligations, any artist could be sure of being in contact with a thoroughly organised field [...]. He had a tradition well established, that he enriched and transmitted'.[102] Yet in light of the change that had swept in new techniques of photographic reproduction and therefore enfeebled the painter's monopoly on representation: 'We have been led out of this contact and this security. How can an artist of today not following the controllable-in-all-points-appearance of nature, be sure of the fertility of his ground?'[103]

Even as he recognised technology's role in discrediting naturalism, Hélion insisted that by inspiring pioneering stylistic tendencies, science could have a constructive influence on art. His understanding was that, by taking an empirical approach to art making, Modernism had provided art with 'a precise meaning and function' that – in the spirit of self-enlightened analysis – had emphasised 'the connections with science, the structure of events, even calculations, geometry, theories of colour and light'.[104] Though Hélion was here voicing a grievance re-garding Modernism's perceived ultra-rationality, he was also analogising between the reductive appearance of modern art and the diagrammatic nature of scientific imagery – in the same way that Nicholson had associated Cubism with relativity physics. This analogy did not imply that artistic Modernism unthinkingly copied scientific imagery but rather that contemporary art had become analytical, like modern science. Form and meaning, Hélion suggested, were inextricably bound together in modern art in the same way that the import of a scientific illustration could not be disentangled from its outward appearance. Indeed, this impression of artistic Modernism becoming more empirical as a response to technological modernity was a perspective shared by Hélion's protégé, Myfanwy Evans, who similarly spoke of Cubism's analytical zeal signifying 'an exhaustive analysis of shapes which, though apparently extreme simplification, was really an elaborate observance of minute differences'.[105]

The subject of artistic Modernism's relationship to science was comprehensively discussed in the book *Circle: International Survey of Constructive Art*. Published to overlap with the exhibition 'Constructive Art' – held at the London Gallery in July 1937 – *Circle* attempted to reconcile the various strands of European abstraction within one coherent international movement. The brainchild of Nicholson, the architect Leslie Martin and the émigré Russian sculptor Naum Gabo, *Circle* aimed to present a consistent theory and practice of artistic Modernism to the British public.[106] While not, in any traditional sense, a manifesto, it nevertheless sought to provide something akin to an internationally unified ideology by bringing the British avant-garde into alliance with the considerable diaspora of European artists who were sheltering from Nazi persecution in Britain at this time.[107] Despite certain conceptual inconsistencies, *Circle*'s genuine commitment to finding a means of social expression for Modernist abstraction allowed several of its contributors to speculate about the rapport between art and science, two subject domains which, it was felt, most clearly impacted upon society and shared a common creative impulse. By marrying a pre-existing British interest in the role science played in stylistic innovation with a European concern for the productive part science performed in generating architectural and industrial design, *Circle* hence represented a high point in the British avant-garde's engagement with modern science.[108]

Circle's introductory essay, 'The Constructive Idea in Art', was penned by Gabo and set the tone for the book's progressive agenda by suggesting that artistic Modernism's social role should involve publicly communicating aspects of the new scientific reality.[109] Gabo had emigrated to England in March 1936 and it is likely that it was during the opening of the 'Abstract and Concrete' exhibition – held in London that same year – that he met Nicholson and Martin and so assisted in formulating the blueprint of what would become *Circle*. At this point Gabo was already recognised as a European sculptor of some repute and his arrival in London served to galvanise the avant-garde community that was settling around the city borough of Hampstead. The fact that just months after arriving in England he had the essay 'Constructive Art' printed in *The Listener* indicates that he had swiftly taken up a prominent role within the London art scene and had formed enough influential contacts to get published in a magazine that was, to all intents and purposes, the acknowledged voice of the establishment.[110]

Gabo had originally trained in the physical sciences and throughout his artistic career maintained a lively intellectual interest in science, drawing upon its concepts and imagery as inspiration for his sculpture and impassioned belief in the responsibility that art and science shared in fashioning modernity.[111] Yet for all his professional engagement with science, aside from brusque treatment in his *The Realistic Manifesto* of 1920 and a 1930 lecture entitled 'Rational and Irrational', 'The Constructive Idea in Art' was Gabo's first major foray into the territory of art and science and, in examining how both disciplines shared similar epistemological trajectories in the history of ideas, went beyond any other contemporary piece of writing in English in arguing for a common ground between the spirit of artistic creativity and scientific discovery.

Noting that something akin to a Hegelian *Zeitgeist* caused 'artistic and scientific activity [to move] at the same time and in the same direction', Gabo drew upon a range of historical precedents to make his case, contending that in the Renaissance 'Copernicus' scientific theory of the world [was] coincident with Raphael's concept in Art'.[112] Elaborating upon this example, he alleged that the blasphemous realism of Raphael's religious paintings was made possible only owing to his belonging 'to the generation which was prepared to abandon the [geocentric] theory of the universe'. Just as Raphael's naturalism repudiated spiritual mysticism, so Copernicus' cosmological theory debunked those numinous medieval theories that positioned the Earth as the divine epicentre of the universe. Using this example as historical support for his contention that the twentieth century had appeared under 'the sign of revolutions and disintegration', Gabo claimed that the modern equivalent to the birth of astronomy and naturalism had been Cubism's fragmentation of the picture-plane, which he likened to that 'which occurred in the world of physics when the new Relativity theory destroyed the borderlines between Matter and Energy, between Space and Time, between the mystery of the world of the atom and the consistent miracle of our galaxy'.[113]

Having experienced the early days of the Soviet project at first hand and having been caught up in the radically left-wing political milieu of the Russian avant-garde, Gabo used 'The Constructive Idea in Art' to present the argument that a common impulse towards reconstruction in art and science had a necessarily ideological aspect.[114] Claiming that the Great War was 'only a natural consequence of a disintegration which started long ago in the depths of the previous civilization', he couched his essay in the apocalyptic language that had become a mainstay of the Depression-era cultural climate, so as to make the utopian vision at the heart of his message appear all the more striking.[115] No revolution, regardless of how far-reaching, could be permanent, he noted, maintaining that 'Constructive' artists remained optimistic, as they saw 'that in the realm of ideas we are now entering on the period of reconstruction', a process of rebuilding that was already being undertaken 'in those two domains of our culture where the revolution has been the most thorough, namely, in Science and in Art'.[116] Indeed, Gabo was keen to proffer the word 'Constructive' as a positive alternative to 'abstract', 'concrete' and even 'Constructivist', which were terms he felt had been foisted onto artists by rapacious critics.[117] While Modernist critics worried over making sense of the superabundance of avant-garde movements during the 1930s, Gabo reminded readers of *The Listener* – in a rehearsal of his *Circle* essay – that 'the word "Constructivism", as most of the other "isms", was invented not by artists themselves, but was given to their work by critics and theorists'.[118]

Also contributing to *Circle*'s debate on art and science was Desmond Bernal, who was, professionally, perhaps better placed than most to comment upon the relationship between art and science. A physicist by occupation, Bernal had been introduced to the sculptor Barbara Hepworth by his wife Margaret Gardiner and, having struck up a lively friendship, was persuaded to write a preface to a 1937 catalogue of her work, as well as to contribute a chapter on 'Art and the Scientist' to *Circle*.[119] Hepworth's appreciation of Bernal's criticism is evinced in a letter to the physicist of 1937 in which she stated that:

> You always seem to be searching for, discovering [and] applying basic laws and principles [...]. Your criticism is most stimulating because you know exactly where a law has been broken [and] can apply the principle so that you can make the solution clear.[120]

Specialising in crystallography, Bernal was attuned to the three-dimensional qualities of Modernist abstraction.[121] Indeed, later in life Hepworth would speak of how Bernal 'had the amazing capacity for comprehending in an instant the nature (and even the formula) of every sculpture, and hours were spent in exciting discussion and drawing'.[122] In addition, Bernal's well documented Marxist political sympathies allowed him to identify with *Circle*'s broader ideological goal of augmenting art's social role. Like his scientific colleagues, Bernal was worried by the mass unemployment triggered by the Great Depression and by the nefarious uses to which science was being applied by European totalitarian regimes. The failure of democratic governments to properly rein in on labour inequities and the worrying distortions of scientific knowledge that fascistic government engendered were seen as ill-omened signs by Bernal that European capitalism was entering into a terminal, though still patently dangerous, state.[123] Thinking about the long-standing causes of this disagreeable state of affairs in relation to art, he observed that:

> One of the features of the civilization out of which we are now passing was its rigid separation of human functions into different spheres. Every man tended to have a job, to be a specialist in something. The great branches of human culture seemed to move further and further apart. In particular, art and science became two entirely separate spheres which did not even touch at any point.[124]

Importantly, Bernal felt that this fractious situation had not always been present. Like his *Circle* colleague Gabo, he recognised in the Renaissance a tradition of interdisciplinary cooperation that had birthed both empiricism in philosophy and engineering in applied science:

> In the great creative periods of science the artists and the scientists worked very closely together and were in many cases the same people. It was to the interest in the visual arts that we owe the birth of accurate observation of nature. It was the problems of architecture that gave rise to the science of mechanics. Leonardo da Vinci, though the greatest, was only typical of whole schools of artist-scientists.[125]

It was the development of the bourgeoisie that caused the Renaissance system of interdisciplinary interchange to collapse as, by separating functional design from pure ornamentation, it had 'used [science] to make money [while] art [was] simply [...] a means of spending it. The result of this separation has been the most incredible mutual ignorance. The scientist totally ignores art [while] the artist works as if science had never existed'.[126] But in spite of this partition, Bernal was confident that the development of artistic Modernism was 'a recapitulation of the progress of mathematical and physical thought over the period from the Renaissance to the present day'. Seeing in Modernist abstraction a rejection of the materialism that had antagonised the gap between the two cultures, Bernal averred that artistic Modernists had tentatively recognised how scientific developments transformed artistic perceptions. The geometrical bias of much Modernist sculpture was the clearest example of this. Though early art was less intricate than the 'hyper-geometries of the modern mathematician', it was nevertheless true that the modern 'artist [had] discovered by intuition and practice many of the stages of this geometrical development. Take for example symmetry. Classical art knew only the simplest bilateral symmetry. Modern art, on the other hand, while ostensibly rejecting symmetry altogether is effectively reintroducing it in more complex forms'.[127] Hepworth's sculptures of geometric solids provided him with the germ of this idea by seeming to echo the forms described by modern mathematics.[128] Similarly, *Balance on Two Points* by Gabo led Bernal to speculate how many 'of the more abstract artists have produced intuitively many of these complex [symmetrical and three-dimensional] rhythms'.[129] Bernal became especially friendly with Gabo after the Russian moved to London and it seems likely that his belief in the possibility of an art that wedded Modernism to science – in ways that moved beyond the 'hackneyed forms' of Socialist Realism – was in some way beholden to his exposure to Gabo's Constructivist ethos.[130]

Interestingly, it was Surrealism that Bernal felt most dexterously incorporated aspects of the New Biology into its oeuvre. Though arguing that its principal point of scientific reference was psychology, he observed that 'Surrealism also significantly draws on a field of content new to art, that of biology'.[131] This judgement was underwritten by Bernal's perception that Surrealism relied upon a sort of psychoanalytically inflected realism which purposefully disfigured the appearance of living things. Nevertheless, in downplaying the importance of the New Biology to Modernist abstraction, Bernal conspicuously overlooked the importance of biologism to many of the leading Constructivist and Bauhaus theorists, artists and architects who had taken up residence in Britain and participated in such radically cooperative projects as the 1934 Penguin Pool at London Zoo – which purposefully brought together architecture, the New Biology and Bauhaus theory.[132] All the same – in positing science as a paradigm for artistic Modernism and questioning how both the artist and the scientist could be socially progressive within a capitalist society which persistently appropriated their efforts[133] – Bernal theorised the parallels between art and science in ways that encapsulated the socio-political conditions in which the New Biology and Modernist sculpture were understood to interact.

Circle's committed defence of Constructivist values represented something of a U-turn in critical approaches to Modernist abstraction which had begun in 1934–35, when the terms of reference for modern art in Britain had started to transform quite drastically. Charles Harrison has proposed that this attitudinal shift was partly down to the increasingly desperate political situation in Europe, which had impressed upon the minds of Modernists the necessity of creating a progressive change in modern art that extended beyond mere technical development.[134] Certainly, by the mid-1930s the term 'revolutionary' had started to acquire a positive value which Modernist critics exploited to support a variety of different, although predominantly left-of-centre, ideological positions.[135] Avant-garde publications such as Betty Rea's politically combative volume *5 on Revolutionary Art* (1935) and *Circle* added to this debate by weighing the contribution that Modernist abstraction could make to society in spite of its apparent esotericism, which many Marxist critics – such as A.L. Lloyd – trenchantly frowned upon as symptomatic of a reactionary bourgeois elitism.[136]

Sympathising with the Marxist sentiments that distinguished avant-garde activity in the mid-1930s, Read consistently felt obliged to defend artistic Modernism's 'revolutionary' principles, claiming – like his colleagues in *Circle* – that it served a functional purpose in a classist society that urgently required rebuilding. Though his conception of revolutionary precepts was somewhat exclusive in privileging the role of the individual artist in fomenting purposive change in society,[137] he was nonetheless consistent in claiming that Modernist abstraction made use of 'certain proportions and rhythms which are inherent in the structure of the universe' and that abstract art thus furnished architecture and industrial design with 'a heightened sensibility to the purity of form'.[138] Fostering a biologistic aesthetic – as we will see in Chapter 4 – Read's interwar criticism extended *Circle*'s scientistic remit by employing morphology to justify the political radicalism of Modernist abstraction.[139]

A late convert to Modernist abstraction, Read only really began championing a biologistic approach to contemporary art in the 1934 Unit 1 manifesto, in which the incipient biocentrism of the British group was couched in terms that would pave the way for a meaningful encounter between Modernist sculpture and the New Biology.[140] Of the majority of Unit 1 members, he wrote, '[though] never for a moment compromising with the discredited notion that art is in any sense a reproduction or simulacrum of nature, they feel that art must at least bear some organic relation to the forms or types of nature'.[141]

Read's first, tentative attempt at theorising the biologistic character of artistic Modernism was laid out in his 1933 introduction to modern art, *Art Now*. Revisiting a long-standing scientistic interest – first expressed in 'Psychoanalysis and the Critic' (1925) – here he proposed a theory of art history which split artistic production into two divergent methodologies. Historically, he asserted, art was divided: on the one hand, by empiricism and, on the other, by a particularly analytical type of scientific attitude. When partaking of an empirical attitude, art was subject to a purely *imitative* impulse which had 'no other aim beyond the reproduction of appearances'. Thus, in its unthinking replication of the external aspect of physical objects, Read negatively likened this approach to philosophical reductionism: 'it is a mechanistic theory of art'.[142] Contrastingly, a scientistic attitude eschewed mere illusionism in favour of a deeper, *analytical* understanding of nature: the artist 'realises that the outward appearance of objects depends on their inner structure: he becomes a geologist, to study the formation of rocks; an anatomist, to study the play of muscles, and the framework of bones'.[143]

While *Art Now* traced these antagonistic attitudes as far back as the Renaissance, Read's praiseful description of aesthetic scientism anticipated the biologism of his later theorisations of Modernist activity – as in his contribution to *5 on Revolutionary Art* and his 1937 monograph *Art and Society* – in which the contemporary artist was seen, in psychobiological terms, to abstractly represent the patterns and forms inherent to nature and organic growth.[144] Resultantly, Read's hypotheses crystallised a variety of biologistic ideas – also current in Constructivist and Surrealist circles – about the Modernist project which abetted the definition of the epistemological parameters in which Modernist sculpture and the New Biology could be seen to profitably interact. In the following chapters, we will see how the New Biology, popular science, aesthetic scientism, neo-romanticism and political ideology collectively created the conditions under which a meaningful encounter between Modernist sculpture and biology could be staged.

Notes

1 See C.A. Jones & P. Galison, 'Introduction: Picturing Science, Producing Art', in C. Jones & P. Galison (eds), *Picturing Science, Producing Art* (New York: Routledge, 1998), p. 2; M. Whitworth, *Einstein's Wake: Relativity, Metaphor and Modernist Literature* (Oxford: Oxford University Press, 2001), pp. 16–17.

2 C.P. Snow, *The Two Cultures* (1959) (Cambridge: Cambridge University Press, 1998), p. 16.

3 Jones & Galison, 'Introduction', p. 3.

4 A. Graciano, 'Introduction', in A. Graciano (ed.), *Visualizing the Unseen, Imagining the Unknown, Perfecting the Natural: Art and Science in the 18th and 19th Centuries* (Newcastle: Cambridge Scholars, 2008), p. xv.

5 Georges Seurat's interest in the chromatic theories of Ogden Rood, Charles Henry and Hermann von Helmholtz is perhaps one of the best-developed aspects of this story. See J. Gage, *Colour and Meaning: Art, Science and Symbolism* (London: Thames & Hudson, 1999), 209–27.

6 Jones & Galison, 'Introduction', p. 3.

7 T.S. Eliot, 'The Idea of a Literary Review', in *The New Criterion*, Vol. IV, No. 1, January 1926, p. 4.

8 Beckett, 'Circle', p. 18.

9 R. Fry, 'Art and Science' (1919), in R. Fry, *Vision and Design* (Harmondsworth; Penguin Books, 1961), pp. 70–1.

10 *Ibid.*, p. 73.

11 Grigson, *Henry Moore*, p. 8.

12 *Ibid.*, p. 7.

13 S.L. Stebbing, *Philosophy and the Physicists* (London: Methuen., 1937), p. ix.

14 M. Whitworth, 'The Clothbound Universe: Popular Physics Books, 1919–39', in *Publishing History*, Vol. XL, 1996, p. 53.

15 *Ibid.*, pp. 67, 71.

16 See Bowler, *Science for All*, pp. 104–5.

17 *Ibid.*, pp. 105–7.

18 L. Daston & P. Galison, *Objectivity* (New York: Zone Books, 2010), pp. 115–90.

19 See E. Juler, 'The Key to a Hidden World: Photomicrography and Close-Up Nature Photography in Interwar Britain', in *History of Photography*, Vol. XXXVI, No. 1, February 2012, pp. 87–98.

20 For more information on the print-runs of *Art Forms in Nature*, see N. Halliday, *More Than a Bookshop: Zwemmer's and Art in the Twentieth Century* (London: Philip Wilson, 1991), p. 76.

21 Whitworth, 'The Clothbound Universe', p. 55.

22 J.W.N. Sullivan, 'Science and Literature', in *Athenaeum*, 13 June 1919, p. 464.

23 H. Newbolt, *The Listener*, Vol. V, No. 114, Supplement No. 12 (English Poetry Today), 18 March 1931, p. viii.

24 Whitworth, 'The Clothbound Universe', p. 56.

25 J.W.N. Sullivan, 'Science: The Entente Cordial', in *Athenaeum*, 9 April 1920, p. 482.

26 See Bowler, *Reconciling Science and Religion*, pp. 160–1.

27 H. Bergson, *Creative Evolution* (1911) (New York: Dover, 1998), pp. 248–9.

28 H. Read, 'The Contrary Experience: Autobiographies' (1963), quoted in Woodcock, *Herbert Read*, p. 195.

29 J.Y. Simpson, 'Science and the Idea of God', in *The Listener*, Vol. IX, No. 217, 8 March 1933, p. 378.

30 J. Jeans, *The Mysterious Universe* (1930) (Cambridge: Cambridge University Press, 1931), unpaginated foreword and p. 111.

31 *Ibid.*, unpaginated foreword.

32 H. Read, 'Books of the Quarter: Review of *Science and the Modern World*', in *The New Criterion*, Vol. IV, No. 3, June 1926, p. 581.

33 Stebbing, *Philosophy and the Physicists*, p. ix.

34 A. Huxley, 'Science – The Double-Edged Tool', in *The Listener*, Vol. VII, No. 158, 20 January 1932, p. 77.

35 *Ibid.*, p. 77.

36 *Ibid.*, p. 78.

37 J.B.S. Haldane, 'Health Before Wealth', in *The Listener*, Vol. VII, No. 161, 10 February 1932, p. 192.

38 Huxley, 'Tissue Culture and Human Habits', p. 953.

39 *Ibid.*, p. 953.

40 *Ibid.*, p. 953.

41 *Ibid.*, p. 953.

42 Haldane, 'Health Before Wealth', p. 190.

43 *Ibid.*, p. 190.

44 B. Russell, *The Scientific Outlook* (London: George Allen & Unwin, 1931), pp. 10, 11.

45 L. Hogben, *Science for the Citizen* (London: George Allen & Unwin, 1943), p. 11.

46 *Ibid.*, p. 11. Original emphasis.

47 Bowler, *Science for All*, p. 107.

48 J. Huxley, H.G. Wells & G.P. Wells, *The Science of Life* (London: Cassell, 1938), p. xxii.

49 *Ibid.*, p. xxii.

50 See S. Collini, *Absent Minds: Intellectuals in Britain* (Oxford: Oxford University Press, 2006), p. 113.

51 Bowler, *Science for All*, esp. pp. 209–11.

52 Margaret Cole, 'A Petition to Scientists', in *The Listener*, Vol. III, No. 64, 2 April 1930, p. 602.

53 *Ibid.*, p. 602.

54 See Whitworth, *Einstein's Wake*, pp. 55–6.

55 Anon., 'Editorial – Broadcasting and Popular Science', in *The Listener*, Vol. I, No. 18, 15 May 1929, p. 676.

56 *Ibid.*, p. 676.

57 Whitworth, *Einstein's Wake*, p. 56.

58 Mary Adams, 'Perspective in Science', in *The Listener*, Vol. III, No. 52, 8 January 1930, p. 64.

59 A.S. Russell, 'Science and Its Popularisation', in *The Listener*, Vol. III, No. 67, 23 April 1930, p. 731.

60 *Ibid.*, p. 731.

61 Gerald Heard, 'Humanizing Science', in *The Listener*, Vol. III, No. 69, 7 May 1930, p. 807.

62 *Ibid.*, p. 808.

63 *Ibid.*, p. 808.

64 H. Read, 'Psychoanalysis and the Critic', in *The Criterion*, Vol. III, No. 10, January 1925, p. 214.

65 See Whitworth, 'The Clothbound Universe', pp. 55–6.

66 Read, 'Psychoanalysis and the Critic', pp. 214–15.

67 On Read's early career and interests see A. Causey, 'Herbert Read and Contemporary Art', in D. Goodway (ed.), *Herbert Read Reassessed* (Liverpool: Liverpool University Press, 1998), pp. 123–6.

68 B. Nicholson, 'Points from Letters: The Art of Picasso', in *The Listener*, Vol. X, No. 238, 2 August 1933, p. 174.

69 On Cubism and relativity physics see L. D. Henderson, 'Editor's Introduction: I. Writing Modern Art and Science – An Overview; II. Cubism, Futurism and Ether Physics in the Early Twentieth Century', in *Science in Context*, Vol. XVII, No. 4, 2004, pp. 455–7; on the initial reception of abstract art in Britain see Causey, 'Herbert Read and Contemporary Art', pp. 127–9.

70 Sir Arthur Eddington quoted in B. Nicholson, Untitled statement, in H. Read (ed.), *Unit 1: The Modern Movement in English Architecture, Painting and Sculpture* (London: Cassell, 1934), p. 89.

71 Botar, *Prolegomena to the Study of Biomorphic Modernism*, p. 47.

72 H. Read, 'The Universal Harmony', in *The Listener*, Vol. XXVIII, No. 708, 6 August 1942, p. 187.

73 For more information on Thompson's importance to interwar artists see: Harrison, *English Art and Modernism*, p. 282; Mundy, *Biomorphism*, pp. 130–1; C. Lichtenstern, *Henry Moore: Work – Theory – Impact* (London: Royal Academy of Arts, 2008), pp. 57, 59; M. Hammer & C. Lodder, *Constructing Modernity: The Art and Career of Naum Gabo* (New Haven: Yale University Press, 2000), pp. 385–7.

74 Read, 'The Universal Harmony', p. 187.

75 Letter to Herbert Read from D'Arcy Wentworth Thompson, dated 28 July 1946, University of St Andrews, D'Arcy Wentworth Thompson Archive, MS 45690. I am grateful to Matthew Jarron for drawing my attention to Thompson's correspondence with Read.

76 Letter to D'Arcy Wentworth Thompson from Herbert Read, undated (1942?), University of St Andrews, D'Arcy Wentworth Thompson Archive, MS 25480.

77 Read, 'The Universal Harmony', p. 187.

78 *Ibid.*, p. 187.

79 See Herbert Read, *Annals of Innocence and Experience* (1940) (London: Faber & Faber, 1946), pp. 226–7.

80 Read, 'The Universal Harmony', p. 187.

81 Read, *Annals of Innocence and Experience*, pp. 202–3.

82 H. Read, 'Books of the Quarter: Review of *Science and the Modern World*', p. 581.

83 I. Jeffrey, *Photography: A Concise History* (London: Thames & Hudson, 1981), p. 116.

84 R. Brielle, 'Laure Albin Guillot ou la Science Féerique', in *Art et Décoration: Revue Mensuelle d'Art Moderne*, Vol. LX, July–December 1931 (Paris), p. 165, quoted in Thomas, 'The Search for Pattern', p. 106.

85 Thomas, 'The Search for Pattern', pp. 105–6; Juler, 'The Key to a Hidden World', pp. 88–90.

86 See: Mundy, *Biomorphism*, p. 113; Botar, *Prolegomena to the Study of Biomorphic Modernism*, p. 509.

87 On the Zwemmer Galleries and the exhibition of Blossfeldt's photographs in Britain, see Halliday, *More Than a Bookshop*, pp. 101–3.

88 See T. Boon, *Films of Fact: A History of Documentary Films and Television* (London: Wallflower Press, 2008), p. 29.

89 For an overview to this topic see J. Mundy, 'Form and Creation: The Impact of the Biological Sciences on Modern Art', in *Creation: Modern Art and Nature* (Edinburgh: Scottish National Gallery of Modern Art, 1984), pp. 16–23.

90 MacColl made his name as a progressive critic in the first couple of decades of the twentieth century. By the 1930s, however, he had become something akin to the voice of the establishment on the BBC, speaking of the ills of abstract art and producing a series of erudite, though hostile,

talks on the impenetrability of the modern movement. The most interesting of these, on the fallacy of the Unit 1 group and the impossibility of transplanting abstraction onto British soil, elicited a vociferous response from Read, who railed against MacColl's 'atrophied' senses. See D.S. MacColl, 'Visual and Vocal Art', in *The Listener*, Vol. XI, No. 278, 9 May 1934, pp. 799–800; Herbert Read, 'Mr MacColl on Abstract Art', in *The Listener*, 16 May 1934, pp. 844–5.

91 See Causey, 'Formalism and the Figurative Tradition in British Painting, pp. 20–2.

92 B. Russell, *Power: A New Social Analysis* (London: Allen & Unwin, 1938), pp. 44–6. On the role of intellectuals in interwar Britain, see Collini, *Absent Minds* esp. pp. 110–33; for an insightful discussion of Russell's quote, see pp. 122–3.

93 On the aesthetic insights offered by photomicrographic imagery to artists see Mundy, 'Form and Creation', pp. 17–18; and on the philosophical grounds for Read's distaste for formalism, see Causey, 'Herbert Read and Contemporary Art', esp. p. 125.

94 See Harrison, *English Art and Modernism*, esp. pp. 232–4.

95 As evinced by MacColl's denunciation of Unit 1 on the grounds of historical precedent and shifts in the cultural economy. See MacColl, 'Visual and Vocal Art', pp. 799–800.

96 Hugh Gordon Porteus, 'The Painter Speaks', in *The Listener*, Vol. XI, No. 273, 4 April 1934, p. 584. It is perhaps notable that Porteus here uses terms strikingly similar to those adopted by campaigners for the reintegration of science into society.

97 *Ibid.*, p. 584.

98 Reginald Wilenski was a key exponent of the role photography had played in transforming artistic values. In *The Modern Movement in Art* he argued that the camera brought about two varieties of 'degraded' naturalism in the second half of the nineteenth century. Wilenski, for example, saw the 'silvery' quality of light that infuses Corot's landscapes as mirroring the metallic sheen present in early photographs. R.H. Wilenski, *The Modern Movement in Art* (London: Faber & Gwyer, 1927), esp. pp. 88–112.

99 M. Evans, 'Dead or Alive', in *Axis*, No. 1, January 1935, p. 3.

100 Hélion set out the potential aims for *Axis* in a letter to Ben Nicholson of 1934: Letter from Jean Hélion to Ben Nicholson, 17 September 1934, Tate Gallery Archive, TGA:8717.1.2.1568.

101 See Maldonado, *Le cercle et l'amibe*, pp. 140–1.

102 J. Hélion, 'From Reduction to Growth', in *Axis*, No. 2, April 1935, p. 19.

103 *Ibid.*, p. 19.

104 *Ibid.*, p. 21.

105 Evans, 'Dead or Alive', p. 3.

106 See J.L. Martin, Ben. Nicholson & N. Gabo (eds), *Circle: International Survey of Constructive Art* (London: Faber & Faber, 1937)

107 See Beckett, 'Circle', p. 11.

108 *Ibid.*, p. 18.

109 M. Hammer & C. Lodder, 'Hepworth and Gabo: A Constructive Dialogue', in D. Thistlewood (ed.), *Barbara Hepworth Reconsidered* (Liverpool: Liverpool University Press, 1996), p. 113. See also N. Gabo, 'The Constructive Idea in Art', in J.L. Martin, B. Nicholson & N. Gabo (eds), *Circle* (London: Faber & Faber 1937), pp. 1–10.

110 Although 'Constructive Art' elaborated many of the themes that Gabo would return to in his *Circle* essay (among them the nature of the 'Constructive' idea itself and the particularity of the early-twentieth-century artistic revolution) on the whole it avoided much discussion of art's correspondence with science and, as such, it was left to Gabo's 1937 paper to lay down the major groundwork for this topic. See N. Gabo, 'Constructive Art', in *The Listener*, Vol. XVI, No. 408, 4 November 1936, pp. 846–8. On the general background to the publication of this essay see M. Hammer & C. Lodder (eds), *Gabo on Gabo* (Forest Row: Artists Bookworks, 2000), p. 87. Given that Gabo's English was quite poor when he first arrived in England, the clarity of language and structure evinced in 'Constructive Art' suggests the assistance of an editor (quite possibly Read).

111 On Gabo's intellectual interest in modern science see Hammer & Lodder, *Constructing Modernity*, pp. 379–401.

112 N. Gabo, 'The Constructive Idea in Art', in J.L. Martin, B. Nicholson & N. Gabo (eds), *Circle: International Survey of Constructive Art* (London: Faber & Faber, 1937), p. 3.

113 *Ibid.*, pp. 4–5.

114 See Hammer & Lodder, *Constructing Modernity*, p. 381.

115 On the pessimism of the interwar years, see R. Overy, *The Morbid Age: Britain Between the Wars* (London: Allen Lane, 2009).

116 Gabo, 'The Constructive Idea in Art', p. 1.

117 Beckett, 'Circle', p. 18.

118 Gabo, 'Constructive Art', p. 846.

119 For some background to Bernal's artistic interests see A. Brown, *J.D. Bernal: The Sage of Science* (Oxford: Oxford University Press, 2005), pp. 153–4.

120 B. Hepworth, undated letter to Desmond Bernal (c.1937), Cambridge University, J.D. Bernal Archive, Add. 8287/J84.

121 Brown, *J.D. Bernal*, p. 154.

122 B. Hepworth, undated letter to E. Bernal (c.1971), Cambridge University, J.D. Bernal Archive, P.6.2.

123 See G. Werskey, *The Visible College* (London: Free Association Books, 1988), pp. 135–6.

124 J.D. Bernal, 'Art and the Scientist', in J.L. Martin, B. Nicholson & N. Gabo (eds), *Circle: International Survey of Constructive Art* (London: Faber & Faber, 1937), p. 119.

125 *Ibid.*, p. 119.

126 *Ibid.*, p. 119.

127 *Ibid.*, p. 120.

128 A.J. Barlow, 'Barbara Hepworth and Science', in D. Thistlewood (ed.), *Barbara Hepworth Reconsidered* (Liverpool: Liverpool University Press, 1996), p. 99.

129 Bernal, 'Art and the Scientist', p. 120.

130 Hammer & Lodder, *Constructing Modernity*, pp. 379–81.

131 Bernal, 'Art and the Scientist', p. 122.

132 See P. Anker, 'The Bauhaus of Nature', MODERNISM/*modernity*, Vol. XII, No. 2, 2005, pp. 229–30, 236–40.

133 Hammer & Lodder, *Constructing Modernity*, p. 381.

134 Harrison, *English Art and Modernism*, p. 303.

135 *Ibid.*, p. 304.

136 See, for example, Lloyd's discussion of Picasso's stylistic multiplicity as being merely a decadent symbol of the ideological crisis that had gripped the bourgeoisie since the onset of the Great Depression: A.L. Lloyd, 'Modern Art and Modern Society', in Betty Rea (ed.), *5 on Revolutionary Art* (London: Wishart, 1935), pp. 67–8.

137 Harrison, *English Art and Modernism*, p. 305.

138 H. Read, *Art and Society* (1937) (London: Faber & Faber, 1956), p. 125.

139 Hammer and Lodder identify *Circle* with the notion 'that contemporary art and science share a common progressive spirit, and that together artist and scientist might contribute positively to the wide process of social construction [...]. Such ideas provided a theoretical framework for justifying abstract art as a radical project'. Hammer & Lodder, 'Hepworth and Gabo', pp. 114–15.

140 On the bioromanticism of Unit 1, see Botar, *Prolegomena to the Study of Biomorphic Modernism*, pp. 47–8.

141 H. Read, 'Introduction', in H. Read (ed.), *Unit 1* (London: Cassell, 1934), p. 16.

142 H. Read, *Art Now: An Introduction to the Theory of Modern Painting and Sculpture* (1933) (London: Faber & Faber, 1948), p. 60.

143 *Ibid.*, p. 61.

144 See Read, *Art and Society*, p. 125.

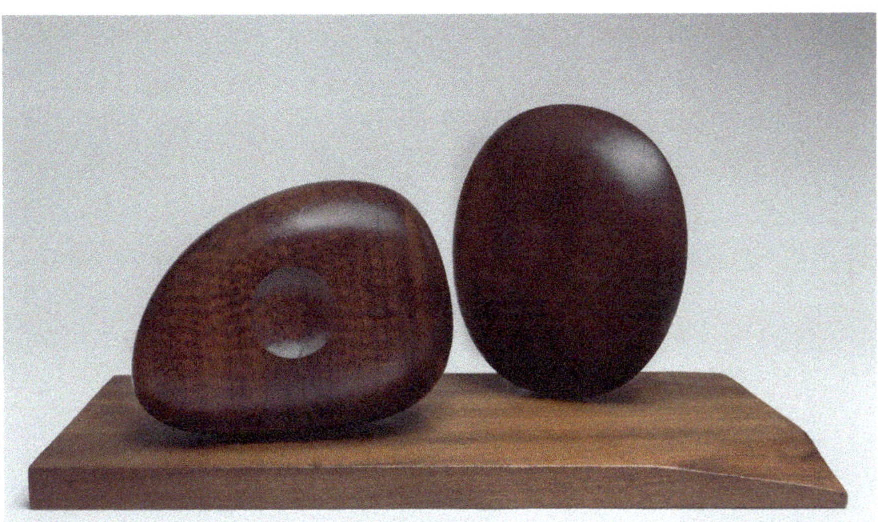

Plate 1

Barbara Hepworth, *Two Forms (Snakewood)*, 1935. Hirshhorn Museum and
Sculpture Garden

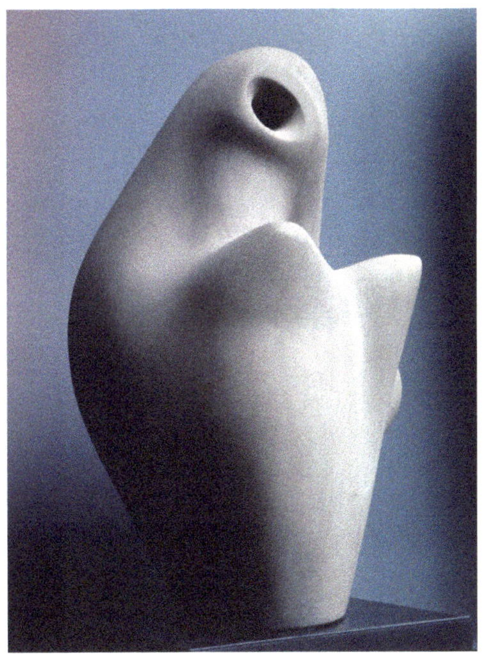

Plate 2

Henry Moore, *Figure*, 1933–34. Private collection

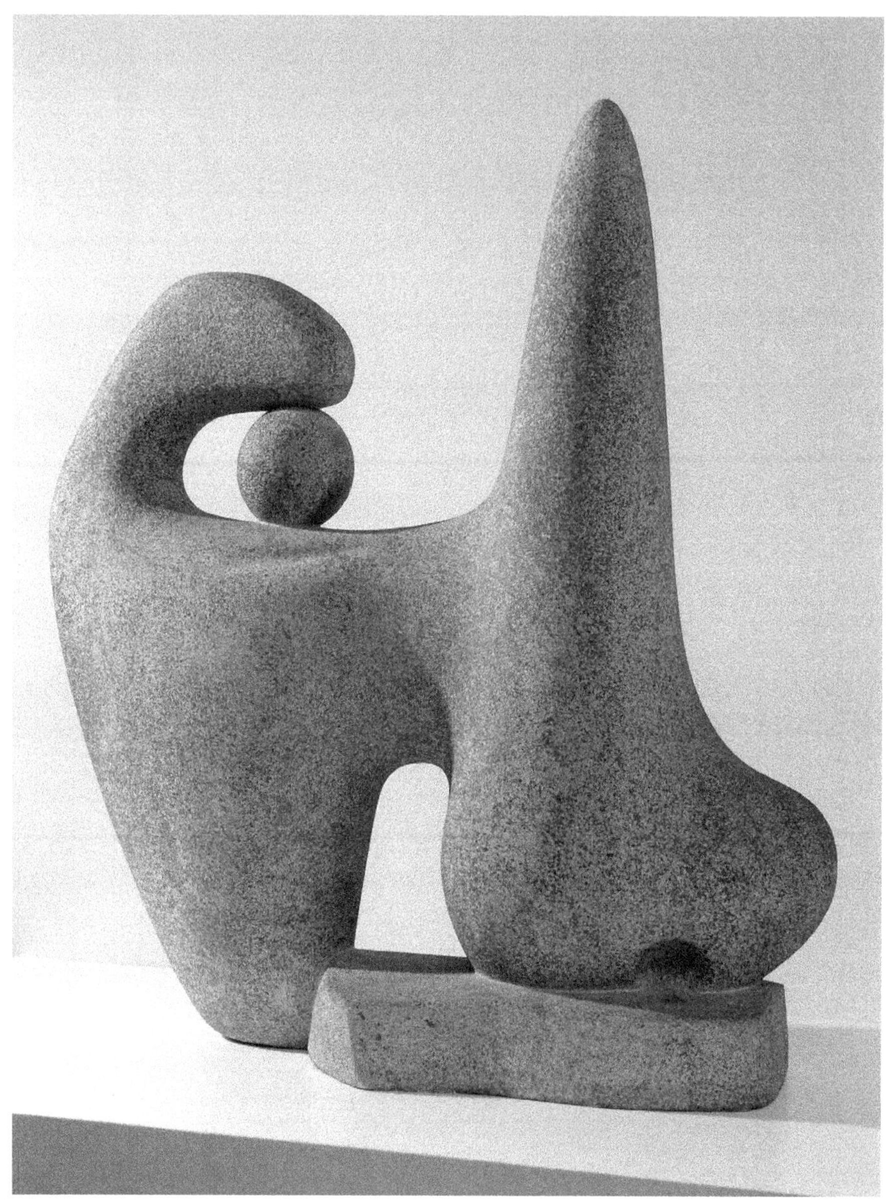

Plate 3

F.E. McWilliam, *Eye, Nose and Cheek*, 1939. Tate, London

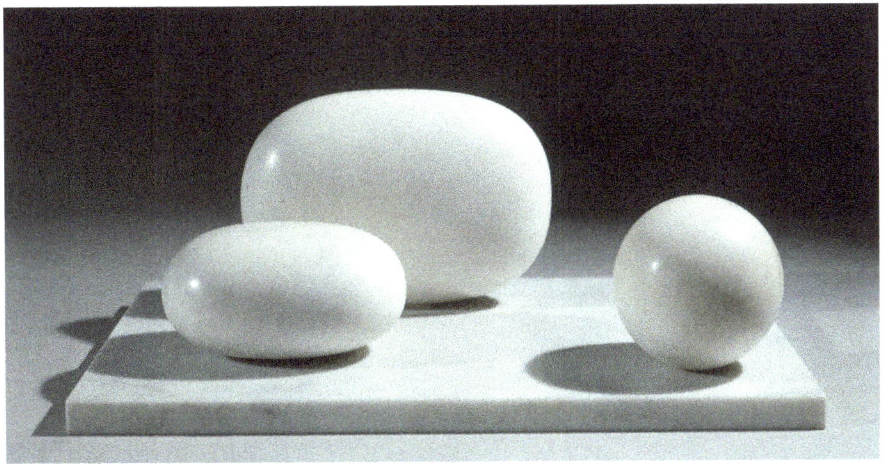

Plate 4

Barbara Hepworth, *Three Forms*, 1935. Tate, London

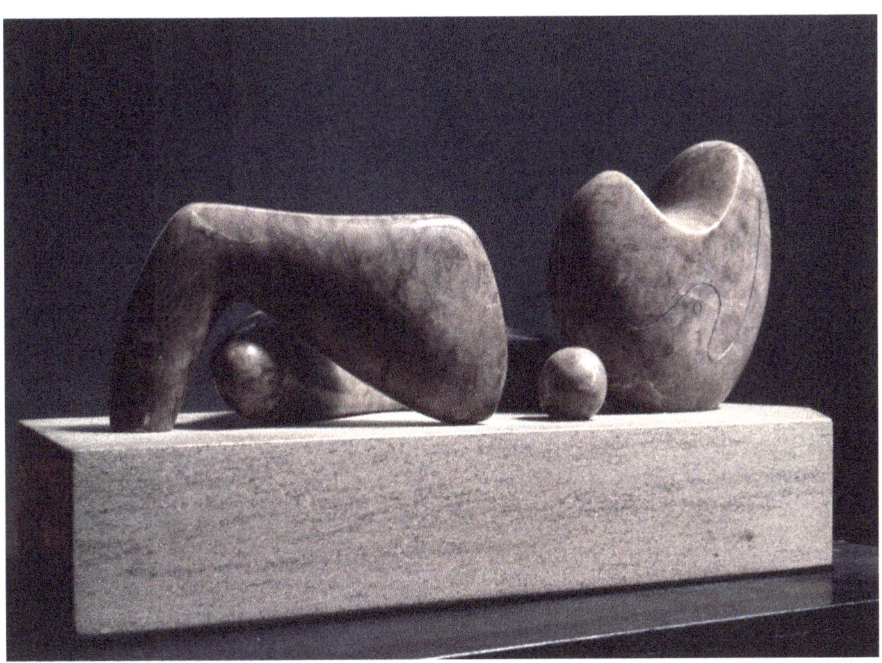

Plate 5

Henry Moore, *Four-Piece Composition: Reclining Figure*, 1934.
Tate, London

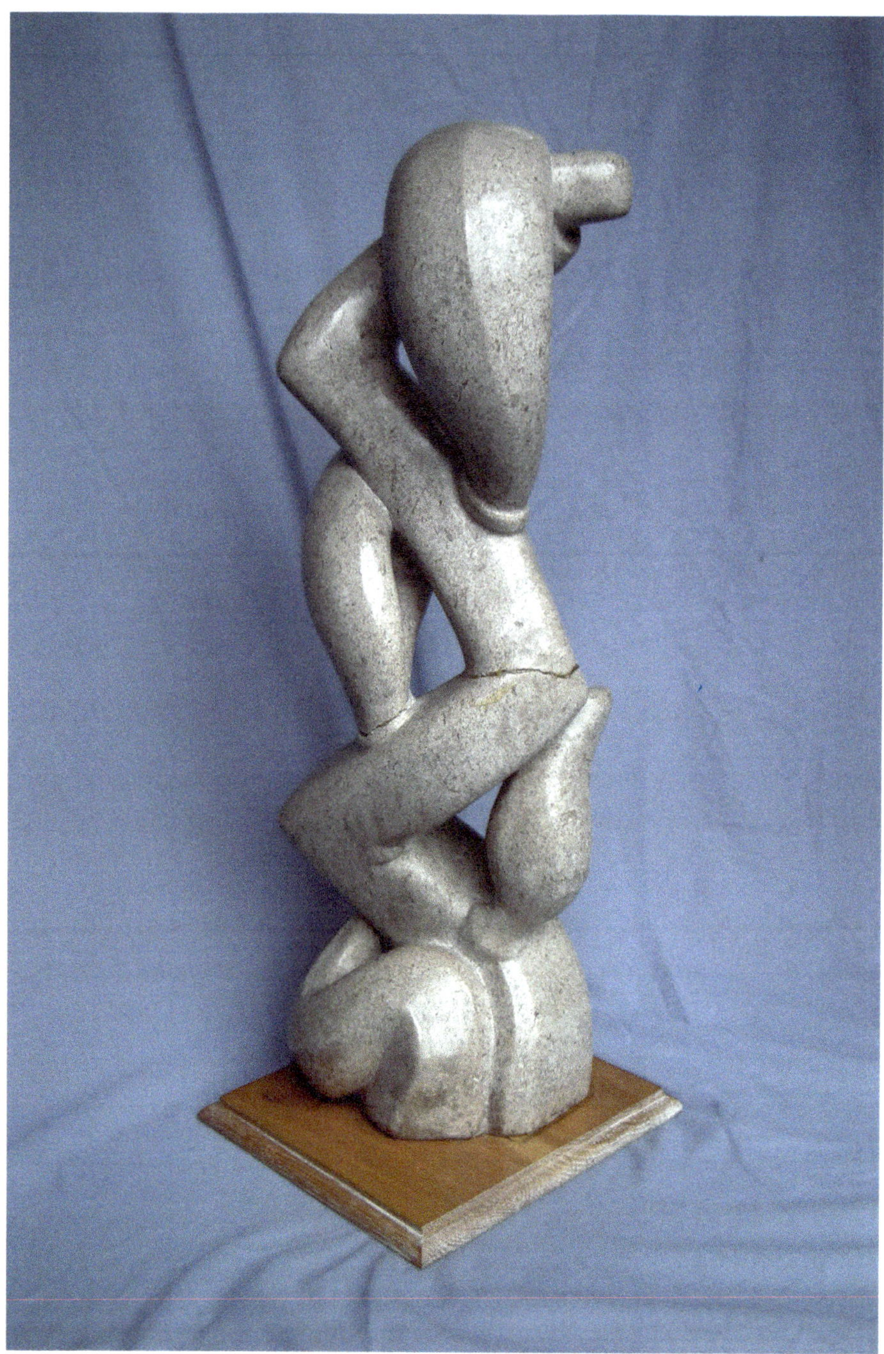

Plate 6
Richard Bedford, *Spring Tree*, c.1932, Mersea Museum

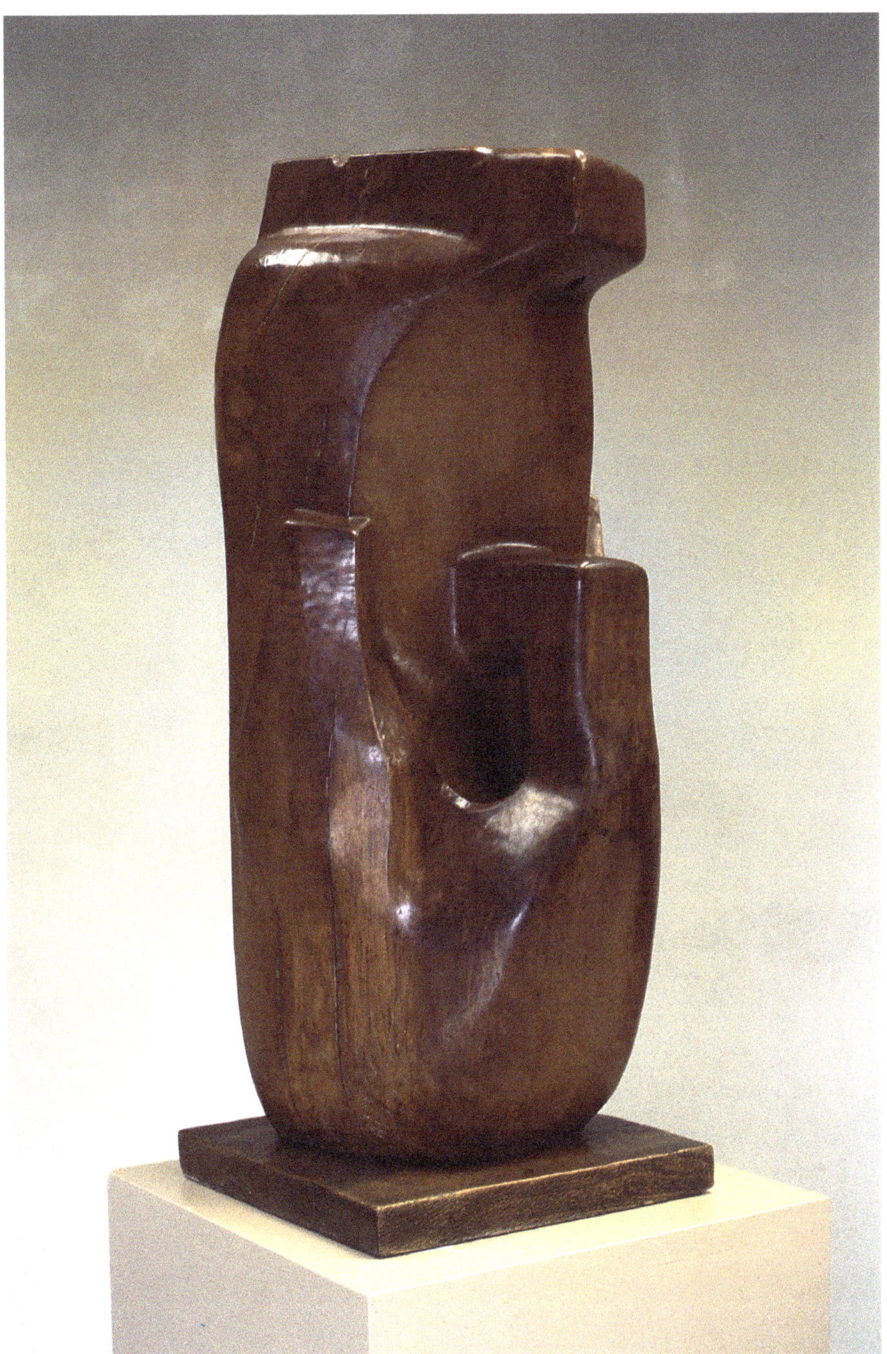

Plate 7

Henry Moore, *Carving*, 1935. Henry Moore Foundation

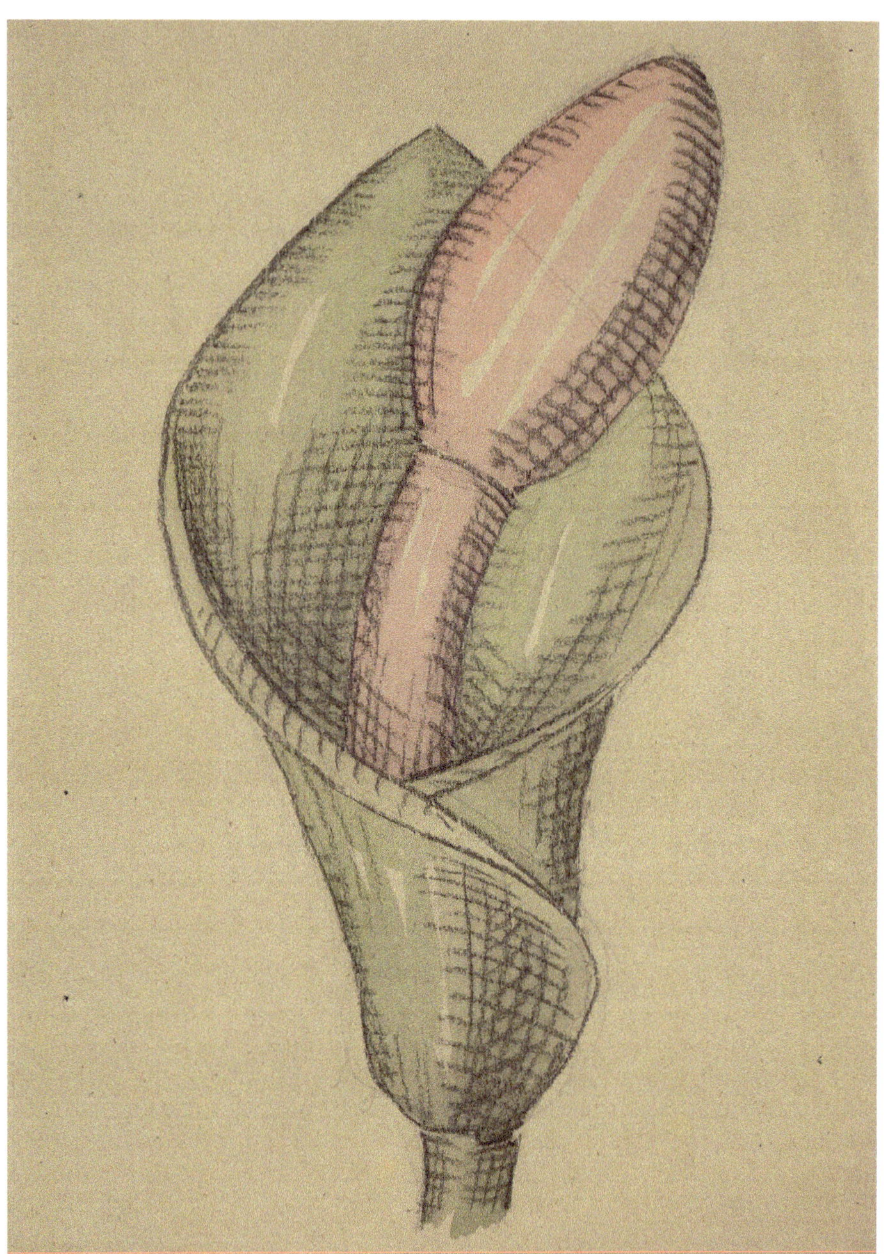

Plate 8
Richard Bedford, *Plant*, c.1938. Victor Batte-Lay Foundation Collection

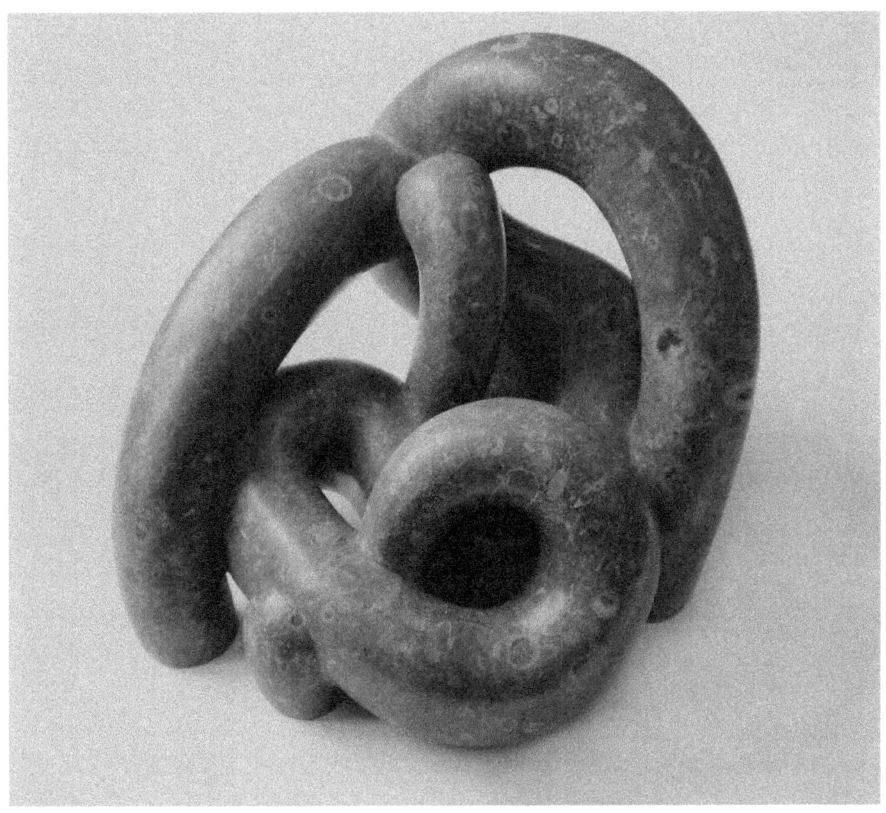

Plate 9

Richard Bedford, *Horned Lizard*, 1936. Mersea Museum

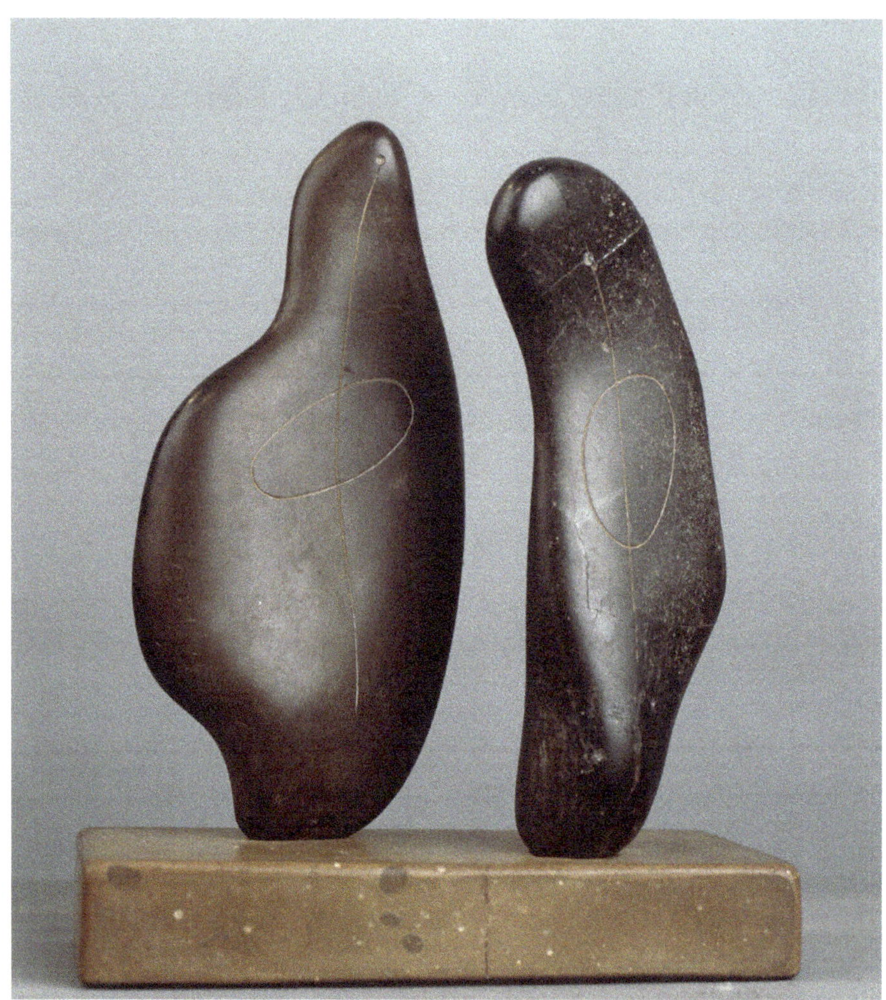

Plate 10

Henry Moore, *Two Forms*, 1934. Private collection

2. Metamorphosis

Beginning around 1930, Henry Moore embarked upon an ambitious series of drawings that would eventually come to be known as his *Transformation Drawings* (c.1930–35). Picturing a range of natural objects in various stages of rotation, these sketches can be readily appreciated as symbolic of the curiosity that Moore directed towards the Surrealist cult of the found object and the expressive capacity of natural forms.[1] Yet, whereas for a Surrealist such as André Breton, certain natural objects possessed an aesthetic appeal precisely because of their ability to conjure up a phantasmagorical plethora of disparate and existentially antithetical entities,[2] Moore's own morphological interests were ostensibly more prosaic and empirically minded.[3] Desiring to unearth the 'principles of form and rhythm from the study of natural objects', Moore hinted that his drawings of lobster claws, tree roots and bones, dexterously imaged from a bewildering variety of viewpoints, were simply studies of the morphological laws that underpinned the development of form in nature.[4] In a turn of phrase that adroitly summed up the visual rationale behind sketches such as his 1932 'Studies of Bones' (Figure 2.1), Moore enthused in the Unit 1 manifesto that: 'Bones have a marvellous structural strength and hard tenseness of form, subtle transition of one shape into the next and great variety in section'.[5] Most evident in the *Transformation Drawings* is the manner in which the angle of vision fundamentally changes the form of the object depicted. For example, one sketch of a pelvic bone illustrates a fragment of the pelvic girdle variously viewed sideways, from the top and bottom. Each shift in viewpoint alters the overall shape of the object – in one sketch it is tall, thin and linear; in another it becomes squat, compressed and twisted, an ossified torque (Figure 2.2).

Moore's acquaintanceship with D'Arcy Wentworth Thompson's *On Growth and Form* is palpable in the way the *Transformation Drawings* chronicle the morphological shifts experienced by a form apprehended in four-dimensional space, with one bony shape seamlessly morphing into another through a simple act of rotation.[6] First published in 1917, Thompson's volume employed mathematics to explain the laws of growth and form in living things and was popular among artistic Modernists who had become drawn to the problems the New Biology posed to art.[7] Germane to Moore's practice was Thompson's chapter on 'The Theory of Transformations', in which the deformation of an organism's structure through

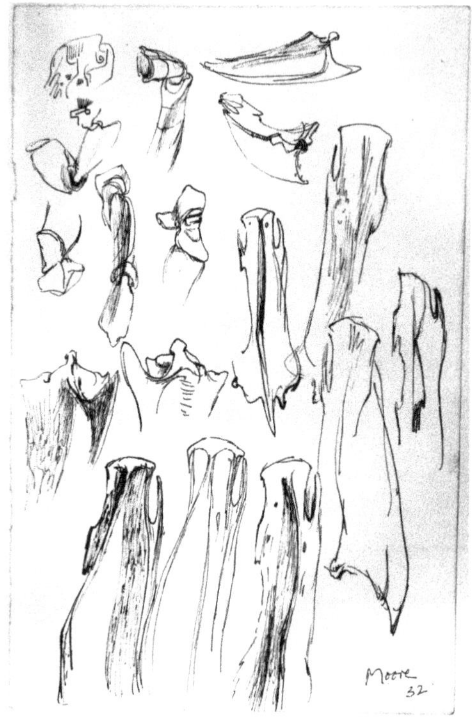

Figure 2.1

Henry Moore, 'Studies of Bones', 1932, HMF936. Current location unknown

the action of physical forces led to it taking on the appearance of an anatomically distinct though morphologically related genus. By plotting the shape of an organism onto a grid-like matrix of mathematical coordinates, Thompson conjectured that any bodily distortion could be geometrically represented so as to exemplify the morphological relatedness between species (Figure 2.3). Importantly, by effecting this method of 'transformation' he noted that it was possible to hypothesise the intermediary but as yet unknown forms through which a particular organism had to pass before it evolved into another genus type:

> [This] process may be especially useful, and will be most obviously legitimate, when we apply it to the particular case of representing intermediate stages between two forms which are actually known to exist, in other words, of reconstructing the transitional stages through which the course of evolution must have successively travelled if it has brought about the change from some ancestral type to its presumed descendant.[8]

Moore's enthusiasm for this morphological project might be surmised by the fact that many of his *Transformation Drawings* served as preparation for his full-scale

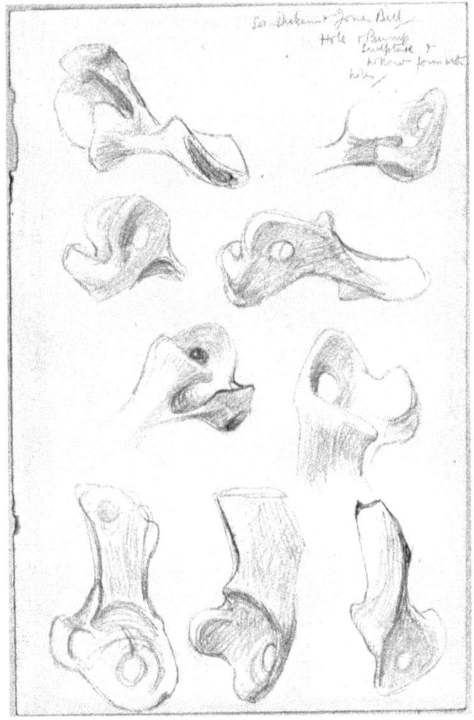

Figure 2.2

Henry Moore, 'Studies of Bones', 1932, HMF937. Henry Moore family collection/
Moore Danowski Trust

sculptures.[9] In one sketch, for instance, a jawbone is swivelled into several different positions, each standpoint being represented in a different dimension of scale, demonstrating 'the subtle transition of one shape into the next' but also allowing Moore to visually ad lib, exaggerating or warping the forms into shapes that are more recognisably sculptural. Notes scribbled across the page signify Moore's wish, *à la* Thompson, to improvise upon the morphology of natural objects so as to envisage *possible* forms, structurally related to the form depicted, but metamorphosed into a new and original sculptural configuration: 'In transferring studies into/stone – harden & tighten, stiffen/taughten [*sic*] them up'.[10] A drawing of a jawbone resting upon its bottom edge is thus transmogrified into a template for a reclining figure (Figure 2.4). That Thompson employed Albrecht Dürer's proportional diagrams of the human figure from his *Four Books on Human Proportion* (1528) as an illustration of his method no doubt helped Moore to appreciate the utility of his morphology to the history of art.[11] Morphological illustrations of this sort powerfully resonated with Herbert Read's scientistic description, in *Art Now*, of artists who enlisted 'the aid of the scientist' in their

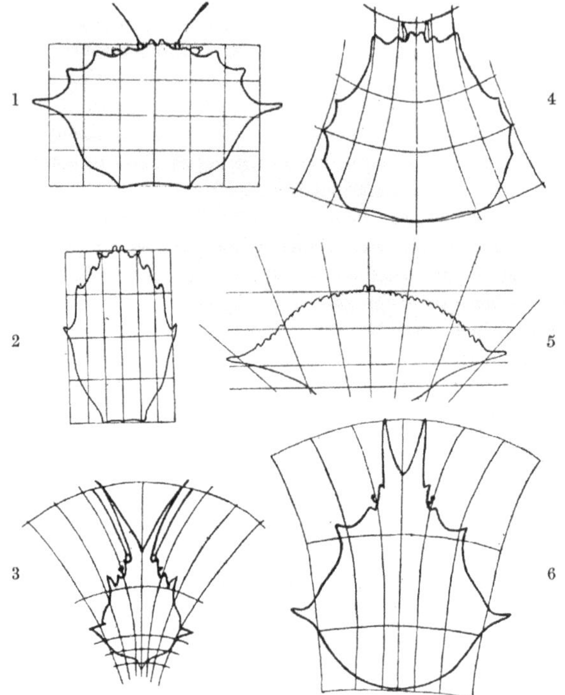

Fig. 513. Carapaces of various crabs. 1, *Geryon*; 2, *Corystes*; 3, *Scyramathia*; 4, *Paralomis*; 5, *Lupa*; 6, *Chorinus*.

Figure 2.3

D'Arcy Wentworth Thompson, 'Carapaces of Various Crabs', Fig. 513 in the second edition of *On Growth and Form* (Cambridge: Cambridge University Press, 1942)

work so as to become surrogate anatomists who studied 'the play of muscles and the framework of bones'.[12] Indeed, as this chapter will demonstrate, Moore was not unique in his appreciation of metamorphosis. Through developments in evolutionary theory, experimental embryology and scientific philosophy – staged within the New Biology – metamorphosis and morphogenesis would emerge as powerful metaphors for creative agency during the 1930s.[13] Under the intellectual auspices of Surrealism and neo-vitalism, Modernists would see the egg shape as a symbol of the illimitability of creation and – in response to the neo-vitalist theories of Hans Driesch and Henri Bergson – the spatio-temporal extension of contemporary sculpture as a token of its burgeoning vitality. Closely linked to Geoffrey Grigson and Alfred Barr's fetishisation of the amoebic form as a turgescent emblem of stylistic energy – as we will see – the purposeful asymmetry of Modernist sculpture was seen to enhance its kinaesthetic properties and thus enliven spectatorial space.

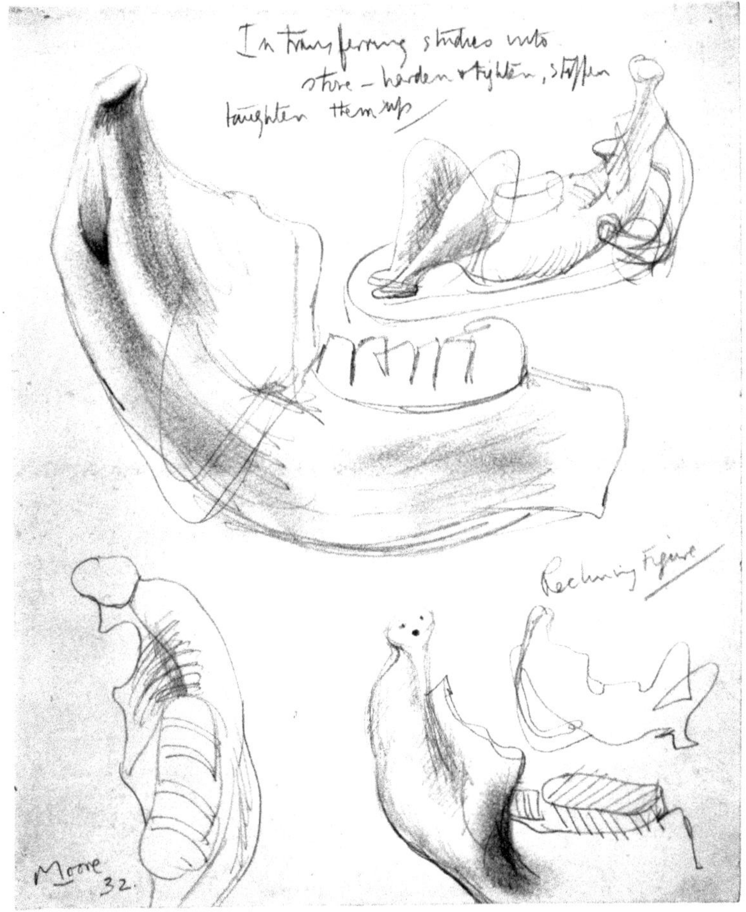

Figure 2.4

Henry Moore, 'Ideas for Sculpture: Transformation of Bones', 1932, HMF941.
The Henry Moore Foundation

The new evolutionary synthesis

The interwar years witnessed renewed scientific interest in evolution as biologists sought to reconcile a neo-Darwinian theory of evolution with Mendelian genetics and classical natural history. This struggle towards a new evolutionary synthesis was in no small part due to the frustrations harboured within sections of the scientific community towards conventional expressions of Darwinism. On the one hand, scientific workers were tired of the second-class status that fields such as experimental morphology had acquired in relation to evolutionary studies. Part of this dissatisfaction stemmed from an irritation with the way in which processes of development and cell differentiation were overlooked in favour of studying embryonic form for purely phylogenetic reasons – effectively abandoning the explanation of function for the study of form.[14] Conventional evolutionary theory, moreover, was seen as highly speculative in petitioning for the role of natural selection without providing a convincing argument as to the origins of variation in species.[15]

The subject that enabled an overhaul of evolutionary theory in the 1920s turned out to be genetics, as it supplied a credible mechanistic explanation for the basis of mutations in natural populations.[16] Yet despite the advances made in the theory of evolution during the early interwar period, residual problems remained, as various schools provided competing interpretations of the new synthesis – not only was there a vast backlog of contentious research that needed reinterpretation in the light of the new theory, but the traditional schism between the mechanist and vitalist factions continued, further problematising scientific interpretation of evolutionary facts.[17] As we shall see, the profitable fusion of evolutionary studies with embryology not only reinvigorated scientific understandings of physico-chemical factors in development, but also compelled thinkers to envisage the growing organism in more holistic terms.[18]

By the 1920s, practitioners of biology had begun to narrow the margins that had existed between hitherto unconnected fields of natural science. The streamlining of scientific knowledge that this process occasioned allowed for a greater understanding of the ways in which all the phenomena of life were subtly related.[19] In many ways, the study of evolution provisioned scientists with the impulse needed to fashion this new biological synthesis – as the complexities of evolutionary change were seen to necessitate the combining of resources from multiple disciplines. For example, in the first of a suite of articles entitled 'Links in the Chain of Life', the embryologist and Professor of Natural History at the University of Glasgow, John Graham Kerr detailed three major fields of scientific interest to the student of evolution. These were, in his opinion: palaeontology, embryology and comparative morphology.[20] To this list Kerr should have added genetics, the impact of which on the theory of natural selection was becoming increasingly evident at this time.[21] Indeed, one of the shortcomings of standard accounts of evolution had been the inability of natural science to explain the basic mechanisms of the theory. Without any means of accounting for the abrupt shifts in form-type made manifest in the fossil record, J.B.S. Haldane claimed that 'many palaeontologists today confine themselves to stating the facts of evolution, and laying down general laws which they obey, rather than attempting to discover the causes underlying those laws'.[22] The principles of evolution were increasingly identified as part and parcel of the hereditary process, which permitted scientists to understand 'each generation [as bound] to the organic semblance of the last'.[23] Although natural selection was still of importance to evolutionary biologists, increasingly its effects were understood as being conditioned by the influence of other biogenetic factors. Out of the three laws of evolution that Julian Huxley forwarded in *Evolution: The Modern Synthesis* – mutation, recombination and selection – two had their basis in genetics.[24]

By the late 1920s, the study of embryology had begun to resolve some of the problems raised by the analysis of the fossil record and therefore reached a key stage in its history. The questions asked by embryologists of the period were, for the most part, responses to the pioneering work undertaken in the late nineteenth century by Wilhelm Roux and Hans Driesch. In looking into the relationship between embryonic development and phylogenetic history, Roux had asked why certain cells in the maturing embryo begin at precise points to differentiate into other cells. To resolve this puzzle, Roux had devised the mosaic theory, which maintained that the hereditary material of each cell was divided unequally during cell division. As cleavage progressed, the potentialities of each cell became increasingly limited, meaning that each cell came to possess just one principal hereditary trait and, subsequently, could belong to only one class of tissue.[25]

The origins of this theory had been the discovery by Roux that two-celled frog embryos failed to develop properly after one of the cells had been killed. Driesch amended this hypothesis after staging an experiment with sea-urchin eggs. Instead of destroying one of the cells, Driesch shook the embryo until the blastomeres separated from each other. Upon allowing each single-celled embryo to develop individually, Driesch observed that each grew into a normal, although rather small, sea-urchin larva. This observation led him to the supposition that the embryo was essentially a collection of cells linked together as a self-adjusting unit – in which 'all parts are equivalent in their potential for producing a whole new organism' – instead of envisaging differentiation, as Roux did, as a consequence of the uneven allocation of hereditary matter.[26]

The controversy that surrounded pre-formationism and epigenesis in the 1930s found initial articulation in these two competing theories of embryology. On the one hand, Roux asserted that the embryo had in its internal constitution all of the elements necessary to stimulate development – thus contending that the embryo is essentially pre-programmed to develop in a certain way. Driesch, on the contrary, emphasised the importance of external stimuli in processes of embryonic growth. Influences as diverse as the mass of the embryo, its contact with the external environment, the position of each cell in relation to the other cells in the embryo, and the effects of gravity were all, in Driesch's view, causal agents in development. He resultantly understood development 'less as a series of rigidly programmed events than as the response of a living entity, the whole embryo, to varying life conditions'.[27]

Driesch's embryological work, in particular, would have significant implications for the philosophy of science in the early twentieth century. Believing that the constant adjustability of the embryo defied materialist explanations, Driesch argued that morphogenesis was the result of a vital force that was independent of physico-chemical factors. Seemingly providing scientific confirmation of an immaterial vital spirit that guided the development of matter, Driesch was an important figure in the revival of vitalism that took place across Europe in the early 1900s. Rather than naively positing a vital substance in order to fit material-ist expectations – as had early proponents of vitalism – 'critical' neo-vitalism focused upon a vital impulse which, as Frederick Burwick and Paul Douglass have explained, operated within the novel 'context of an ontology of energy and idea. German *Naturphilosophie* was joined by the critical vitalisms of Bergson and Driesch, and the aesthetic and social vitalism of Nietzsche. There can be no doubt that, as the century closed, the vitalist *tradition* was being powerfully reinterpreted by some of the most celebrated intellectuals in the West'.[28]

Like Driesch, Henri Bergson was a key player in the revival of vitalism's fortunes in the early years of the twentieth century. Published in 1907 (and translated into English in 1911), Bergson's neo-vitalist thesis *Creative Evolution* attracted much attention in the British Isles and occasioned a successful lecture tour of England in 1911. Bergson's vision of a haphazard though unrelenting life force – the *élan vital* – was widely recognised as a reaction against materialism and an effort to rejuvenate a certain strand of philosophical conservatism that was more in tune with religion.[29] Certainly, Modernist critics were inspired by the teleological implications of Bergson's neo-vitalist cogitations and, even as late as 1955, Herbert Read still claimed that 'the inspiration I continue to receive from the only metaphysics that is based on biological science [is] the metaphysics of Henri Bergson'.[30]

While, by the 1930s, outright neo-vitalism no longer enjoyed much influence in biology (a point illustrated by Driesch's well publicised defection to philosophy),[31] its philosophical connection to organicism, nature-centrism and anti-mechanism meant that it foreshadowed many of the developments that would intellectu-ally cohere in the formation of the New Biology.[32] Indeed, the fact that many biologists had lingering doubts about mechanism's ability to fully describe a living system meant that neo-vitalism continued to enjoy influence in scientific philosophy – mainly as a source of inspiration for the philosophies of organicism and holism – even as it seemingly eschewed epistemological coherency.[33] In terms of the arts – as George Rousseau has shown – the philosophical controversies over the nature of life that emerged in embryology between 1900 and 1930 inflamed the passions of a whole generation of British writers, artists and philosophers who looked to neo-vitalist principles so as to develop a meaningful philosophy of life.[34] For sure, the popularity of neo-vitalism as a rhetorical device through which an understanding of Modernist sculpture could be obtained lay in its opposition to materialism and 'the bleak rationalism' that troubled Read.[35] The philosopher Cyril Edwin Mitchinson Joad penned a neo-vitalist apologia in 1928 – titled *The Meaning of Life* – in which he critiqued materialism's philosophical reductiveness. Like many neo-vitalists, Joad argued that relativity physics had destroyed 'the principle of conservation of substance, [so far as] matter is concerned' through new ideas about radioactivity and spatio-temporal positionality and had made a mockery of materialism's founding tenets.[36] Related work in biology on the dynamic interconnectivity of living matter had similarly led him to believe in a neo-vitalist 'non-material principle of life which [manifested] itself in matter [and gave] rise to the phenomena we call organisms'.[37]

These neo-vitalist musings on the New Biology would find expression in Modernist art theory as a profoundly anti-reductivist biologism. Philosophically connected to organicism, critical neo-vitalism defended the irreducibility of the artwork by equating compositional rhythm to the spatio-temporal quintessence of artistic intentionality. Jean Hélion thus expounded in an *Axis* essay of 1935:

> It is necessary to remove taboos of any sort, though keeping an axis of progression; to constitute paintings, like beings whose term can develop indefinitely, where everything remains possible, instead of paintings where more terms become forbidden as we go on. The work considered as an organism in growth.[38]

Here Hélion used embryological terms to convey the spatio-temporal extension of the artwork; how, like a living thing, it actively grew in complexity through morphogenetic development. As we will see, developments in experimental embryology and neo-vitalist philosophy combined to produce an environment in which Modernists could interpret certain types of sculpture as 'vital' signs of creativity and kinaesthetic intricacy.

The life of forms: metamorphosis, Focillon and art theory

In an essay printed in *The Listener* in 1933, bearing the melodramatic title of 'The Age of Change', Julian Huxley reminded his audience that the 'basic idea of the present age is the idea of change' and that this made it distinct to other phases of human development, in which change was regarded as either not occurring or as catastrophic and hence undesirable. To start with, the blistering pace of technological development had highlighted the transformation of daily life through inventions such as the airplane and wireless technology. Yet in 'the sphere of general thought [it was] the idea of evolution [that had] played a preponderating role [in bringing the concept of change to the masses], by introducing us to the spectacle of continuous change in every branch of life, and of steady progress in some branches'.[39]

The impression that a new awareness of evolution had forever transformed humanity's perception of change – substituting old ideas about the universe existing in a steady state with notions of the world being in perpetual flux – had a deep impact upon how Modernist historians and critics theorised the development of art, both modern and ancient, in the 1930s. Although Eric Fernie maintains that evolutionary metaphors fell out of fashion in art history during the twentieth century as historians started to doubt whether art evolved solely according to material conditions or even that it advanced *progressively*,[40] the new evolutionary synthesis nonetheless created the conditions under which fresh ideas about evolution were promoted, many of which found expression in art theory to explain stylistic development. For example, the professor of art history at the Sorbonne, Henri Focillon, produced a neo-vitalist study of artistic form, entitled *The Life of Forms in Art* (1934), which explicitly used 'the methods employed by biology'.[41] Notably, this biologistic treatise drew upon the Bergsonian concept of an *élan vital* to theorise the development of form in art. Contending that an independent creative principle purposefully drove the evolution of art forms, Focillon described an artwork as an *activity* that had its counterpart in the agency of the vital force that produced form in nature:

> Life is form, and form is the modality of life. The relationships that bind forms together in nature cannot be pure chance, and what we call 'natural life' is in effect a relationship between forms, so inexorable that without it this natural life could not exist. So it is with art as well [...], forms constitute an order of existence and [...] this order has the motion and the breath of life.[42]

Focillon's notion of a fluid continuity binding together the various manifestations of art over the centuries bears an uncanny similarity to Bergson's conception of evolutionary mutability.[43] Just as Focillon recognised form in art to be endlessly changing, so Bergson saw form as inherently fluxional, arguing that it was a philosophical redundancy to speak of form as being in any way static:

> [Life] is an evolution. We concentrate a period of this evolution in a stable view which we call a form, and, when the change has become considerable enough to overcome the fortunate inertia of our perception, we say that the body has changed its form. But in reality the body is changing form at every moment; or rather, there is no form, since form is immobile and reality is movement. What is real is the continual *change of* form. Therefore, here again, our perception manages to solidify into discontinuous images the fluid continuity of the real.[44]

In a passage on Raphael's painting, Focillon's emphasis upon the variability of forms, on how Raphael ceaselessly adjusted the morphology of the human body to differing compositional standards, echoes Bergson's neo-vitalist focus upon metamorphosis as a condition of life:

> [The] models of nature may themselves be regarded as the stem and support of metamorphoses. The body of man and the body of woman can remain virtually constant, but the ciphers capable of being written with the bodies of men and women are inexhaustibly various, and this variety works on, activates and inspires all works of art [...]. [It is] in those compositions by Raphael that are laden with whole garlands of human bodies that we can best comprehend the genius for harmonic variations that combines over and over again those shapes wherein the life of forms has absolutely no aim other than itself and its renewal [...]. Here the metamorphosis of shapes does not alter the factors of life, but it does compose a new life [...]. Form may, it is true, become a formula and canon; in other words, it may be abruptly frozen into a normative type. But form is primarily a mobile life in a changing world. Its metamorphoses endlessly begin anew, and it is by the principle of *style* that they are all coordinated and stabilized.[45]

While Focillon's stress upon the fluxional character of form underscored his philosophical indebtedness to a Bergsonian model of neo-vitalism, his portrayal of metamorphosis as an aspect of life also owed something to the New Biology and its embryological hypotheses.[46] Though not necessarily neo-vitalist in philosophical orientation, many biologists agreed that the new evolutionary synthesis had proved the importance of environmental change in evolutionary development in ways that preceding theories of evolution had not anticipated. For example, John Graham Kerr observed in *The Listener* in 1930 that just as organisms were subject to fluctuations of form during their life cycles, so too, '[t]he environment is in a constant state of flux'.[47] In explaining how species adapted to their surroundings, he felt that biology had traditionally laid too much emphasis on the *teleological* character of natural selection. In reality, however, 'the elusive environment is all the while itself changing, so that the evolving organism, while continually approaching perfect adaptation, never actually attains it. In this way, evolution by natural selection never comes to an end, but remains ceaselessly progressing'.[48] Focillon's thoughts on the role of the environment in shaping the forms of art thus corresponded closely with the epigenetic interactivity presupposed by the New Biology:

> [Forms] mingle with life, whence they come; they translate into space certain movements of the mind. But a definite style is not merely a state in the life of forms, nor is it that life itself: it is a homogeneous, coherent, formal environment [...]. It is, too, an environment that may move from place to place *en bloc*. We find in Gothic art imported in such wise into northern Spain, into England, into Germany, where it lived with varying degrees of energy and with a rhythm sufficiently rapid to permit an occasional incorporation of older forms that, although they had become localized, were never vitally essential to the environment.[49]

Written in the autumn of 1933 and published in 1934, *The Life of Forms in Art* provided a powerful neo-vitalist interpretation of the epigenetic properties of art forms, presenting Modernists with an authoritative historical model through which to theorise upon the internal coherency of an artwork and its compositional vitality.[50] Certainly, echoes of Focillon's thinking can be found in the writings of Jean Hélion and Moore; and Barbara Hepworth, in particular, was captivated by his assertion that vivacity in art was conveyed by a dynamic interrelationship between form and space.[51] Nevertheless, despite its applicability to biologistic art theory, *The Life of Forms in Art* was translated into English only in 1948 and Focillon's influence upon Modernist sculpture in Britain must be seen as part of a wider current of interest in the new evolutionary synthesis and its impact upon the arts. For instance, in 1932 the critic Stephen Haden-Guest penned a biologistic theory of stylistic development which was clearly indebted to popular, neo-Darwinian accounts of evolutionary change:

> The arts most clearly change from age to age to tally with the changing needs and desires and capacities of mankind. There is change in evolution, new species come into

existence supplanting old ones, and peculiar qualities become alternatively emphasised and diminished. What will be the chief forms of the arts, the most important drives and impulses behind their creation in this magnificently rich and troubled world of 1932, and in the quarter of a century or so ahead of us?[52]

Like Huxley, Haden-Guest viewed this emphasis upon change through the prism of the rapid pace of contemporary social development. The prodigious rapidity of socio-cultural change in the twentieth century – technological, political and sociological – had not only dramatically altered the material conditions of human-kind but in so doing had impressed upon humankind's psyche the prevalence of metamorphosis:

> The world today is rushing forward at a tremendous and ever-increasing speed – to what? None of us knows: we know only two things. First that the actual *pace* of change has increased almost incredibly since the eighteenth century, when there was no modern machinery, mass production, communications, or political, educational, or economic democracy; and then second, we do know that the material of the human race [...] is basically the same as in the past, though we may fairly assume it to be in the rapid process of adaptation to new conditions.[53]

For Haden-Guest, this unparalleled pace of change had led to artists adapting their practice to the brave new world in which they found themselves: 'exploring, mapping, providing new countries of the mind' and producing abstract artworks which 'are as definite as machinery in its colours, which in juxtaposition are more emphatic than anything in nature'.[54] Although Haden-Guest's critique lacked the depth of Focillon's historical analysis, it nonetheless saw metamorphosis as the driving force of artistic change and, like *The Life of Forms in Art*, used a biologistic methodology to explain the stylistic transformations to which art was subject over time. As Focillon remarked, the benefit of exploiting 'the methods employed by biology' was precisely that they allowed for an analysis of the tangible phenomena – space and matter – of which form was comprised far more effectively than any psychological study of artistic intentionality. While he agreed that the process of inception occurred in the mind, the world of art forms 'exists in space and matter, its measures and laws are no longer those of the mind in general, but measures and laws that are highly specialized [...]. Form cannot withdraw from matter and space [...] even before it takes possession of matter and space, it already exists within them'.[55] The physicality of the artwork, its corporeal spatio-temporal extension, hence justified the use of biological method as a strategy of interpretation as, like a living thing, it possessed 'properties, movements, measures, metamorphoses' that, in line with neo-vitalist philosophies, evinced an *élan vital* which purposefully drove its formal development.[56] Yet if the new evolutionary synthesis presented Modernist theoreticians with a prescient model to explain the historical emergence of form in art, it was of no less importance to modern artists who, inspired by the fluxional character of developing form, drew upon embryological science to explore the genesis of form in contemporary art.[57]

The fruit of the womb:
embryology, Surrealism and the 'egg shape'

'A prophet can never claim to certainty', wrote Julian Huxley in an analysis of the future of humankind in the age of experimental embryology. Summing up the extent to which improvements in embryology had encouraged prophecies regarding the impending development of the human race, Huxley was convinced that the 'Biological Revolution' would lead to a comprehensive transformation of human habits and biology: 'But if we cannot say for sure that one day human infants will achieve their embryonic stages in a bottle we can at least assert that it is no more impossible than a great many inventions that are accomplished facts today'.[58] Evidence that a biological revolution was by now well underway was provided by improved biomedical knowledge and the advent of birth-control technologies, both of which had deeply impacted upon the general health of the public, extending life spans, increasing the average weight and height of the populace as well as regulating demographic growth in those regions of the world where overpopulation was becoming an increasing socio-political burden. Recent advances in embryology, however, augured the most significant influence upon the evolution of humankind by offering 'the possibility of babies in bottles, motherhood without birth'.[59] In this way, Huxley's embryological speculations underscored the role that the new evolutionary synthesis played in heightening popular fantasies and misgivings about the future development of the human race, to the extent whereby the philosopher C.E.M. Joad openly worried whether humankind would continue to purposefully evolve or whether it was 'doomed to failure and extinction' like so many other species held by the fossil record.[60]

Popular interest in embryology found expression in the writings of Modernist critics, artists and philosophers who recognised in the discipline evocative metaphors through which to hypothesise upon the purposiveness of creative agency in modern art. The poet and critic Adrian Stokes, for example, used gestatory analogies in a review of Hepworth's sculpture – published in *The Spectator* in 1933 – to explain how a sculptor conceptualised form by bringing forth 'stone-shapes' embryonic within the block. 'The true carver attacks his material', he wrote, 'so that it may bear a vivifying or plastic fruit. He woos the block'.[61]

Like his contemporaries, Stokes depended on gender stereotypes that were a commonplace of Modernist criticism, using metaphors of procreativity to designate the quintessential *femininity* of women artists as opposed to the *masculine* artworks produced by artists of the opposite sex.[62] Warming to his theme, he continued thus: 'These stones are inhabited with feeling, even if, in common with the majority of "advanced" carvers, Miss Hepworth has felt not only the block, but also its potential fruit, to be always feminine'.[63] As Anne Wagner has persuasively contended, it was undoubtedly the, now lost, alabaster sculpture *Figure: Mother and Child* of 1933 (Figure 2.5) which provoked Stokes' embryological speculations.[64] The swollen, pebble-like form that the principal figure cradles

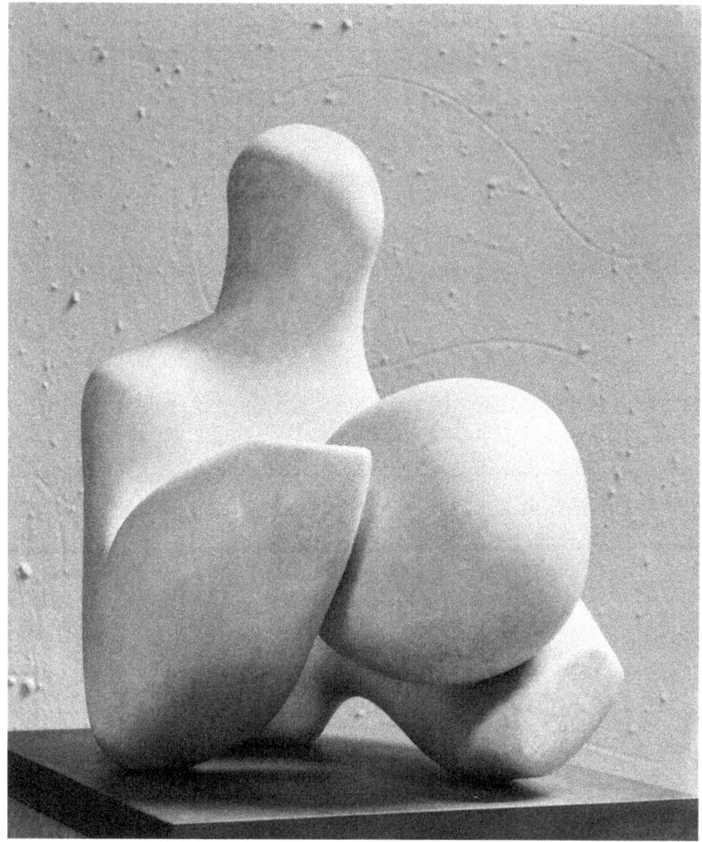

Figure 2.5

Barbara Hepworth, *Figure: Mother and Child*, 1933, destroyed

emphasises an unresolved tension within the sculpture, whereby its separation from the central form appeared to Stokes only partial:

> The stone is beautifully rubbed: it is continuous as an enlarging snowball on the run; yet part of the matrix is detached as a subtly flattened pebble. This is the child which the mother owns with all her weight, a child that is of the block yet separate, beyond her womb yet of her being [...] Miss Hepworth's stone *is* a mother, her huge pebble its child.[65]

Composed of the same stone as the larger form, the 'child' underscores its material origins in the 'mother': a point reinforced by the way the principal figure appears structurally to 'give anchorage to the child'.[66] Indeed, the roundedness of the pebble formally stresses the interdependent nature of this familial relationship; though physically separate, its plump curviness – sat squarely in the abdominal region of the mother – mimics the form of a pregnant belly, highlighting the child's genesis in the mother's womb yet eventual estrangement from it. While the highly charged language of closeness, separation and loss that Stokes utilised in his writing no doubt bore witness to his ongoing fascination with the psychoanalysis of Melanie Klein,[67] the biologistic undertones of 'Miss Hepworth's Carving' signposts Stokes' awareness of the potency of embryological jargon and its applicability to contemporary debates regarding creative agency.

Gendered fantasies about artistic creativity have a history that stretch back to pre-Enlightenment accounts of sexual anatomy, in which the female body was imagined as a negative imprint of the male and thus operated as a passive vehicle for male procreative power.[68] All the same, Stokes' gestatory suppositions appeared at a time when sexualised models of creative agency were inventively re-entering the lexicon of modern art through the philosophical and poetical endeavours of the Surrealist movement. Certainly – even as it championed female emancipation from the patriarchal demands of bourgeois society – Surrealism extolled the poetic qualities of femininity using a terminology that remained codified by clichéd stereotypes of procreativity and repressed eroticism; and it seems probable that such an understanding of female creativity re-entered the lexicon of British art criticism through the agency of poets and artists familiar with Surrealist theory in the 1930s.[69]

Historically, 11 June 1936 is given as the day on which Surrealism officially arrived in Britain, under the auspices of the first International Surrealist Exhibition at the New Burlington Galleries, London, although, in reality, British Modernists – such as Edward Wadsworth and Paul Nash – had for years Channel-hopped, visiting the studios of the Parisian avant-garde and gaining access to French-language Surrealist periodicals through regional outlets such as Zwemmer's bookstore in London.[70]

To appeal to the British public's sense of tradition and therefore underscore Surrealism's relevance to British artistic sensibility, the International Surrealist Exhibition catalogue highlighted the movement's intellectual roots in romantic literature.[71] Indeed, André Breton's celebrated description of Surrealism as 'an excessively prehensile tail' of romanticism served to emphatically underscore its philosophical debt to the nineteenth-century movement.[72] As Jennifer Mundy has remarked, this intellectual obligation to romanticism's philosophical and artistic ideologies was signalled by Surrealism's open fetishisation of outlandish flora and fauna and general enthralment to the power of metamorphosis.[73] For example, in 'Beauty Will Be Convulsive' (1934), Breton marvelled at the ability of coral – a polypean organism which until the eighteenth century had been widely viewed as a plant – to symbolise, by virtue of its protean cycles of sprouting and decay, the evolutionary forces of nature:

> Life, in the constancy of its processes of formation and destruction, does not seem to me more concretely encompassed for the human eye than by the Australian 'Great Barrier Reef's' aragonite hedges of blue-tits and bridges of treasure.[74]

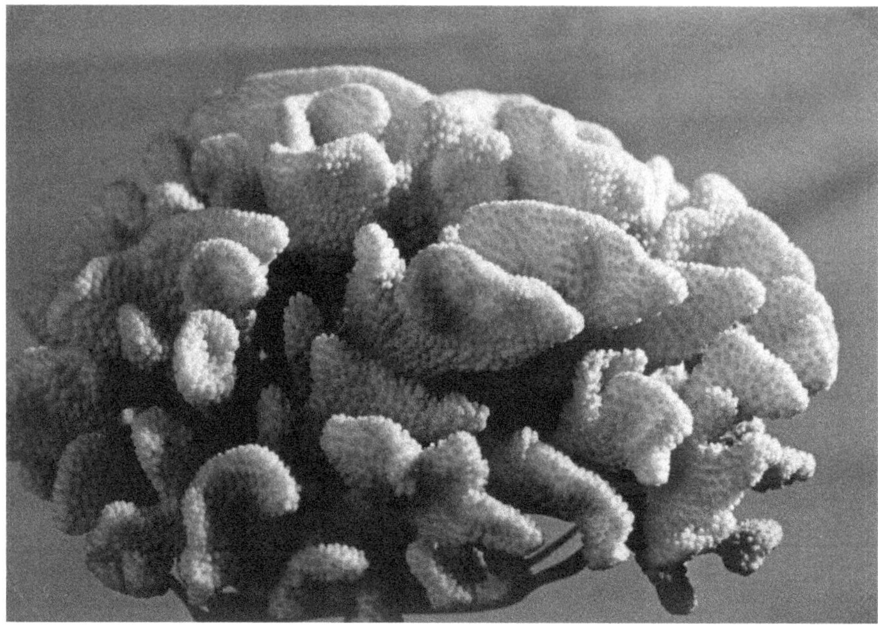

Figure 2.6

Brassaï (Gyula Halász), *Madrépores*, 1930. Centre Pompidou, Paris

Breton, in this paean to life's dynamic properties, regarded coral forests as a paradigm for a type of spontaneous creation which he understood to be the apogee of beauty in both nature and art.[75] Although it was the solidity of the crystalline that most inspired Breton's meditations on artistic perfection,[76] the capacity of corallaceous beings to mimic inorganic structures *even as* they abided by life's evolutionary laws encouraged him to eulogise the fruticant properties of such primitive organisms, seeing in their life cycle a vigorous prototype for Surrealist agency.[77]

Gestatory imagery formed a powerful undercurrent in Surrealist aesthetics in the 1930s, especially in the genre of Surrealist photography, which drew upon precedents in applied and scientific photography to create purposely disorienting close-ups of natural forms in embryonic or bourgeoning states.[78] To illustrate Breton's 1934 article in *Minotaure*, for instance, the Hungarian photo-journalist and photographer Brassaï (the pseudonym of Gyula Halász) chose a selection of photographs of *Madrepore* coral in which the dense wrinkles of the organism are seen to proffer a 'mysterious extension' whereby identical forms appear at every level of the structure,[79] budding endlessly outwards like a fractal in mathematics (Figure 2.6). The self-similarity of form thus exhibited within the object,

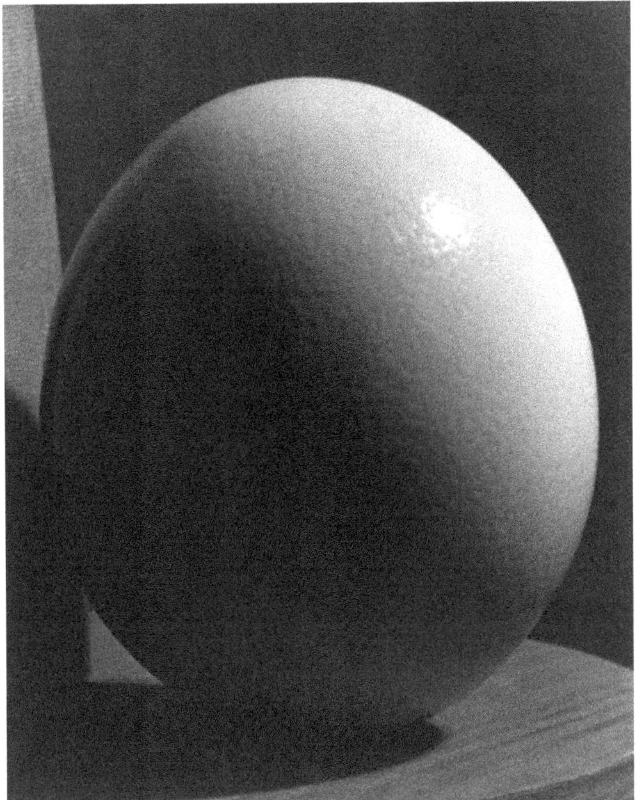

Figure 2.7

Man Ray, *Ostrich Egg*, 1944

irrespective of depth of stratum, offered a compelling representation of coralline growth's ability to evoke the continual process of formation and destruction in life.

The potential of scientific photographs of developing forms to act as templates for Surrealist agency was exploited to flamboyant effect by the American photographer Man Ray. His *Ostrich Egg* of 1944 (Figure 2.7) shows a large egg, set atop a wooden plinth and dramatically side-lit in a manner that lampooned the stylistic foibles of the New Objectivity school of photography in Germany – a contemporary movement in art photography whose appreciation of the analytical precision of scientific photography extended even to the reproduction of photomicrographs, X-rays and other scientific photographs in its yearbooks.[80] Like Brassaï's image of *Madrepore* coral, Man Ray's photograph coolly mimics the conventions of scientific photography even as it purposefully defamiliarises our conception of an egg: magnified and harshly lit, so that each pockmark assumes startling dimensions, the object more closely resembles a cratered lunar surface

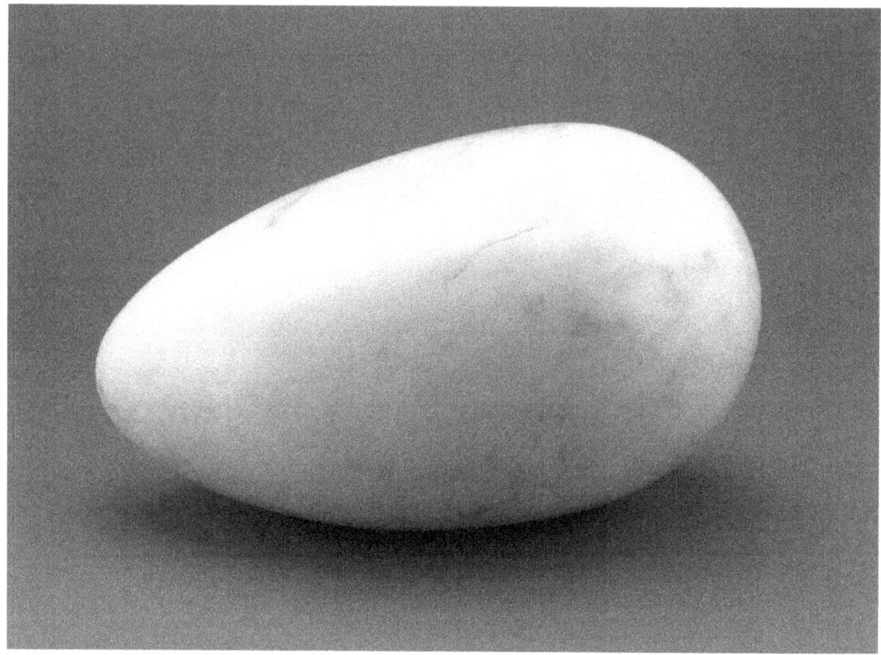

Figure 2.8

Constantin Brancusi, *Sculpture for the Blind [I]*, c.1920. Philadelphia Museum of Art, Louise and Walter Arensberg Collection

than a bird's egg.[81] Photographed like so, the egg ontologically vacillates between its latent potential as a life-giving, self-generating object and its photographic presence as a stony, lifeless entity, unsettling the expectations of the viewer while simultaneously embodying Breton's call for artworks which more closely exemplified life's formative processes.[82]

With its stagy emphasis upon the smooth-edged form of a bird's ovum, Man Ray's *Ostrich Egg* thematically related to the highly charged current of artistic interest that was directed towards the egg as a symbol of aesthetic purity during the late 1920s and 1930s.[83] Drawn to the processes of concretion and structural simplification which produced the egg's perfectly ovoid form, artists as diverse as the Bauhaus photographer Hans Finsler and the Romanian Modernist sculptor Constantin Brancusi experimented with the object's formal properties and materiality, transforming its distinctively ellipsoid shape into a veritable leitmotif of Modernist practice. Brancusi, in particular – through such works as his *Sculpture for the Blind [1]* (c.1920) (Figure 2.8) – became synonymous with the egg-like figure, which was seen by many as emblematic of the Modernist artistic project as a whole. Henry Moore, for example, in a 1962 interview, extended a line of argument he had first put forward in 1937, specifying that Brancusi 'got people to look at shape again for its own sake, and made a martyr of himself [...], for a single form, for the egg, or the egg form, as the basis of a sculpture'.[84] In 'The Sculptor Speaks' (1937), Moore underlined the importance of this type of formal essentialism to his own practice: '[The sensitive observer of sculpture] must, for example, perceive an egg as a simple single solid shape, quite apart from its significance as food, or from the literary idea that it will become a bird'.[85] Equally, Hepworth's form experiments of the mid-1930s testify to Brancusi's influence. Thinking about the embryological connotations of sculptures such as *Two Forms (Snakewood)* of 1935 (Plate 1), for instance, the critic Hugh Gordon Porteus remarked, somewhat derisively, upon the formal similarities between Hepworth's new work and Brancusi's egg-like forms: 'Take away from Barbara Hepworth all that she owes Moore, and nothing would remain but a solitary clutch of Brancusi eggs'.[86] However, it was not simply that the egg represented the apogee of Modernist formal simplicity that artists found appealing; it was also that, in its boundless capacity for growth, the ovum offered a potent metaphor for the illimitability of the creative process.[87]

Especially for the Surrealists, the developmental potential of the fertilised egg far outstripped the imaginative powers of the human being, making it the ideal template for a freethinking artistic consciousness. The physician and writer Pierre Mabille expounded upon the ovum's creative aptitudes in a 1935 essay for the Surrealist journal *Minotaure*: 'The spermatozoon meets the ovule in the womb of the woman. From this minute, a construction begins, which exceeds the human frame for merging vast natural elaborations'.[88] As this process of intrauterine construction led inexorably to embryonic development, the iconography of the foetus was also an important aspect of this biologistic discourse.[89]

The photograph *Portrait d'Ubu* (1936) by the French Surrealist artist Dora Maar (the pseudonym of Henriette Dora Markovitch) – while imitating the stylistic principles of scientific photography – hideously distorts the scale of the object,[90] emphasising the monstrosity of a creature whose squashed physiognomy and delicate paws disturbingly echo the anatomy of a human foetus (Figure 2.9). The physiognomic overlap between the grotesque appearance of *Portrait d'Ubu* and an embryo at an early stage in its development suggests that Maar was passingly familiar with aspects of the theory of recapitulation, at least as they were summarised by her Surrealist fellow traveller Mabille in 'Preface to a Eulogy on Popular Prejudices'. Reprising an age-old evolutionary idea that had been popularly reappraised by the nineteenth-century evolutionary biologist Ernst Haeckel, Mabille thus summarised recapitulation:

> The [embryo] will be informed by species of animals and their progeny extending right across the geological ages. All the species living or extinct do not repeat themselves in formation in the embryo as one might think. Individually, or in a latent state, the formative tendencies found in periods distant from their emergence in the forms of higher animals reappear, roughly hewn, in the foetus.[91]

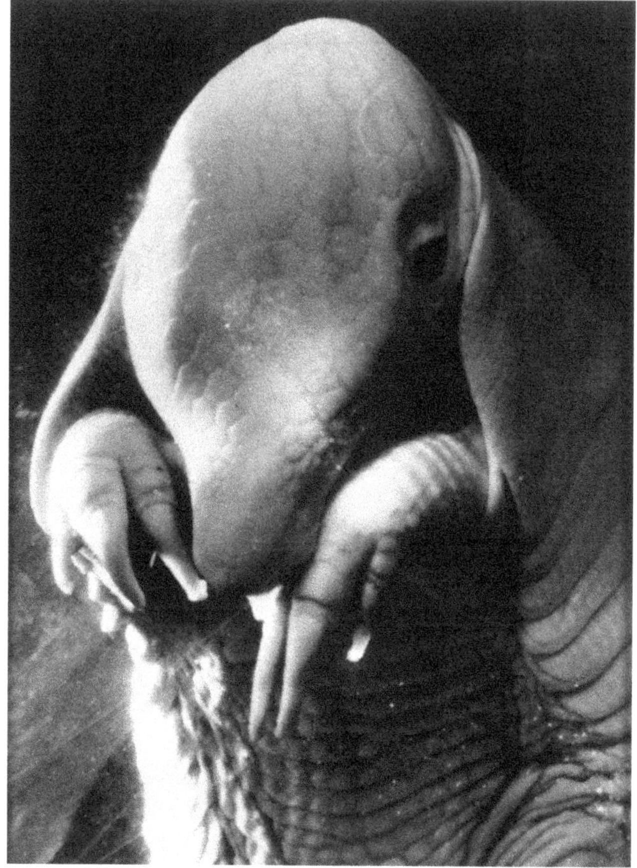

Figure 2.9
Dora Maar (Henriette Dora Markovitch), *Portrait d'Ubu*, 1936.
Centre Pompidou, Paris

Taking the principle that ontogeny (the development of the embryo) recapitulated phylogeny (the development of the species) to an appositely psychoanalytically informed extreme, Mabille postulated that the ancestral physical characteristics which emerged during embryonic development also impacted upon the psychological development of the individual: '[The unconscious] is made from the mass of things acquired in the current of the ages [...] which were [at one point] conscious and which have since, by a method of diffusion, entered into oblivion'.[92] If, as Dawn Ades convincingly argues, Maar's photograph was a mordant portrait of the dictator Ubu (a prodigious anti-hero created by the playwright Alfred Jarry), whose misrule was typified by inhuman acts of barbarism,[93] then its embryological iconography is suggestively apt: Ubu's physical degeneracy therefore reflects a psychological atavism, the 'vast wellspring of social conservation that [like the embryonic template] only increases slowly and changes little'.[94]

Perhaps no other contemporary artist was more closely associated with the iconography of embryology than the German-French sculptor and poet Hans Arp, whose sculptural vocabulary of nebulous, egg-like shapes eloquently conveyed the aesthetic principles that defined biomorphic abstraction.[95] Allied to Surrealism, but never a central member, Arp was drawn to the imagery of transformation and flux as a means of by-passing the intellectual straightjacket of rationalism.[96] Certainly, the slippery, metaphorical form-language of his artworks openly problematised the question of iconographic categorisation – much to the delight of his Surrealist devotees; as the writer Michel Leiris noted in a review of Arp's work at the Galerie Goemans in 1929: 'He buckles forms and, systematically, making everything nearly resemble everything else, disrupts the illusory classifications and even the scale of the things created'.[97] Nevertheless, the unstable, metamorphic quality of Arp's compositions was neither straightforwardly anti-rationalist nor was it an oblique return by the ex-Dadaist to figuration; rather it aimed at creating a novel, open-ended type of artistic morphology. The titles of works such as *Leaves and Navels* (1929), which appeared to undercut the coherency of representation by suggesting that Arp's equivocal forms were clearly identifiable, were really a 'play on the morphological possibilities of forms, of a formal language and its permutations'.[98] It was thus the analogical relationship struck *between* objects – the possibility of one form approaching the morphology of another – that psychologically animated Arp's compositions.[99]

Like many European artists who adopted a biocentric viewpoint while rejecting the principles of rationalism, Arp was attracted to the philosophy of neo-vitalism and purposively borrowed metaphors from biology so as to emphasise the autonomous and vital properties of his art.[100] Metaphors of gestation were of enormous importance to him, as they signalled a commitment to the spontaneous development of the artwork; like a germinating seed or developing ovum, the sculpture or collage was seen to be the product of dynamic internal forces affecting the object's materiality.[101] Unpremeditated yet brimming with morphological possibilities, the amorphous form of sculptures such as *Head with Annoying Objects* (1930) (Figure 2.10) played upon the *potential* of Arp's creations not only to signify a multiplicity of different things – navels, faces, leaves, rocks – but also to structurally develop, like a maturing embryo, into the forms that they suggested. Consequently, embryological imagery performed an important role in Arp's biologistic philosophy of art. For example, in his 'Notes from a Diary' he remarked that 'art is a fruit growing out of man like the fruit out of a plant like the child out of the mother'.[102] Published in the influential English-language, Paris-based literary journal *transition*, in 1932, this essay was an important vehicle through which Arp's gestatory notions of creative agency were conveyed to British audiences in the 1930s. Indeed, Stokes' 1933 discussion of the sculptor's block bearing 'vivifying or plastic fruit' (see above) bears the unmistakable watermark

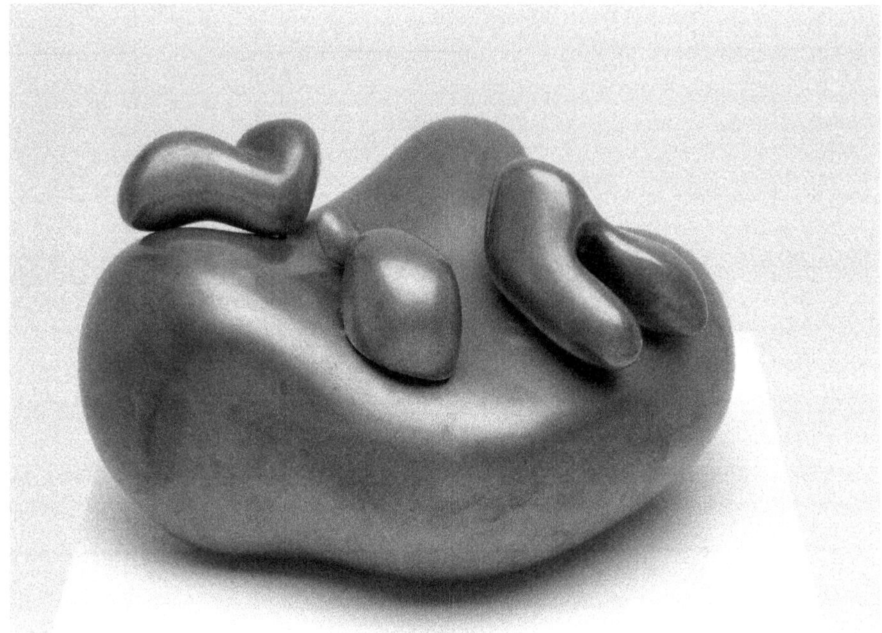

Figure 2.10

Hans Arp, *Head with Annoying Objects*, 1930, cast 3/5 (1970).
Stiftung Hans Arp und Sophie Taeuber-Arp e. V. Rolandswerth

of Arp's philosophy and both Moore and Hepworth testified to the determining influence that Arp had on their practice.[103] As we will see, the predilection towards embryological imagery and metaphor displayed by Surrealist artists fed into a Modernist fascination with neo-vitalist philosophy and procreative energies. Stimulated by Bergson's concept of an *élan vital*, artistic Modernists openly analogised between the artist as creator and life's creative energies, and conflated the spatial projection of the new sculpture with Driesch's notion of a dynamic, self-adjusting entelechy.

The core of origins: embryology, entelechy and sculptural dynamism

Of those artists involved in Unit 1 between 1933 and 1934, none was perhaps more influential in bringing Surrealist ideas pertaining to procreativity and embryology to a British audience than the painter, writer and photographer Paul Nash. Inspired by the visionary qualities of the empty cityscapes produced by the

Italian painter Giorgio de Chirico, Nash's own painterly and photographic experiments in Surrealism – stylistically influenced by the pseudo-scientific qualities of the New Objectivity photography – endeavoured to dramatise the physical and metaphysical relationships between objects.[104] Like many within the Modernist community who found philosophical solace within neo-vitalism and aspects of the New Natural Theology movement, Nash was professionally attracted to biomorphic imagery and framed his biologism through an incipient neo-vitalism.[105] Perhaps Nash's most eloquent disquisition on neo-vitalism in contemporary art, 'The Life of the Inanimate Object', was published in *Country Life* in 1937 and explained the psychological motivations which drove the Surrealists to select certain types of *objet trouvé*. Throughout this essay, Nash emphasised the nature-centric character of his practice, stressing that 'I am treating only of natural phenomena. The object, therefore, will belong to one of the three kingdoms: animal, vegetable, or mineral'.[106] Yet unlike the Surrealists, Nash highlighted how he wished to retain a level of conscious control over the object, as '[the] more the object is studied from the point of view of its animation the more incalculable it becomes in its variations'.[107] Importantly, Nash envisaged this *animation* to stem from the object's internal life, a kind of vital energy that infused certain types of natural entity, providing them with imaginative potential:

> To attain personal distinction, an object must show in its lineaments a veritable personality of its own. It may be a stone which looks like a bloodhound, as [with the example of] the Avebury megalith [...]; but it must not have only an amusingly canine look: it must be a *thing* which is an embodiment and most surely possesses power.[108]

The pantheistic undertones of this statement demonstrate just how closely artistic neo-vitalism was related to the emergence of neo-romanticism in Britain during the middle years of the 1930s.[109] Certainly, Nash's emphasis upon the associative meanings of landscape and acknowledged fondness for the nature philosophies of William Wordsworth and William Blake prove that his engagement with neo-vitalism was, at least partly, driven by a contemporary resurgence of interest in romantic *Lebensphilosophie*.[110]

Unsurprisingly, given Nash's commitment to the Surrealist project and preoccupation with neo-vitalism, an iconography of egg-like forms appears, in a slightly understated way, in his painting and photography. Positioned next to a hawk, the ovoid form in *Landscape from a Dream* (1936–38) (Figure 2.11), for example, plays upon both the Modernist status of the egg as a paradigmatic form-type and also Surrealist interest in the creative agency of embryological development. In 'The Nest of Wild Stones' – an essay printed in the neo-romantic publication *The Painter's Object* (1937) – Nash wittily drew upon Brancusi's well established penchant for ovoid forms while considering the ontological relationship between stones and eggs. Both objects are seen by Nash as replete with morphological possibility: a stone, through the agency of the sculptor, has the potential to be changed into all manner of forms – just as an egg, through

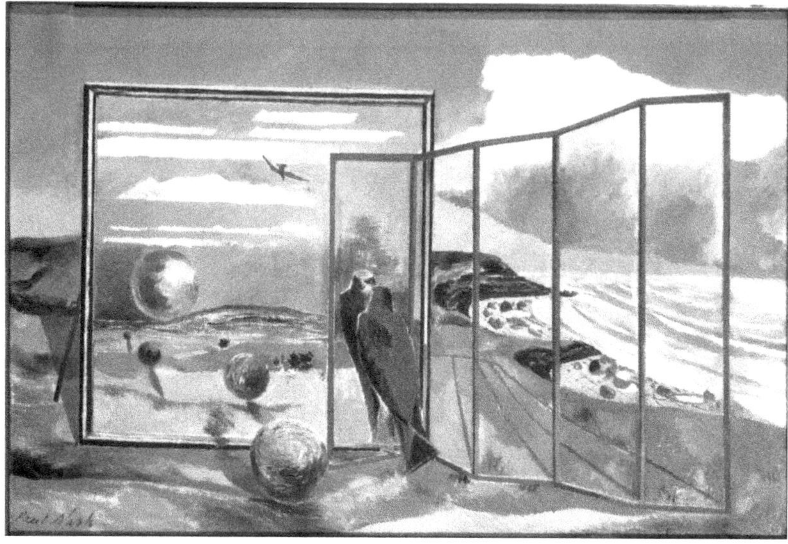

Figure 2.11

Paul Nash, *Landscape from a Dream*, 1936–38. Tate, London

embryological development, is able to transform into any type of bird. Yet a stone left untouched by the artist *retains* its morphological potential, bestowing it with the kind of psychological animation that Nash identified in 'The Life of the Inanimate Object':

> If stones are eggs they are birds, too. Not even grosbeaks always, or comic birds like toucans, but partridges and landrails much more, and pretty little quails. Larks even. All birds of the furrow and the down. Sculptors knock birds out of stones. By the time they have done with them they are neither birds nor stones. Except Brancusi's. But the stone birds of the field are always both. They do not insist. Perhaps, when they are lying on the ground they look like stones, and when they stand up they are birds, but, thank God, they never look like stone birds.[111]

It is significant that Nash chose sculpture as a metaphor through which to express his understanding of the morphological possibilities of certain types of natural object. As early as 1932, an exhibition of recent work by Hepworth and Nicholson at the Arthur Tooth Galleries in London saw him use gestatory metaphors as a means of elucidating the novelty of artistic Modernism. Here, the conscious control which Nash sought to impose over his interpretations of objects is seemingly emphasised by a process of rational embryonic development in which the artwork appears to expand steadily over time – rather than the more disorderly appreciation of temporality which Rosalind Krauss has identified within Surrealist practice.[112]

Embryology, in this sense, provided Nash with a conceptual template for a type of creativity which animated physical space through feigning morphogenesis:

> These works, whether carvings or paintings, suggest, as few are able to do, a process of exteriorization. The sculptor has not imposed, but has inhabited the stone and grown within, outwards, as a child grows in the womb. The painter has not applied, but builds up from the plane of his canvas outwards, a horizontal edifice of infinitely subtle balance. This is why in both cases we are given a sense of projection which in turn attracts our interest inward to the core of its origin.[113]

Antedating slightly Stokes' gendered model of creative agency, it is significant that Nash utilised embryological metaphors to account for Hepworth's carvings while, at the same time, employing a sexually neutral language of 'significant form' to describe Nicholson's paintings. Such interpretations of artistic agency unwittingly served to reinforce contemporary gender hierarchies by associating female creativity with procreativity and thus relegating, in some way, femininity to a sphere of nature onto which any number of sexual, maternal or erotic fantasies could be projected and sustained.[114]

Defining femininity as beholden to its gestatory potential, the appeal to Modernists of embryological metaphors lay partly in the socio-political machinations of the eugenics movement, whose efforts to spread the message of efficacious birth control acquired considerable public interest in the years between the wars. While eugenicists ardently believed that by organising childbirth more effectively society could severely limit the number of unwanted births and hence improve both maternal health and the general lot of children, at the heart of the birth-control movement was the Malthusian belief that the material demands of an expanding British population were rapidly outstripping supply and that those sectors of society that were currently experiencing a demographic boom were precisely those that contributed very little to the nation's socio-economic wealth.[115] Prominent eugenicists such as Marie Stopes were not only instrumental in establishing clinics that handed out information on birth control and contraception without legal constraint (Stopes herself set up the first such clinic in Holloway in 1921) but were also active in publicising the broader objectives of the eugenics movement through town-hall meetings, publications and promotional flyers. Yet, as Stopes' arch rival Margaret Sanger made clear in her *The Pivot of Civilization* of 1923, the defining issue for eugenicists was not simply to augment the wellbeing of society by preventing an unnecessary drain on its resources but to actively restrict the breeding of 'undesirables' so as to ease the burden of social 'defectives'. 'It is not without significance', wrote Sanger, 'that the moron and the imbecile set the pace in living up to [the doctrine of glorious and divine fertility], and that the intellectuals, the educators, the archbishops [etc.] who are most insistent on it, are the staunchest adherents in their own lives of celibacy and non-fertility'.[116]

As the eugenicists' circle of influence expanded to include many famous thinkers during the 1930s (such as John Maynard Keynes, Julian Huxley and George Bernard Shaw), so, too, an ever-increasing range of articles and radio talks promoting the prudency of birth control were submitted to the public sphere by institutions with a national audience, such as the BBC. Out of the morass of pro-eugenics writing at this time, the science writer Mary Adams stands apart for her vociferousness in calling, in the pages of *The Listener*, for the government to adopt a national policy of eugenics to counter the growing threat to social cohesion mounted by the 'feebleminded' from the swelling ranks of the lower classes.[117] By politicising motherhood and making selective use of cutting-edge genetic research, the eugenics movement exploited contemporary fears about imperial and, hence, racial decline that had proliferated following the Great Depression and responded to a half-acknowledged perception of Britain's shrinking political importance in world affairs.[118] Not all scientists, it should be said, were publicly convinced of the veracity of the eugenicists' scientific arguments (favouring the role of environment over that of genetics) and the challenge to democracy posed by a programme of eugenic socio-political intervention was famously satirised by Aldous Huxley in *Brave New World* (1932); yet, all the same, the public marketing of birth control did much to promote certain key issues then affecting the New Biology, most especially those involving the study of heredity and clinical embryology.

Hence, the focus upon embryos, fecundity and womanhood that typified critical responses to Modernist sculpture and, most especially, the act of carving in the early 1930s functioned as an indirect reaction to the pressures impressed on society by the eugenics movement in the politically turbulent years between the wars. From objects such as Leon Underwood's roughly hewn *Untitled (Foetus)* of 1924–25 (Figure 2.12) to Moore's foetal *Figure* of 1933–34 (Plate 2), sculptural production of the interwar period demonstrated an emphatic rejoinder to the fixation upon maintaining high birth rates and promoting maternal health that characterised socio-political discourse in the years following the Great War. Having children of her own no doubt drew Hepworth's attention to the way in which maternity had become radically politicised by the competing claims of the suffragettes and the birth-control movement, both of which increasingly drew a stark line between, on the one hand, choosing emancipation over the social shackles of motherhood and, on the other, undertaking a national duty by engaging in procreation.[119] By weaving the act of carving into a broader discussion of the relationship between creativity and *pro*creativity, critics such as Stokes and Nash inadvertently stepped into a battle being fought over women's rights and responsibilities by implicitly linking femininity to natural cycles and generative energies.

Yet within the impassioned battleground of interwar gender politics, embryology still had a more literal set of meanings pertaining to intrauterine development that were publicly communicated by conduits of the New Biology such as the layman's biological bible, *The Science of Life* (see Chapter 1). Such texts provided vivid, up-to-date accounts of embryonic growth from conception to birth, laying particular stress upon the subtle interplay between genetic and epigenetic factors that epitomised cellular differentiation in the developing organism.[120] Although by no means impervious to the politicised debates that were raging, a range of

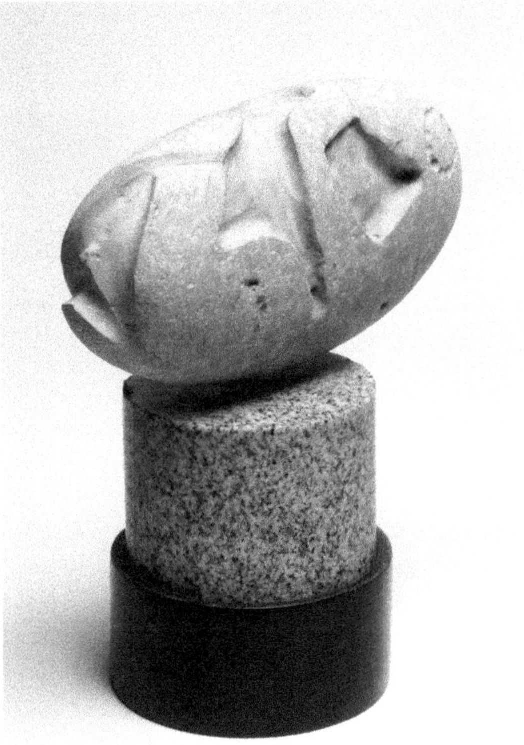

Figure 2.12

Leon Underwood, *Untitled (Foetus)*, 1924–25. Sherwin Collection, Leeds

popular publications sought to portray the 'facts of life' in as dispassionate a manner as possible, using technical terms that strikingly foreshadowed the gestatory metaphors taken up by artists and critics in the years leading up to the Second World War. That such texts strengthened cultural prejudices about the natural bonds existing between a mother and child by accentuating the embryo's 'embedment' in the womb was secondary to their expressive portrayals of embryonic growth in the animal kingdom, in which mitosis was seen to progressively delimit the organs of the body so that the organism could gradually regain possession over the hitherto undifferentiated mass of its cells.[121]

Recognising in embryology a ready set of metaphors through which to convey the rationale behind Modernist abstraction, artists drew upon the theme of cell division to suggest how an artwork developed in time and space without necessarily obeying any preconceived plan of action or blueprint of design.[122] A trailblazer in terms of tailoring the phraseology of the New Biology to the needs of modern art, the painter Jean Hélion couched an essay in *Axis* in embryological jargon so as to explain how a non-figurative composition incrementally progressed. 'It is not enough', he wrote, 'to develop an element or a zone; every element, taken as an end in itself, soon reaches an end, and fattens and degenerates; only groups of elements can grow'.[123] Though individually each pictorial element could present only a limited range of options to the painter, these could be multiplied by 'evolving' the connections between the different parts of the picture-plane:

> All the parts of a living organism [...] cannot reach the same degree of importance: they must keep connected in the process of development, but do not develop the same. There are simple cells, and all degrees presenting a different figure, a different stage.[124]

Conceiving of the canvas as an integrated field of activity that, like a growing embryo, differentiated itself into a patchwork of interlocking parts, Hélion drew upon popular accounts of mitosis to give meaning to Modernist abstraction. Like an embryo, whose form was seen to embody functional time,[125] the artwork came to be understood as somehow picturing its own development, each mark wrought upon its surface being emblematic of its making, in the same way that an embryo's structure kept traces of its growth cycle.

Hélion's organismal philosophy provided Modernists with a template for a type of creative agency which took developmental embryology as its paradigmatic metaphor. Naturally, the philosophical imperatives of neo-vitalism – which increasingly dominated modern art theory during the 1930s – encouraged a biologistic conception of artistry which explicitly analogised between the artist as creator and the creative powers of an incorporeal life force which wilfully shaped organic matter.[126] Hans Driesch, through his experiments with sea-urchin eggs, had become an influential proponent of neo-vitalism, arguing that the life force (or 'entelechy') was an active, purposeful entity that could not be described in physico-chemical terms.[127] Morphogenesis, he claimed, was unequivocally '*not* a specialised arrangement of inorganic events [...]: life is something apart, and biology is an independent science'.[128] Invited to give the Gifford Lectures in 1907–08 – which he published as *Science and Philosophy of the Organism* (1908) – Driesch became a key ally of the New Natural Theology movement in Britain, defending the moral implications of vitalism while pointing to the presence of 'an original primary entelechy', a divine spirit that had created the universe and was responsible for the existence of harmony in nature.[129] Importantly, nature's predisposition towards harmony was recognised by Driesch as an aesthetic quality which, in terms of entelechy, was akin to an artistic sensibility. The 'original primary entelechy' had produced harmony in nature, 'just as the artist makes an object of art'.[130]

The dynamic, self-organising character of Driesch's entelechy resonated among artistic Modernists who looked to the new science as a means of philosophically reasserting the irreducibility of the artwork.[131] Ewald Wasmuth – in response to a question posed in the periodical *transition* on the future of individuality in an age of collectivism – noted that 'all new theories, be they cosmological, physical or biological, point to the methodical basic concept of totality'.[132] For Wasmuth, Driesch had proved instrumental in the struggle against the straightjacket of materialism by using the concept of entelechy to posit a purposeful, dynamic interconnectivity between the separate parts of an organism:

> Driesch founded neo-vitalism. He coined the concept of the *causality of the whole* which premises methodically the idea of totality. The self-dividing eggs of sea urchins had a revolutionary effect [...]. The collectivity, as an organism, is [thus] not only to be understood because of the linear evolution, but we must also consider the 'field organization' which was discovered as a result of these experiments.[133]

Even though neo-vitalism had gradually been losing credibility as a biological theory,[134] the discovery by Hans Spemann of a mysterious substance, termed the 'organiser', which caused embryonic tissue to differentiate as part of a morpho-genetic field had encouraged those – Wasmuth included – who supported Driesch's philosophical teachings.[135] While embryologists such as Julian Huxley insisted that the organiser was simply a sophisticated chemical substance and so implicitly materialistic,[136] hazy knowledge of the organiser's properties and its similarity to Driesch's conception of the embryo as a 'harmonious equipotential system' meant that neo-vitalist sympathisers found plenty of philosophical ammunition with which to arm themselves during the 1930s.[137]

Conceived as a dynamic entity that impinged upon the development of matter, entelechy was understood to have a particular relationship with space, projecting its will – via matter – into the empty void of unrealised spatial extension. In his wide-ranging 1912 critique of Driesch's work, the philosopher C. Lloyd Morgan described entelechy's rapport with space thus:

> Entelechy [...] is affected by and acts upon spatial causality as if it came out of an ultra-spatial dimension; it does not act in space, it acts into space; it is not space, it only has points of manifestation in space. So, too, it is not in the material organism but only 'manifests' itself in this material.[138]

Since the romantic period, vitalism had appealed to artists and writers as diverse as William Blake and Samuel Taylor Coleridge in its capacity to express mystical states-of-being and the unity of time and space;[139] yet by the 1930s, the purposeful implications of entelechy were increasingly used by Modernists of a biocentric persuasion to explain the spatial dynamism of contemporary sculpture.[140] It was this conception of entelechy's dynamic spatial properties that led Hepworth to envisage the 'vitality' of sculpture as being the product of the artist's will purposively directing the development of form in inert material: 'The idea – the imaginative concept – actually *is* the giving of life and vitality to material [...]. Vitality is not a physical, organic attribute of sculpture – it is inner force and energy'.[141] Certainly, the recognition of a temporal rhythm within artistic design had been a fundamental aspect of Modernist responses to neo-vitalism in the early 1900s and, more specifically, Henri Bergson's concept of duration. The decorative colour arrangements of the Fauvist John Duncan Fergusson, for example, were viewed as a metaphor for a 'pantheistic life force within artistic creativity' as well as the *élan vital* of the artist's own psychology.[142]

Clearly, the equipotential, self-adjusting nature of Driesch's entelechy provided Modernists with a potent metaphor through which to conceptualise the irreducibility and compositional dynamism of the artwork. Envisaging the pictographic elements of a canvas as parts of a morphogenetic field, Hélion thus explained the 'rhythm' of an artwork as a complex entity, an intricate tapestry of self-organising elements whose signification constantly changed in relation to one another and the vacillating attentiveness of the viewer. Citing the enduring appeal of Nicolas Poussin's painting *Eliezer and Rebecca* (1648) as evidence for his biologistic argument, he hypothesised that:

> Poussin's *Eliezer and Rebecca* is still legible today because faces, gowns, landscapes, were used as bearers of rapports, multiplying the relations, developing their quality, intensity, oppositions to the maximum. Not only the maximum of one or two opposi-tions, but the maximum of all conceived oppositions. This is where concentration takes its full meaning. Concentration of a being in growth [...]. The same simplicity has no meaning except when applied to a complex organism. Simplification must be obtained on the way to growth, as a stage of that growth, instead of being obtained on the way to reduction, by suppression.[143]

Like Hélion, who here used the concept of a morphogenetic field to signify the stages through which a painting evolved until it attained rhythmic complexity, artistic Modernists drew upon embryological metaphors to explain the com-positional dynamism of the new sculpture, how it enlivened haptic and visual sensibilities by physically interrupting the spectatorial field. As Moore explained in biologistic terms in 1930: 'The sculpture which moves me most is full-blooded and self-supporting, fully in the round, that is its component forms are completely realised and work as masses in opposition, not merely being indicated by surface cutting in relief'.[144] This neo-vitalist rejection of 'surface cutting in relief' repre-sented a distinctly Modernist position which sought to question the reductionist shortcomings of formalism by synchronously destabilising the reductive principles of contemporary exhibitory practice.[145]

As Alex Potts has shown, until the early twentieth century, ways of beholding sculpture were restricted to an exhibitory paradigm that was founded upon the two-dimensionality of bas-reliefs – from which a principal viewpoint was believed to be most clearly established.[146] Originating in the writings of the nineteenth-century art critic Adolf Hildebrand, this philosophy rejected the visual confusion prompted by having to constantly move around a sculpture to gauge its overall disposition, and instead aimed at finding the viewpoint from which the object's primary form could be apprehended with the greatest clarity.[147] Only by adopting such an approach did Hildebrand feel that the viewer would be guaranteed to see the sculpture as a pleasing whole that was not disrupted by the act of traipsing from viewpoint to viewpoint.[148]

Therefore, by making use of a suite of neo-vitalist and embryological metaphors to explain the 'exteriorisation' of Modernist sculpture, both Moore and Nash drew attention to just how strangely these objects challenged contemporary cultures of display by projecting intrusively into the spectatorial field. Under these terms, the impression of turgescency that pervaded Moore's work led Geoffrey Grigson to think of ballooning, engorged forms which dynamically inhabited space like a gestating foetus: 'Analogies are at once obvious: the darkness of the womb, and shapes swelling and thrusting from it'.[149] Embryology here metaphorically supplies a sense of corporeal connectivity between the artist's creative agency and its spatio-temporal effects. At the same time, the purposiveness that Grigson attributes to the artist's dynamic will – a consciousness that manifests itself *in* material and acts *upon* space – is beholden to Bergson's understanding of the creative impulse and Driesch's neo-vitalist conception of entelechy. As we will see, the dynamic compositional character of Modernist sculpture was directly linked to its asymmetrical appearance – a morphological quality that was believed to kinaesthetically enliven the artwork. Combined with a neo-vitalist philosophy, this biologistic aesthetic likened the new sculpture to living things precisely because it appeared to *act* spatio-temporally, obliging the viewer to appreciate its form in four-dimensional terms.

Vital signs:
asymmetry and the eclipse of fixed-point perspective

> For me a work must first have a vitality of its own. I do not mean a reflection of the vitality of life, of movement, physical action, frisking, dancing figures and so on, but that a work can have in it a pent-up energy, an intense life of its own, independent of the object it may represent.[150]

Henry Moore's above synopsis, in *Unit 1*, of the expressive powers of sculpture reveals the extent to which his biologistic aesthetic was shaped by the metaphysical priorities of neo-vitalist philosophy and the neo-romantic belief that the contemporary artist did not simply reproduce objects, but, rather, added novel forms to nature's inventory.[151] This neo-vitalist conception of the artwork was connected, in the minds of critics and artists, to the peculiar spatial disposition of Modernist sculpture, which appeared – like Driesch's self-adapting entelechy – to spatio-temporally modify itself to the phenomenological capacity of the viewer. In *The Meaning of Art*, for example, Herbert Read presented an unabashedly Bergsonian, kinaesthetic model of the sculptural object, in which the perception of a sculpture was an explicitly spatio-temporal experience:

> We cannot see all round a cubic mass; the sculptor therefore tends to walk round his mass of stone and endeavours to make it satisfactory from every point of view. He can thus go a long way towards success, but he cannot be so successful as the sculptor whose act of creation is, as it were, a four-dimensional process growing out of a conception which inheres in the mass itself.[152]

To speak of physical development of any kind necessitates invoking the transition of matter from one form into another.[153] Naturally, sculpture's tactile properties mean that it conveys a subtle, phenomenological interplay between *materiality* and *spatiality*. Yet biologistic emphases upon morphogenesis show not only how Modernist sculpture dynamically inhabited the spectatorial field but also actively *shaped* viewers' kinaesthetic experience by forcing them to move around the artwork so as to fully apprehend the sculptural volume.[154]

The radical, kinaesthetic character of the new sculpture was associated with a break from fixed-point perspectival systems. Indeed, Hildebrand – while discussing Auguste Rodin's work in 1917 – anticipated this eventuality, bemoaning that Rodin's works 'fell apart' precisely because they denied modern sculpture coherency and stasis of form.[155] Considering Rodin's legacy, the critic Anatole Jakovski, in an *Axis* article of 1935, highlighted the debt Modernist sculptors owed to his wilful disintegration of sculptural volume:

> Rodin's touch, his little cells of shadow and light, disintegrate the volume, in the same way as spectral divisionism – the decomposition of the old 'white' by [Impressionist] painters [...] killed the modeling, exhausted the perspective and consequently destroyed the whole weight of the picture.[156]

Whereas Moore claimed that 'surface cutting in relief' implied a flat surface with only one principal viewpoint,[157] the vitality of Modernist sculpture was linked to its asymmetrical character – a quality that was seen to significantly augment the number of perspectives from which it could be apprehended. Hepworth claimed in *Circle* that: 'It does not matter whether a sculpture is asymmetrical or symmetrical – it does not lose or gain by being either: [but] it can be said that an asymmetrical sculpture has more points of view'.[158] While sidestepping any clear-cut position on asymmetrical composition (no doubt because symmetry still formed an important part of her sculptural production), Hepworth nonetheless spoke fluently of its phenomenological qualities, of how asymmetry transformed the perceptual relationship between viewer and object by opening up multiple perspectives, impelling the active circumnavigation of the sculpture: 'asymmetry can be found in the tension, balance, inner vital impact with space and in the scale [of the sculpture]'.[159] Likewise, in *Unit 1* Moore wrote of how carving in the round allowed for the proliferation of viewpoints:

> Sculpture fully in the round has no two points of view alike. The desire for form completely realised is connected with asymmetry. For a symmetrical mass being the same from both sides cannot have more than half the number of different points of view possessed by a non-symmetrical mass.[160]

Aside from possessing dynamic phenomenological qualities, asymmetry was recognised as being a property inherent to living things and was discussed by sculptors in developmental terms as symbolic of the formative processes of nature.[161] Moore thus spoke of how asymmetry abstractly represented natural forces shaping organic material: 'Asymmetry is connected with the desire for the organic rather than the geometric. Organic forms though they may be symmetrical in their main disposition, in their reaction to environment, growth and gravity, lose their perfect symmetry'.[162] Such impressions of asymmetry stemmed from experimental morphology, in which physical energies were seen to impress themselves upon organic material, forcing biological systems to undergo endless cycles of change or bodily deformation. In *On Growth and Form*, D'Arcy Wentworth Thompson wrote at length of how the emphatic curvature of organic structures was a consequence of surface tension and the contractility of protoplasm under the pressure of gravitational or thermodynamic forces. The form of the amoeba, in particular, was explained to be 'the very negation of rest or of equilibrium [as the] creature [was] always moving, from one protean configuration to another; its surface tension [being] never constant, but continually [varying] from here to there'.[163] The constantly changing tensity of the amoeba's protoplasmic surface meant that it could never acquire symmetry, forcing it to endlessly fluctuate from one vaguely spherical form to another. Thompson maintained that the equilibrium sought by most organisms was generally absent, resulting in forms that 'are in continual flux and movement, each portion of the surface constantly changing its form, passing from one phase to another of an equilibrium which is never stable for more than a moment'.[164] That said, Thompson was sceptical that asymmetry was *quintessential* to living things, arguing that '[in] all cases where the principle of maxima and minima comes into play, as it conspicuously does in films at rest under surface-tension, the configurations so produced are characterized by obvious and remarkable *symmetry*. Such symmetry is highly characteristic of organic forms and is rarely absent in living things'.[165]

Asymmetry was, of course, a formal quality that was much feted in biomorphic abstraction during the 1930s.[166] In his essay 'Painting and Sculpture Today' of 1935, for instance, Geoffrey Grigson contended that Constantin Brancusi's 'polished, unicellular forms' had provided the kernel for Moore's experimentations with biomorphic Modernism.[167] And writing in the catalogue of the 'Cubism and Abstract Art' exhibition – held at the Museum of Modern Art, New York, in 1936 – the American curator, Alfred Barr, depicted biomorphism as 'the silhouette of an amoeba' and 'a soft, irregular, curving silhouette half-way between a circle and the object represented'.[168]

As we will see in Chapter 4, curvature of form had long been associated with nature's generative energies – most especially within romantic nature philosophy.[169] In *The Seven Lamps of Architecture* (1849), for example, the nineteenth-century art theorist John Ruskin pronounced that 'all perfectly beautiful forms must be composed of curves; since there is hardly any common natural form in which it is possible to discover a straight line'.[170] Yet for artistic Modernists, informed by the morphogenetic suppositions of the New Biology, it was the capacity of an irregular, protuberant line to abstractly convey nature's developmental energies that gave it particular currency in representing the neo-vitalist irreducibility of

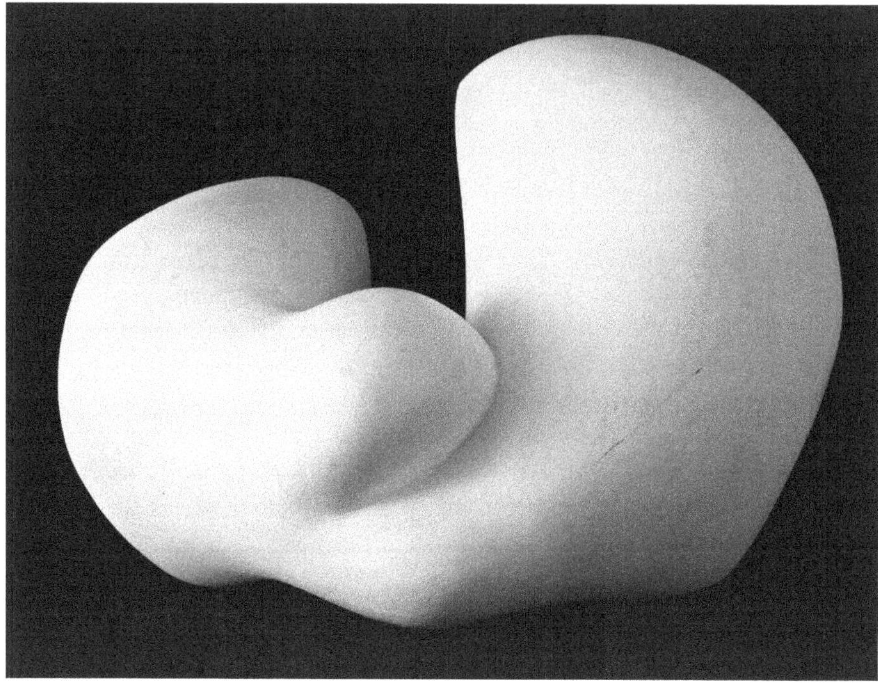

Figure 2.13

Hans Arp, *Human Concretion*, 1935. Museum of Modern Art (MoMA), New York

the artwork. Indeed, whereas a logarithmic curve or spiral form, though indicative of vital energies, represented a perfect geometry that appealed to a neo-romantic, idealist morphology, an amoebic or irregularly shaped circular form suggested, by virtue of its linear deformation, a vital process of metamorphosis or epigenesis.[171] Bergson, for one, spoke of the spontaneity and unpredictability of the amoeba's form as embodying the quintessence of the *élan vital*: 'The amoeba deforms itself in varying directions; its entire mass does what the differentiation of parts will localize in a sensori-motor system in a developed animal'.[172]

Tellingly, it was Arp's move away from a frieze-style mode of composition that inspired Barr's speculations on asymmetrical form: 'Recently, Arp has turned away from stratified relief to sculpture in the round. His *Human Concretion* is a kind of sculptural protoplasm, half-organic, half the water-worn white stone'.[173] Along these lines, the undulating, polymorphous nature of *Human Concretion* (1935) (Figure 2.13) ardently communicates the impression of a protoplasmic organism existing in a state of flux.[174] The irregular swells and distensions of Arp's sculpture imply a kind of volatility of surface which expands centrifugally into the spectatorial field. As Rosalind Krauss has discerned: 'Projecting a sense of

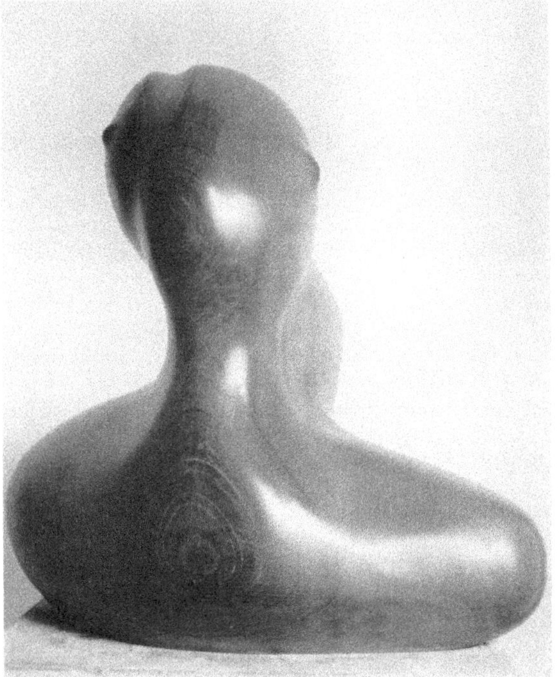

Figure 2.14

Henry Moore, *Composition*, 1932. Tate, London

fluid pressure from within the volumetric container, Arp's stressed surfaces become images of variability and change'.[175] Dynamic properties similar to those present in *Human Concretion* are equally evinced in objects such as Moore's *Composition* (1932) (Figure 2.14), in which a liquid volume bulges sinuously outwards, kinaesthetically enlivening space through a miscellany of irregular protrusions. Conceived as four-dimensional entities, such asymmetrical volumes thus demand the active circumlocution of the sculptural object as, jutting lopsidedly into the spectatorial field, their *complete* spatial disposition can be fully apprehended only through a 360-degree mode of viewing (unlike a 'stratified relief', which can, under Hildebrand's phenomenology, be completely realised only from a single, static viewpoint). Thompson's supposition that the '*conformation* of the organism' was nothing more than the 'interaction or balance of forces'[176] helped in the formulation of this aesthetic by providing Modernists with an effective metaphor through which a sculpture could be seen as an adaptable, metamorphic system that responded to the fluctuating intentionality of the artist and the vacillating position of the viewer. Hence, by envisaging the sculptural object as a morphogenetic field that kinaesthetically animated space through its asymmetry and spatio-temporality, British Modernists could conceive of it, in neo-vitalist terms, as embodying the 'rhythm and growth of life'.[177]

Notes

1 Christa Lichtenstern has argued that Moore began producing these drawings at the same time as he was exposed to Surrealist ideology and began working in a biomorphic style. See Lichtenstern, *Henry Moore*, p. 53.

2 Breton, for example, speaks in 'La beauté sera convulsive' of the imaginative potential of crystal and geological forms: 'The inanimate touches here so closely upon the animated that the imagination is free to play upon the infinity of the forms of appearance of any mineral, reproducing in its being the reasoning which consists in recognising a nest, a bouquet held by a petrified fountain'. ('L'inanimé touche ici de si près l'animé que l'imagination est libre de se jouer à l'infini sur ces formes d'apparence toute minérale, de reproduire à leur sujet la démarche qui consiste à reconnaître un nid, une grappe retirés d'une fontaine pétrifiante'.) A. Breton, 'La beauté sera convulsive', in *Minotaure*, No. 5, 1934, p. 13.

3 As Lichtenstern has shown, while an interest in morphogenesis was central to the Surrealist project, Moore generally ignored the psychological revelation afforded by Surrealism in favour of a systematic exploration of the possibilities of form. While it has been more recently shown by scholars such as Andrew Causey that Moore was deeply interested in Surrealism's engagement with Freud, this shift towards Surrealism largely took place around 1936, so confirming Lichtenstern's argument for the period in which the *Transformation Drawings* were produced. See C. Lichtenstern, 'Henry Moore and Surrealism', in *The Burlington Magazine*, Vol. CXXIII, No. 944, November 1981, p. 651.

4 See H. Moore, Untitled statement, in H. Read (ed.), *Unit 1: The Modern Movement in English Architecture, Painting and Sculpture* (London: Cassell, 1934), p. 30.

5 *Ibid.*, p. 30.

6 See Lichtenstern, *Henry Moore*, p. 57.

7 See Harrison, *English Art and Modernism*, p. 282; M. Jarron & C. Caudwell, *D'Arcy Thompson and His Zoology Museum in Dundee* (Dundee: University of Dundee Museum Services & The College of the Life Sciences, 2010), p. 53.

8 D'Arcy Wentworth Thompson, *On Growth and Form* (1917) (New York: Dover, 1992), pp. 1069–70.

9 Lichtenstern, 'Henry Moore and Surrealism', p. 651.

10 H. Moore, notes on 'Ideas for Sculpture: Transformation of Bones' (1932), AG 32.52 HMF941, reproduced in A. Garrould (ed.), *Henry Moore, Volume II: Complete Drawings, 1930–39* (London: Henry Moore Foundation, 1998), p. 70.

11 Thompson, *On Growth and Form*, p. 1053.

12 Read, *Art Now*, p. 61.

13 Maldonado, *Le cercle et l'amibe*, pp. 181–2.

14 See Allen, *Life Science in the Twentieth Century*, p. 9.

15 *Ibid.*, pp. 8–19.

16 See Bowler, *Reconciling Science and Religion*, pp. 147–59; Allen, *Life Science in the Twentieth Century*, pp. 144–5.

17 Bowler, *Reconciling Science and Religion*, pp. 147–8.

18 Allen, *Life Science in the Twentieth Century*, p. 145.

19 *Ibid.*, p. 113.

20 J.G. Kerr, 'Links in the Chain of Life: I. The Amoeba, an Old-Fashioned Creature', in *The Listener*, Vol. III, No. 55, 29 January 1930, p. 188.

21 See Bowler, *Reconciling Science and Religion*, p. 147.

22 J.B.S. Haldane, *The Causes of Evolution* (London: Longmans, Green & Co., 1932), pp. 28–9.

23 See Huxley, Wells & Wells, *The Science of Life*, p. 350.

24 J. Huxley, *Evolution: The Modern Synthesis* (London: George Allen & Unwin, 1942), p. 29.

25 See Allen, *Life Science in the Twentieth Century*, pp. 27, 29.

26 *Ibid.*, p. 30.

27 *Ibid.*, p. 33.

28 F. Burwick & P. Douglass, 'Introduction', in F. Burwick & P. Douglass (eds), *The Crisis in Modernism: Bergson and the Vitalist Controversy* (Cambridge: Cambridge University Press, 1992), p. 1.

29 Bowler, *Reconciling Science and Religion*, p. 378.

30 H. Read, *Icon and Idea: The Function of Art in the Development of Human Consciousness* (London: Faber & Faber, 1955), p. 19.

31 See Bowler, *Reconciling Science and Religion*, p. 167.

32 Botar, 'Defining Biocentrism', p. 27.

33 Bowler, *Reconciling Science and Religion*, p. 167; Botar, *Prolegomena to the Study of Biomorphic Modernism*, p. 190.

34 G. Rousseau, 'The Perpetual Crises of Modernism and the Traditions of Enlightenment Vitalism: With a Note on Mikhail Bakhtin', in F. Burwick & P. Douglass (eds), *The Crisis in Modernism: Bergson and the Vitalist Controversy* (Cambridge: Cambridge University Press, 1992), p. 57.

35 H. Read quoted in Woodcock, *Herbert Read*, p. 195.

36 C.E.M. Joad, *The Meaning of Life* (London: Watts & Co., 1928), p. 14.

37 *Ibid.*, p. 22.

38 Hélion, 'From Reduction to Growth', p. 24.

39 J. Huxley, 'The Age of Change', in *The Listener*, Vol. IX, No. 222, 12 April 1933, p. 590.

40 E. Fernie, *Art History and Its Methods* (London: Phaidon, 1995), p. 337.

41 H. Focillon, *The Life of Forms in Art* (1934) (New York: Zone Books, 1989), p. 117.

42 *Ibid.*, pp. 33, 41.

43 See Maldonado, *Le cercle et l'amibe*, p. 119.

44 Bergson, *Creative Evolution*, p. 302.

45 Focillon, *The Life of Forms in Art*, pp. 42–4.

46 See Maldonado, *Le cercle et l'amibe*, p. 120.

47 J.G. Kerr, 'Links in the Chain of Life: VI – The Animal World of Today', in *The Listener*, Vol. III, No. 60, 5 March 1930, p. 421.

48 *Ibid.*, p. 421.

49 Focillon, *The Life of Forms in Art*, p. 60.

50 Maldonado, *Le cercle et l'amibe*, p. 120.

51 Even though Hepworth did not own a copy of Focillon's book until at least late 1944, from the tone of her statements and her correspondence with Herbert Read, it is clear that she possessed an understanding of its aesthetic precepts. See Barlow, 'Barbara Hepworth and Science', p. 105.

52 S. Haden-Guest, 'A New Energy in the Arts', in *The Listener*, Vol. VII, No. 156, 6 January 1932, p. 19.

53 *Ibid.*, p. 19.

54 *Ibid.*, p. 20.

55 Focillon, *The Life of Forms in Art*, pp. 119–20.

56 *Ibid.*, p. 117.

57 See Maldonado, *Le cercle et l'amibe*, p. 119.

58 Huxley, 'Tissue Culture and Human Habits', p. 952.

59 *Ibid.*, p. 953.

60 C.E.M. Joad, 'The Confusion of Modern Thought' in *The Listener*, Vol. X, No. 240, 16 August 1933, p. 229.

61 A. Stokes, 'Miss Hepworth's Carving', in *The Spectator*, November 1933, reprinted in *The Critical Writings of Adrian Stokes: Vol. I, 1930–1937* (London: Thames & Hudson, 1978), p. 309.

62 For additional information on this topic, see K. Deepwell, 'Hepworth and Her Critics', in D. Thistlewood (ed.), *Barbara Hepworth Reconsidered* (Liverpool: Liverpool University Press, 1996), esp. pp. 75–86.

63 Stokes, 'Miss Hepworth's Carving', p. 310.

64 A.M. Wagner, 'Miss Hepworth's Stone *Is* a Mother', in D. Thistlewood (ed.), *Barbara Hepworth Reconsidered* (Liverpool: Liverpool University Press, 1996), p. 57.

65 Stokes, 'Miss Hepworth's Carving', p. 310.

66 A. Potts, 'Carving and the Engendering of Sculpture: Adrian Stokes on Barbara Hepworth', in D. Thistlewood (ed.), *Barbara Hepworth Reconsidered* (Liverpool: Liverpool University Press, 1996), p. 48.

67 See Wagner, 'Miss Hepworth's Stone *Is* a Mother', esp. pp. 59–67.

68 Potts, 'Carving and the Engendering of Sculpture', p. 47.

69 For a discussion of this subject see W. Chadwick, *Women, Art, and Society* (London: Thames & Hudson, 1990), pp. 309–10.

70 Remy, *Surrealism in Britain*, pp. 62–7. Despite Moore's stylistic bipartisanship, which led to his remarkable claim that 'the violent quarrel between the abstractionists and the surrealists seems [...] quite unnecessary', 1936 witnessed a perceptible change in his work, especially in his drawings, which began to incorporate darker, more claustrophobic elements into their compass, using Surrealism as a filter through which to channel a personal interpretation of Freud's notion of the uncanny. Andrew Causey has described this transformation thus: 'There is a shift from the smooth natural forms of Moore's sculptures to the sinister, morbid and occasionally even grotesque character of the drawings'. A. Causey, 'Henry Moore and the Uncanny', in J. Beckett & F. Russell (eds), *Henry Moore: Critical Essays* (Aldershot: Ashgate, 2003), pp. 83–4.

71 See K. Marwood, 'Shadows of Femininity: Women, Surrealism and the Gothic', in *re.bus*, No. 4, autumn/winter 2009, pp. 1–2. Marwood notes that this tendency to view Surrealism as dependent upon romantic and Gothic precedents was particularly evident in critiques of female Surrealists, whose work was negatively perceived as wantonly anachronistic and thus unoriginal. Indeed, certain figures within Surrealism were increasingly tired of the negative connotations stemming from the movement being viewed as something akin to romanticism's last hurrah. See also Harrison, *English Art and Modernism*, p. 312.

72 André Breton, 'What Is Surrealism?', reprinted in Franklin Rosemont (ed.), *André Breton: What Is Surrealism? Selected Writings* (London: Pluto Press, 1978), p. 132.

73 Mundy, *Biomorphism*, p. 154.

74 'La vie dans la constance de son processus de formation et de destruction, ne me semble pour l'oeil humain pouvoir être concrètement mieux enclose qu'entre les haies de mésanges blues de l'aragonite et le pond de trésors de la "grande barrière" australienne'. A. Breton, *L'amour fou* (1937) (Paris: Éditions Gallimard, 2010), p. 18. This text was first published as: André Breton, 'La beauté sera convulsive', in *Minotaure*, No. 5, 1934, p. 13.

75 See D. Ades, 'Little Things: Close-Up in Photo and Film 1839–1963', in D. Ades & S. Baker (eds), *Close-Up: Proximity and Defamiliarisation in Art, Film and Photography* (Edinburgh: Fruitmarket Gallery, 2008), p. 48.

76 See Breton, *L'amour fou*, pp. 16–17.

77 Mundy claims that Breton was enamoured of 'a crystalline and not an organic model of artistic form' and states that 'Breton took not a living organism [as a symbol of life] but the inorganic structures of coral'. Mundy, *Biomorphism*, p. 154. Nonetheless, it was well known by the 1930s that coral was a living organism and not a crystal or stone. Indeed, Breton differentiates here between the crystalline as a model for the artwork and the corallaceous as a model for *life* – suggesting that it was precisely coral's vital properties (its susceptibility to both growth *and* decay) that made it appropriate as a template for artistic agency even as, in its apparent solidity or petrification, it seemed to embody crystalline qualities. It is thus unlikely that Breton would have chosen such a complex organism (one prone to calcification as well as budding) simply to illustrate a point about crystals – or indeed that he was unaware of the distinction.

78 For information on the role scientific photography played in Surrealist aesthetics, see M. Warehime, *Brassaï: Images of Culture and the Surrealist Observer* (Baton Rouge: Louisiana State University Press, 1996), pp. 9–12.

79 The photographs of coral were displayed alongside images of crystalline forms, thus emphasising the self-similarity of form evident in both types of natural object. See also Ades, 'Little Things', p. 48.

80 Botar, *Prolegomena to the Study of Biomorphic Modernism*, p. 505.

81 The image begs comparison, for example, with James Nasmyth's astronomical photographs of the moon from the late 1800s.

82 See Ades, 'Little Things', p. 52.

83 On this subject, see Maldonado, *Le cercle et l'amibe*, p. 82.

84 Henry Moore interviewed by John and Vera Russell, 'Moore Explains His Universal Shapes', in *New York Times Magazine*, 11 November 1962, reprinted in P. James (ed.), *Henry Moore on Sculpture* (London: MacDonald, 1966), p. 198.

85   H. Moore, 'The Sculptor Speaks', in *The Listener*, Vol. XVIII, No. 449, 18 August 1937, p. 338.

86   H.G. Porteus, 'New Planets', in *Axis*, No. 3, July 1935, p. 22.

87   Maldonado, *Le cercle et l'amibe*, pp. 83–4.

88   'Le spermatozoïde a rencontré l'ovule dans la matrice de la femme. Dès cette minute, une construction commence, qui dépasse le cadre humain pour rejoindre les vastes élaborations naturelles'. P. Mabille, 'Préface à l'éloge des prejugés populaires', in *Minotaure*, No. 6, 1935, p. 1.

89   Maldonado, *Le cercle et l'amibe*, p. 84. As Vivian Endicott Barnett has demonstrated, Wassily Kandinsky drew upon scientific diagrams to depict embryos in various stages of development in his paintings. Barnett, 'Kandinsky and Science', p. 216.

90   Ades, 'Little Things', p. 52.

91   'Les échafaudages seront fournis par les espèces animales et les fils tendus au travers des âges géologiques. Toutes les séries vivantes ou éteintes ne se répètent pas dans l'embryon en formation comme on pourrait trop simplement le penser. Seules, et a l'état d'intentions, les tendances d'édification qui ont trouvé en des périodes lointaines leur réalisation dans les formes des animaux supérieurs reparaissent ébauchées dans le fœtus'. Mabille, 'Préface à l'éloge des prejuges populaires', p. 1.

92   '[L'inconscient] est faite de la masse des choses apprises au courant des âges [...], qui furent conscientes et qui par diffusion sont entrées dans l'oubli'. *Ibid.*, p. 2.

93   Ades, 'Little Things', p. 52.

94   'Vaste fond de conservation sociale qui ne s'augmente que lentement et se modifie peu'. Mabille, 'Préface à l'éloge des prejuges populaires', p. 2.

95   Maldonado, *Le cercle et l'amibe*, pp. 82–3.

96   Mundy, *Biomorphism*, pp. 129–30.

97   'Il fait se gondolier les formes et, systématiquement, rendant tout presque semblable à tout, bouleverse les classifications illusoires et l'échelle même les choses créées'. M. Leiris, 'Exposition Hans Arp', in *Documents*, No. 6, 1929, p. 340.

98   B. Fer, *On Abstract Art* (New Haven: Yale University Press, 1997), p. 76.

99   Krauss, *Passages in Modern Sculpture*, p. 138.

100 Mundy, *Biomorphism*, p. 121; Botar & Wünsche, 'Introduction', pp. 3–4. On vitalism in the history of the life sciences, see Allen, *Life Science in the Twentieth Century*, p. xxii.

101 Maldonado, *Le cercle et l'amibe*, pp. 187–9.

102 H. Arp, 'Notes from a Diary', in *transition*, No. 21, 1932, p. 191.

103 Both Moore and Hepworth visited Arp's studio in the 1930s. As Sally Festing points out, Hepworth was transfixed by the way in which Arp fused landscape and human forms in his art. S. Festing, *Barbara Hepworth: A Life of Forms* (London: Viking, 1995), p. 98.

104 Remy, *Surrealism in Britain*, pp. 36–43.

105 Botar, *Prolegomena to the Study of Biomorphic Modernism*, pp. 47–8.

106 P. Nash, 'The Life of the Inanimate Object', in *Country Life*, 1 May 1937, pp. 496–7, reprinted in A. Causey (ed.), *Paul Nash: Writings on Art* (Oxford: Oxford University Press, 2000), pp. 138–9.

107 *Ibid.*, p. 139.

108 *Ibid.*, p. 139.

109 Botar, *Prolegomena to the Study of Biomorphic Modernism*, p. 47.

110 See Harrison, *English Art and Modernism*, pp. 317–26. For a discussion of the philosophical relationship between neo-vitalism and neo-romanticism, see Botar, 'Defining Biocentrism', pp. 15–17, 23.

111 P. Nash, 'The Nest of Wild Stones', in M. Evans (ed.), *The Painter's Object* (London: Gerald Howe, 1937), p. 38.

112 In this sense, Nash's writing can be linked to Arp's practice, which, as Rosalind Krauss has shown, similarly used *growth* as a metaphor through which to indicate a sense of rational, stage-by-stage development in his art. See Krauss, *Passages in Modern Sculpture*, p. 140.

113 P. Nash, 'A Painter and a Sculptor', in *Weekend Review*, 19 November 1932, p. 613.

114 Chadwick, *Women, Art, and Society*, pp. 310–11.

115 For an overview of this topic see Overy, *The Morbid Age*, pp. 93–135.

116 M. Sanger, *The Pivot of Civilization* (London: Jonathon Cape, 1923), pp. 32–3.

117 In the early 1930s Adams regularly produced broadcasts and articles on this subject. To gauge a measure of their strident flavour the reader is directed to: M. Adams, 'What Do We Owe Our

Forefathers?', in *The Listener*, Vol. IV, No. 97, 19 November 1930, pp. 838–9; M. Adams, 'Has Human Nature Ever Changed?', in *The Listener*, Vol. IV, No. 98, 26 November 1930, pp. 883–4; M. Adams, 'Shall the Unfit Survive?', in *The Listener*, Vol. IV, No. 99, 26 November 1930, p. 933.

118 Overy, *The Morbid Age*, pp. 99–101.

119 These ideas are more fully discussed by Anne Wagner in her canonical study of interwar British sculpture: A.M. Wagner, *Mother Stone: The Vitality of Modern British Sculpture* (New Haven: Yale University Press, 2005), esp. pp. 6–8, 45–9.

120 See, for example, the discussion of embryonic development provided by Huxley, Wells & Wells, in *The Science of Life*, esp. pp. 93–5.

121 *Ibid.*, pp. 314–30.

122 Maldonado, *Le cercle et l'amibe*, pp. 187–9.

123 Hélion, 'From Reduction to Growth', p. 22.

124 *Ibid.*, p. 23.

125 See Canguilhem, *Ideology and Rationality*, p. 108.

126 See Krauss, *Passages in Modern Sculpture*, p. 140.

127 Bowler, *Reconciling Science and Religion*, pp. 166–7.

128 H. Driesch, *Science and Philosophy of the Organism* (Aberdeen: Aberdeen University Press, 1908), Vol. I, p. 142.

129 Bowler, *Reconciling Science and Religion*, p. 167.

130 Driesch, *Science and Philosophy of the Organism*, Vol. II, p. 370.

131 See Mundy, *Biomorphism*, p. 121.

132 Ewald Wasmuth, untitled response to the question of how the individual and metaphysics will develop in a collectivist epoch, in *transition*, No. 21, March 1932, p. 142.

133 *Ibid.*, p. 143. Original emphasis.

134 Bowler, *Reconciling Science and Religion*, p. 167.

135 Allen, *Life Science in the Twentieth Century*, p. 122.

136 See J. Huxley, 'Organisers in Animal Development', in *The Listener*, Vol. X, No. 234, 5 July 1933, pp. 28–9.

137 The ambiguity between the organicist and neo-vitalist positions – both of which stressed the irreducibility of living matter – did much to sustain popular faith in neo-vitalism, even as prominent organicists like J.S. Haldane took pains to stress the differences between the two worldviews.

138 C. Lloyd Morgan, *Instinct and Experience* (London: Methuen, 1912), p. 157.

139 Rousseau, 'The Perpetual Crises of Modernism', p. 46.

140 Krauss, *Passages in Modern Sculpture*, pp. 140–1.

141 B. Hepworth, 'Sculpture', in J.L. Martin, B. Nicholson & N. Gabo (eds), *Circle: International Survey of Constructive Art* (London: Faber & Faber, 1937), p. 113.

142 See M. Antliff, *Inventing Bergson: Cultural Politics and the Parisian Avant-Garde* (Princeton: Princeton University Press, 1993), pp. 80–2.

143 Hélion, 'From Reduction to Growth', p. 23.

144 H. Moore, 'A View of Sculpture', in *Architectural Association Journal*, Vol. XLV, May 1930, p. 408. Concomitant with a preoccupation with direct carving, this difference resulted from a contemporary interest in 'carving in the round', which, simply put, encouraged sculptors to think of a sculptural idea as something that was contingent on the particular form of the block set before them and thus inherently three dimensional. See Read's description of this idea in H. Read, *The Meaning of Art* (London, Faber & Faber, 1931), pp. 151–3.

145 As we will see in Chapter 3, the holistic attitude that such neo-vitalist opinions augured has been viewed by scholars as 'a revolt against formalism'. See C. Lawrence & G. Weisz, 'Medical Holism: The Context', in C. Lawrence & G. Weisz (eds), *'Greater than the Parts': Holism in Biomedicine, 1920–1950* (Oxford: Oxford University Press, 1998), p. 6.

146 A. Potts, *The Sculptural Imagination: Figurative, Modernist, Minimalist* (New Haven: Yale University Press, 2000), pp. 124–7.

147 See A. Hildebrand, *The Problem of Form in Painting and Sculpture* (London: G.E. Stechert & Co., 1932).

148 See also Potts, *The Sculptural Imagination*, p. 125.

149 Grigson, *Henry Moore*, p. 10.

150 H. Moore, Untitled statement, p. 30.

151 See Krauss, *Passages in Modern Sculpture*, p. 140.

152 Read, *The Meaning of Art*, p. 153. The template here for Read's biologistic musings was undoubtedly Henri Bergson's neo-vitalist tome *Creative Evolution*. Specifying the nature of the vital impetus, Bergson discriminated between the concepts of manufacturing and organisation – seeing the former as an *artificial* construct which imposed itself, in a piecemeal fashion, upon matter from without, whereas organisation was perceived as a *natural* unfolding of creative will.

153 See G. Beer, *Darwin's Plots: Evolutionary Narrative in Darwin, George Eliot and Nineteenth-Century Fiction* (Cambridge: Cambridge University Press, 2000), pp. 100–1.

154 See Potts, *The Sculptural Imagination*, esp. pp. 1–9.

155 A. Hildebrand quoted in P. Curtis, *Sculpture: 1900–1945* (Oxford: Oxford University Press, 1999), p. 77.

156 A. Jakovski, 'Brancusi', in *Axis*, No. 3, July 1935, p. 3.

157 Moore, 'A View of Sculpture', p. 408.

158 Hepworth, 'Sculpture', pp. 113–14.

159 *Ibid.*, pp. 113–14.

160 Moore, Untitled statement, p. 29.

161 See Mundy, *Biomorphism*, esp. pp. 128–35.

162 Moore, Untitled statement, p. 29.

163 Thompson, *On Growth and Form*, pp. 359–60.

164 *Ibid.*, pp. 404–5.

165 *Ibid.*, p. 357.

166 Mundy, *Biomorphism*, pp. 127–9. Although, given the ambivalence of professional morphologists over whether asymmetry was an essential trait of living beings, it is telling that most art critics chose only those single-cell organisms that suffered most from surface-tension forces as examples of the life forms that biomorphic art most clearly resembled.

167 G. Grigson, 'Painting and Sculpture Today', in G. Grigson (ed.), *The Arts Today* (London: John Lane, 1935), pp. 89–90.

168 A. Barr, *Cubism and Abstract Art* (1936) (London: Secker & Warburg, 1964), pp. 19, 186.

169 See also Mundy, *Biomorphism*, p. 134.

170 J. Ruskin, *The Seven Lamps of Architecture* (1849), reprinted in E.T. Cook & A. Wedderburn (eds), *The Works of John Ruskin*, (London: George Allen & Unwin, 1903), Vol. VIII, pp. 144–5.

171 Maldonado, *Le cercle et l'amibe*, pp. 187–9.

172 Bergson, *Creative Evolution*, p. 96.

173 Barr, *Cubism and Abstract Art*, p. 186.

174 On Arp's concretions, see E. Robertson, *Arp: Painter, Poet, Sculptor* (New Haven: Yale University Press, 2006), pp. 110–12. Eric Robertson views Arp's concretions as representing the very quintessence of metamorphosis by utilising a vocabulary of growth and decay – on the one hand, suggesting bourgeoning cells or sprouting buds and, on the other, nuggets of stone worn down by the elements.

175 Krauss, *Passages in Modern Sculpture*, p. 141.

176 Thompson, *On Growth and Form*, pp. 404–5.

177 Barbara Hepworth, Interview, in *The Studio*, Vol. CIV, July–December 1932, p. 332.

3. Organismal composition

The significant sculpture of the last five years is distinguished from all previous sculpture by a novel departure – it has *split itself up*, as mercury splits up, or as certain primitive organisms, or the fish *poi* found in the South Seas, split up to propagate their kind.[1]
Hugh Gordon Porteus, 'New Planets', 1935

As the poet Hugh Gordon Porteus' biologistic synopsis of the new sculpture suggests, as early as 1935, the dissolution of the object – which took place in the 1930s under the creative auspices of Surrealism – led to the emergence of a new, *organismal* vocabulary through which Modernist critics conceptualised sculptural practice. Grounded, lacking a plinth and composed of seemingly separate components, artworks such as Barbara Hepworth's *Two Forms* (1933) (Figure 3.1), Henry Moore's *Composition* (1934) (Figure 3.2) and F.E. McWilliam's *Eye, Nose and Cheek* (1939) (Plate 3) problematised the impression of unity of design within the sculptural object, making *composition* – a concept hitherto more readily associated with painting – a topic of central importance to Modernist sculpture.[2] So as to theorise the rationale behind 'multipart' sculpture, Modernist critics drew upon the theory of organicism, an ancient idea in the history of aesthetics which hypothesised a sense of unity between the different parts of a composition.[3] In the words of the painter Jean Hélion – who would become a key Modernist exponent of organicism as an aesthetic strategy – the artwork, newly conceived of as a diagram of dynamic interconnectivity, thus evinced a relationship between the parts and the whole which mirrored those present in 'a complex organism'.[4]

Connected to the biocentric currents which elicited a biomorphic conception of Modernist abstraction, sculptural organicism was nonetheless *distinct* in its emphasis upon part-to-whole relationships and the internal arrangement of a composition as opposed to any outward semblance to living things. As Jennifer Mundy has argued: 'Strictly speaking, the term organic refers not to a resemblance to the shapes of nature but to form principles seen to underlie natural form'.[5] Nonetheless, like biomorphism, sculptural organicism drew upon the biologistic currents which informed Modernist art theory, finding in the New Biology an organicist sensibility through which to appreciate the structural integrity of a multipart composition. In a review of Alfred North Whitehead's philosophy of

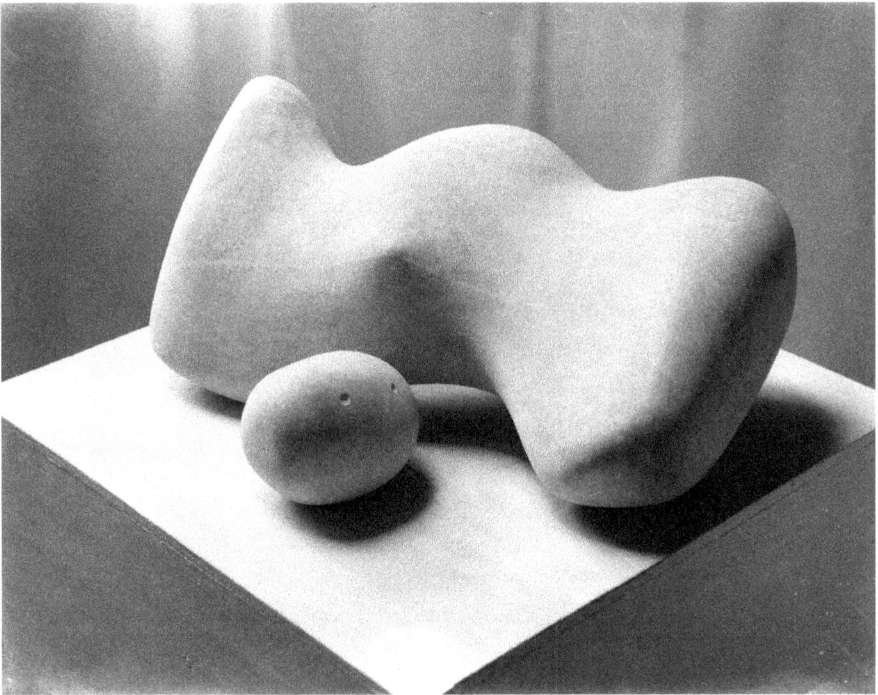

Figure 3.1

Barbara Hepworth, *Two Forms*, 1933

organicism, for example, Herbert Read praised its applicability to art theory, extolling Whitehead's conception of the artwork as a pattern of creative unity. Beyond its prescience to art theory, Read recognised Whitehead's philosophy as a 'doctrine of the interconnection of everything', which embodied 'the material of a revolution in our whole concept of life or being, and seeks to reinterpret, not only the categories of science and philosophy, but even those of religion and art'.[6] In particular, he applauded Whitehead's disapproval of intellectual callousness in society, claiming that a brutal lack of sensitivity in the world, manifest in overly reductive analyses of nature, accounted 'for all the vagaries of modern art'.[7] The deeply anti-reductionist prejudice of this standpoint highlights organicism's intellectual relationship to the wider movement of holism, which emerged in Europe in the late nineteenth and early twentieth century as a profound 'revolt against positivism [and] formalism'.[8] Like holism, organicism would seek to combat the fragmentation and brutality of modern, mechanistic thought by hypothesising a sense of 'organic' wholeness between objects, modes of thought and states of

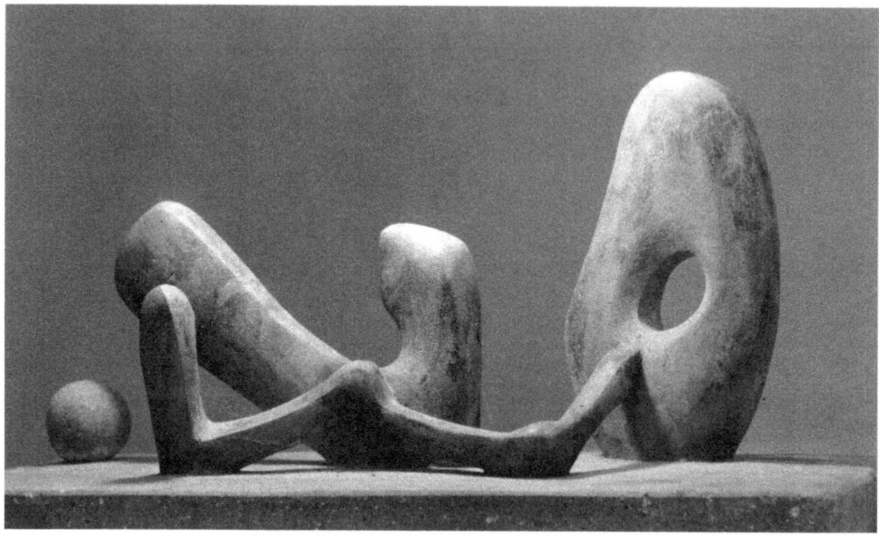

Figure 3.2

Henry Moore, *Composition*, 1934. Henry Moore Foundation

being that would otherwise have been deemed separate.[9] This chapter will thus explore the emergence of an organicist sensibility in Modernist sculpture, firstly by examining its roots in the New Biology and social theory, and then by assessing its impact upon new conceptions of the sculptural object in the mid-1930s.

Between the quick and the dead: organicism and the mechanism–vitalism controversy

In an article on 'The Origin of Life' of 1933, A.S. Russell hailed the passing of the mechanism–vitalism controversy, noting that 'vitalism [had] been whittled down and mechanism [had] become a bigger thing'.[10] In place of vitalism there now appeared the '"organismal" view', which maintained 'that many happenings can only be explained in terms of the organism as a whole'. Tellingly, in an earlier essay, of 1930, Russell had quipped that such disputes were potentially an issue of metaphor, noting that while 'it is unwise to deny the inorganic the power to fashion the organic [...] "mechanism" is an absurd word to describe the process'.[11] Yet what had previously been an issue of dogma in 1930 had, by 1933, become principally an issue of technique; the dividing line in biology had seemingly been redrawn over whether the organism was an indivisible biochemical whole or simply the sum of its inorganic components.

Russell's interest in publicising the reputed outcome of the mechanism–vitalism dispute stemmed from a genuine excitement, then keenly felt, that a philosophical dispute that had been ongoing since the late eighteenth century and which ultimately affected the core principles of science was, once and for all, nearing resolution. The longevity of the controversy was due, in part, not only to the inherently conflicting worldviews held by the mechanists and vitalists but also to the magnitude of the topic, which, at its heart, concerned the very origins of life itself.

Historically, *vitalism* originated in a seventeenth-century doctrine devised by Robert Fludd and J.B. van Helmont (among others), who posited the existence of life-bearing seeds or spirits, speculating that a non-physical 'vital' force animated living matter.[12] Contrastingly, *mechanism* stemmed from classical ideas about the primacy of matter that had gained momentum through seventeenth-century studies of mechanics, leading to the belief that living matter was underpinned by distinctive physico-chemical laws.[13]

In many ways, the nineteenth century presaged the slow eclipse of vitalism as an increasingly materialistic science – galvanised, on the one hand, by Darwin's theory of natural selection and, on the other, by the study of artificial partheno-genesis – seemed to confirm the mechanistic belief that all of life could be understood according to physico-chemical laws. Nonetheless, the fruits of the mechanistic approach, conversely, kept alive important aspects of the vitalist argument. For those who opposed mechanism on theological or idealistic grounds, the nineteenth-century transition from a matter-based physics to one founded on energy paved the way for the introduction of 'critical vitalism', a creed that abandoned previous attempts to posit an animating substance or vital fluid (which broadly tied in with a matter-based ontology) in favour of a vital impetus that more closely matched contemporary visions of an energy-based universe.[14] Equally, the more materialistic aspects of evolutionary theory were artfully subverted by a vitalistic teleology – promoted through the writings of the physicist Oliver Lodge – that conjectured that life was involved in a process of spiritual evolution of which 'mind', in some form, was the ultimate goal. Henri Bergson's *Creative Evolution*, translated into English in 1911, was highly influential in publicising this view.[15] His vision of an indeterminate yet unyielding vital spirit that was able to 'beat down every resistance [...] perhaps even death',[16] tempted those biologists who wished to see the integrated activity of a living body as something more than the sum of knowable mechanical principles.[17]

Yet, in reality, while strongholds of vitalist philosophy still stood firm in the 1920s and 1930s, it was becoming increasingly difficult to sustain the myth of an enlivening *élan vital* when it had been proven that the organisation of most regulatory processes was achieved through the complex interaction of a variety of physico-chemical elements.[18] Indeed, many who sympathised with the vitalist cause had long sought to accommodate a belief in the 'higher-level' integrated activity of an organism with an acceptance of the fundamental laws of physics.

They consented that any 'vital force' was ultimately chimerical while, at the same time, testifying that it made little sense to explain all of the properties of a living creature in starkly materialistic terms.[19]

This strongly anti-mechanistic outlook took the view that while vitalism was too much under the sway of mysticism to be of much practicable use, mechanism itself was far too reductive and had failed to account for 'the persistent and specific coordination of structure, activity and environment which [showed] itself in the lives of organisms'.[20] In place of these old philosophies, reformists posited an 'organismal' viewpoint that attended more closely to the *organisation* manifest between the parts of a living being, extending the 'concept of organism' to the quest of looking 'for organizing relations at all the levels, coarse and fine, of the living structure'.[21] Interestingly, in breaking down the certainties of nineteenth-century science, it was physics that led the way – via the newly hypothesised relativity theory and quantum mechanics – in rebutting the conventionally reductive principles of pure mechanism.[22]

The publication, in 1926, of *Science and the Modern World* by the philosopher Alfred North Whitehead did much to promote the view that the findings of modern physics had relevance for the New Biology, not least in Whitehead's efforts to abandon materialism in favour of 'an alternative doctrine of organism'.[23] By all accounts a popular success,[24] *Science and the Modern World* provided a historical critique of mechanism and sought to introduce an entirely new approach to nature by applying essentially *organic* categories to the study of the universe.[25] Inspired by Bergson and developments in field physics, Whitehead argued that contemporary notions of energy transfer and gravitational fields of force had destroyed the Newtonian cosmos of abstract physical entities – 'simple locations' – that existed as static units in barren space, unconnected to 'other regions of space and to other durations of time'.[26] His theory of 'organic mechanism' registered how the behaviour of atoms and electrons was affected by the plan of the larger unit of which they were part, regardless of whether it was a living body or nature as an entirety.[27] In this worldview, the fundamental unit of science was termed the 'event', which consisted of all the material qualities existing in any one place and time, and its spatio-temporal relationship to 'all other events'.[28] Indeed, by forwarding a radically holistic reading of the physical world, 'organic mechanism' offered a third way to those biologists who, on the one hand, wished to escape from the philosophical stranglehold of mechanism yet, on the other, opposed the 'mystical' undertones of outright vitalism.[29] As Oliver Botar has noted, Whitehead, in his nature-centric endeavour to redefine human knowledge as simply an experiential subset of nature, was a potent influence on British neo-romanticism and its organicist offshoots.[30]

At heart, *organicism*, as a biological term, hypothesised a particular sense of unity between the parts of a living body and the whole. While vitalists, such as Hans Driesch and Henri Bergson, had all along argued that the unique wholeness of an organism provided indisputable evidence of its underlying vitality, and mechanists, such as Jacques Loeb, had never really denied that, at some level, the study of organ and tissue interactions could offer a description of the complete organism,[31] 'organicists' differed in both denying the existence of any express vital force while, at the same time, focusing attention on how the properties of the whole differed *qualitatively* from the properties of the parts, extending this concept to include the physico-chemical environment of the organism. In *The Philosophical Basis of Biology* (1931) John Scott Haldane explained this quality as being akin to a kind of existential interdependency in which 'the spatial relations of the parts [did] not imply their separate existence from one another', as it was impossible to 'define them as existing separately when their very existence expresses coordination with one another'. He saw this coordination extending 'over the surrounding environment, and the spatial relations of parts and environment [expressing] unity, not separation'.[32] Indeed, the idea of coordinated interactivity being a primary quality of life itself was integral to Whitehead's conception of 'organism' and his vision of different grades of matter blending indissolubly into a single unity formed a convincing template over which organicists could map their own holistic philosophy.[33]

Although many materialists, such as the Marxist biologist Lancelot Hogben, believed that organicism represented nothing more than a repackaging of neo-vitalist dogma,[34] its emphasis upon dynamic interconnectivity as a primary function of life was philosophically broad enough to appeal to scientists on both sides of the mechanism–vitalism divide who wished to establish definitive principles for the subject of biology. For those artistic Modernists who looked to the New Biology to satisfy their biologistic leanings, neo-vitalism and organicism appeared as complementary worldviews, serving to buttress the pantheistic, biologistic and anti-reductionist footings of their incipient neo-romanticism.[35] As we will see, the organicist conception of the artwork as a cipher of dynamic interdependency reinforced neo-vitalist perceptions of the artwork as a morphogenetic field, while simultaneously providing a sense of unity to a problematically fragmented object. In turn, the impulse Modernists felt towards organicist sensibilities in art was nourished by the sustained 'organic' critique of mechanism, materialism and reductionism that increasingly dominated socio-cultural discourse throughout the 1920s and 1930s.

Organicism in British culture between the wars

Organicism exerted a considerable pull on the contemporary imagination at this time. By the twentieth century, 'organic' concepts had widely percolated into social thought – especially of a conservative kind – and were frequently employed to signify the dissimilarity between 'organised' statehood and the more 'organic' societies of pre-industrial Britain.[36] Views of this kind – such as those espoused by F.R. Leavis and Denys Thompson in *Culture and Environment* of

1933 – explicitly contrasted the '"organized" modern state' with the 'Old England [...] of the organic community'[37] by evoking an image of organised statehood that was 'mechanistic' in the original sense meant by the political philosophers of the seventeenth century.[38] Such comparisons owed much to popular reactions against the cataclysmic effects of the Great War, which in being the world's first mechanised conflict was largely discussed in negatively mechanistic terms. The narrator of George Orwell's novel *Coming Up for Air* (1939) speaks of the war as having been 'like an enormous machine that had got hold of you' in which '[you] had no sense of acting of your own free will, and at the same time no notion of trying to resist'.[39] Indeed, during the interwar years, mechanistic notions of technology and statehood were typically identified with Germanic culture and contrasted with an organic conception of English libertarianism and pastoralism.[40] Thus for conservative thinkers the term 'organic' increasingly came to symbolise a 'slowly adapting society and tradition'.[41] In an essay on 'Law' for the rightist anthology *The English Genius* of 1939, for example, E.S.P. Haynes quoted with approval that the English legal system was 'far more elastic and organic in its nature' than the 'despotic' regimes of continental Europe.[42] By the same token, the rapid pace of industrialisation in the interwar years generated a backlash among neo-romantic conservatives, who viewed modernisation as little short of an aesthetic and social disaster.[43] As a result, neo-romantic publications such as *England and the Octopus* (1928) by Clough Williams-Ellis posited a pastoral and Arcadian image of Englishness as a cathartic defence against the perceived disorder and ugliness of industrial Britain. Such publications hypothesised a unity between the body politic and nature that, in the modern era, industrialisation had allegedly begun to destroy.[44]

Interwar debates regarding the 'synthetic' nature of modernity fed into a broader critique of post-Cartesian rationalism, in which empirical philosophy was blamed for having destroyed 'the unique reality of an object and reduce[d] it to a mere instance in a series of instances'.[45] The critique of materialist philosophy that Whitehead forwarded in *Science and the Modern World* was by far the most successful exposition of this view.[46] In a 1926 appraisal of this tome, Read confirmed that it was Whitehead's rebuttal of age-old materialist certainties, achieved through the findings of the new science, which made his philosophy relevant to the neo-romantic, holistic concerns of Modernist intellectuals:

> [The] most important effect [of Whitehead's philosophy] is to get rid of the idea of an inert, valueless matter independent of mind. The assumption of modern science, first made explicit by Descartes, is that bodies and minds are independent substances, each existing in its own right apart from any necessary reference to each other. The consequences of this assumption, both for art and religion have been nothing less than disastrous.[47]

The mind–matter split hence enshrined by materialist philosophy had led to a 'mechanism entirely valueless', which, in Read's opinion, had directly culminated in the moral bankruptcy of 'modern industrialism and a world war'.[48]

The refrain that post-Cartesian science had diminished the spiritual value of life and, in its place, erected an unprincipled system of reductive, mechanistic thought was repeated by Modernists across the neo-romantic spectrum, the majority of whom beseeched modern science to adopt a more holistic perspective that would symbolically repair the damage wrought by an incomplete, materialist knowledge of life. The religious writer Hugh I'A. Fausset was a prominent representative of this holistic standpoint, bewailing science's attempt to reduce the 'human mind to a sensitive machine' and claiming, in neo-vitalist terms, that proponents of mechanism were 'cut off from the deep rhythm of life and [were] condemned to a sterile service'.[49] The reductionist attitude of science was, he contended, 'fatal to the growth of the finer spiritual qualities. When, however, it is recognised to *be* partial, it may prove of real service to men in the task of making themselves and their world whole'.[50] Deeply connected to the organicist sensibility that challenged the convictions of mechanistic thought, this holistic approach – like organicism – identified fragmentation (be it social, scientific or psychological) as a negative condition of modernity and correspondingly sought to overcome it by conjecturing metaphors of wholeness, equilibrium and organicism as a philosophical solution.[51]

Breaking the mould: multipart sculpture and sculptural organicism

The New Biology's resolution of the mechanism–vitalism controversy, through the establishment of an organicist paradigm, in effect represented the crowning moment of a holistic tendency in European culture that had been inaugurated in the early nineteenth century and which rejected the philosophical insensitivity and socio-cultural fragmentation that proceeded from an unbridled materialism. Evolving in fields as disparate as history, politics and philosophy, holism thus acted as a critique of both reductionist philosophies and the discordant culture they engendered by accentuating the quality of relations between objects within any given system (be it biological, cultural or otherwise) and by holding that every system was, like Driesch's self-adjusting embryo, a self-regulating equilibrium.[52] Although aesthetic organicism would frequently be presented as a palliative to the atomistic mind-set of mechanistic science, in reality, nature-centric Modernists, such as Paul Klee and Hans Arp, were drawn to the organicist sensibilities of the New Biology and from it acquired an understanding 'of principles of emergent order serving to distinguish [organic] form from inert structures'.[53] This emergent sensitivity to organicist philosophies within Modernist circles arose in tandem with other biologistic ideologies, although aesthetic organicism's particular attentiveness to the interrelationships between parts, concepts of wholeness and the internal organisation of form – while generally congruent with biophilic philosophies of art – distanced it from the principles of similitude that characterised

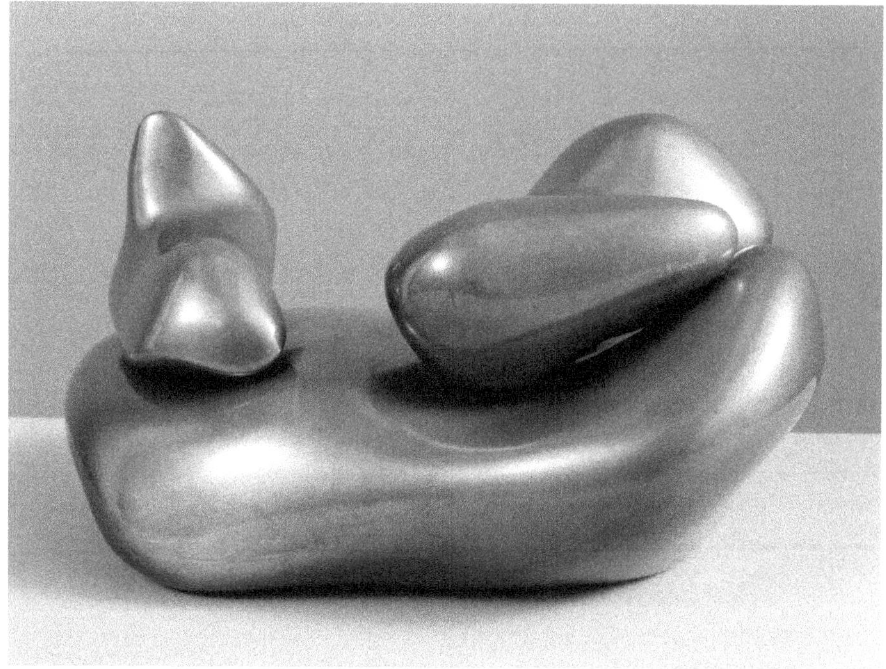

Figure 3.3

Hans Arp, *Sculpture to be Lost in the Forest*, 1932. Tate, London

biomorphic artworks.[54] As Jack Burnham has stated on the subject of organicism in sculpture:

> Visual biological metaphors exist on many levels besides the obvious total configuration of an animal or human. The aesthetics of true organicism [...] is not grounded in the appearances of natural forms and their carryover into sculptural materials, but is concerned with organization of processes and interacting systems.[55]

Philip Ritterbush has spoken of Arshile Gorky, Roberto Matta and Hans Arp as producing compositions which echoed the principles of aesthetic organicism: the 'interdependence of parts, and differentiation directed from within. Their compositions reinterpret the basic principles of organic form. They do not [...] represent organisms directly'.[56] Arp's inclusion in this analysis is apposite as his part-objects of the 1930s, such as *Sculpture to be Lost in the Forest* (1932) (Figure 3.3), by virtue of their composite makeup, did much to inspire an organicist tendency in Modernist sculpture. For example, Hugh Gordon Porteus, in his article on sculptural Modernism, acknowledged Arp's centrality to what he termed the project of the 'new sculpture', adopting holistic metaphors to explain how

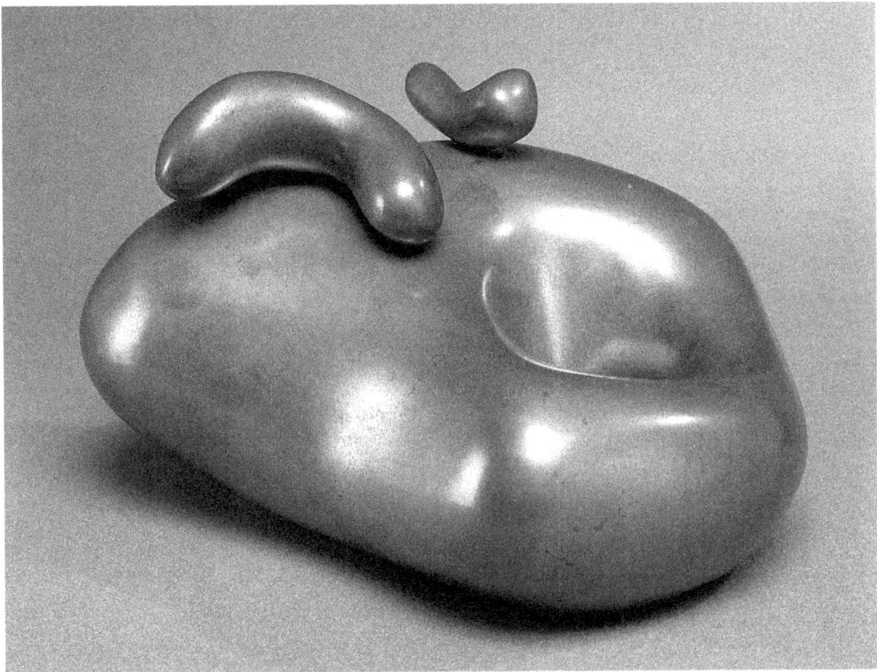

Figure 3.4

Hans Arp, *Navel and Two Thoughts*, 1932. Israel Museum, Jerusalem

the multiple components of this idiom were united not by resemblance, but by underlying structural affinities or bonds of association:

> Sculpture has hitherto concerned itself, on the plastic side, with relations between surfaces of a single object. The new sculpture, without neglecting the claims of plastic motion within or upon the object itself, attempts to make relations between itself and other separate objects of the same kin, or between dislocated bits of itself.[57]

Penelope Curtis has described this development in Modernist sculpture as a 'grounding' of the object: 'The object gave way to sculpture as the plinth gave way to a ground-plane which was part of the sculpture itself. No one part was more important than another, even in pieces which invited the viewer's involvement'.[58] Originating in Surrealist experiments with involuntary and participatory sculpture, the multipart compositions of, say, Arp and Giacometti shared a disregard for the role of the pedestal in sculptural display, suppressing the elevation it provided by rendering it horizontal and thus fashioning it as though it were a compositional element in its own right.[59] Indeed, in an artwork such as *Navel and Two Thoughts* (1932) (Figure 3.4), Arp carefully modelled the pedestal into an abdominal form that functioned as a compositional counterpart to the two smaller biomorphs – rather than something as hieratic as a plinth. This act of grounding therefore places the sculpture into an ambivalent relationship with its surroundings: on the one hand, it becomes indistinguishable from its environs, literally spilling out into the space that surrounds it; 'yet it is also, paradoxically, a way of asserting the sculpture as being distinct from its surroundings, as a material object in its own right, existing in real space, and therefore requiring no plinth to mark it off as belonging to the representational space of sculpture'.[60] This staging of groundedness can, in turn, be related to the description of 'formlessness' by the Surrealist philosopher Georges Bataille, in his critical dictionary, as a total levelling of structure, something which, he claimed, would 'knock form off its pedestal and bring it down in the world'.[61] Certainly, for Modernist critics, like Porteus, who were perhaps more accustomed to the exigencies of single-block carving, the fragmentation of the sculptural object, accompanied by its symbolic grounding, represented a polemic that promised the total dematerialisation of the artwork.[62] In her *Modern Plastic Art: Elements of Reality, Volume and Disintegration* (1937), the art historian Carola Giedion-Welcker explained that Modernist sculpture had become ever more concerned with 'disintegration' as problems of physics overtook previously more 'literary' representational interests:

> The problems of statics and dynamics, as of the disintegration of mass and the space–time interrelation of volumes, are bound to become a new plastic medium once their divorce from literary and psychological suggestion allows a return to first principles.[63]

This process of volumetric disbanding was staged by Giedion-Welcker as a confrontation between 'static' and 'dynamic' conceptions of sculpture in which a neo-classical attitude was transformed through the 'surface complexities' of Impressionism, which – through the agency of Cubism, Dada, Surrealism and Constructivism – ultimately led to the 'disintegration of material solidity' and 'the shaping of volumes perceived in terms of space'.[64] Similarly, in a piece of writing published in *Axis*, the critic Anatole Jakovski hypothesised a novel history of art in which, from the age of Rodin onwards, the sculptural object was seen to have become ever more fragmentary and ethereal.[65] In particular, Modernist sculpture was presented as a symptom of those new theories of space-time that had materialised in the ashes of the nineteenth century and which had fundamentally changed contemporary understandings of spatiality.[66] Renaissance conceptions of space 'were to last until mankind's new age', wrote Jakovski, 'until the new theories of space of Riemann, Lobachevski and Einstein. Rotation, gyration, cyclical movement [served] as a framework for all artistic creation'.[67] Conditioned by a new comprehension of space imparted by modern physics, Modernist sculpture thus represented the breakdown of the last stable elements of sculptural volume:

> Everything scatters, turns to dust, vanishes. The roundness of nudes that would once have delighted the Impressionists lengthened and shrivelled. Spirals no longer

make us dizzy. We are intoxicated with the straight line, that which shows the way to another limitless space, where numerous light-giving dynamos are already in action. Aero-dynamics affirms its power on all sides. Brancusi keeps the last traces of roundedness [...].[68]

While Brancusi was understood to have retained an impression of volume in his work that derived from the imperatives of single-block carving, the fragmentary and grounded aesthetic of the new generation of sculptors had opened out Modernist sculpture to an unbounded sense of space:

> [Brancusi's] successors – most of the younger generation – Giacometti [...], Calder, work in geometrics. They calculate the infinite. They work in counterpoint. They build, but their works have no boundaries, as a photo has no margins. Like energy, or light, which we cannot yet isolate.[69]

In his 1937 apologia 'The Sculptor Speaks', Henry Moore hinted that these developments had their origins, like holism itself, in a disillusionment with Modernism and the formalist conventions of Modernist practice.[70] Noting that since 'the Gothic, European sculpture had become overgrown with moss, weeds – all sorts of surface excrescences which completely sealed shape', he claimed that, as a response, sculptural Modernism had travelled too far in the opposite direction, purging itself of all complexity by adopting a reductionist approach to form:

> It has been Brancusi's special mission to get rid of this overgrowth, and to make us once more shape-conscious. To do this he has had to concentrate on very simple direct shapes, to keep his sculpture, as it were, one-cylindered, to refine and polish a single shape to a degree almost too precious [...]. But it may now be no longer necessary to close down and restrict sculpture to the single (static) form unit.[71]

Here, Moore's comments on the evolution of multipart sculpture are instructive, as it was in Britain that the possibilities of a composite aesthetic particularly resonated. While remarking upon the Surrealist origins of this form of Modernist practice, Penelope Curtis has observed: 'A striking example is [Yves] Tanguy's *The Certitude of the Never Seen* (1933), which has a carved wooden frame with forms which appear in the sculpture of the French Surrealists and, more especially, the British sculptors Moore, Hepworth, and [F.E.] McWilliam'.[72] Correspondingly, Charles Harrison has discussed the revelation that the examples of Arp and Giacometti afforded Hepworth thus: 'Arp's influence was [evident in] the device of relating various forms on a flat surface [and] mediated by a recent acquaintance with works by Giacometti [...] in which more simply geometric forms were similarly deployed'.[73] Certainly, Arp was a major influence on Hepworth's practice, a 1933 visit to his Paris studio precipitating her own experiments with multipart composition.[74] What particularly appealed to Hepworth was 'the way Arp had fused landscape with the human form in so extraordinary a manner',[75] although, as Eric Robertson has discerned, it was Arp's grounded part-objects in particular which inspired Hepworth to produce her own multipart sculptures during the mid-1930s.[76]

In a 1933 review of Hepworth's sculpture at the Lefèvre Galleries, the critic Adrian Stokes identified the roots of Hepworth's multipart compositions in her concern with the theme of the mother and child: the composite makeup of *Figure: Mother and Child* (1933) (Figure 2.5, p. 62), in particular, representing maternity more by virtue of the intimacy suggested by the interdependency of its separate parts than by any principle of resemblance:

> This is the child which the mother owns with all her weight, a child that is of the block yet separate, beyond her womb yet of her being. So poignant are these shapes of stone, that in spite of the degree in which a more representational aim and treatment have been avoided, no one could mistake the underlying subject of the group. In this case at least the abstractions employed enforce a vast certainty. It is not a matter of a mother and child group represented in stone. Miss Hepworth's stone *is* a mother, her huge pebble its child.[77]

Stokes' emphasis upon the mutual dependency of the separate components in Hepworth's sculpture – though indebted to a Kleinian vocabulary of object relations – had its philosophical origins in holism and anticipated many of the ways in which organicism would be employed by critics as a means of cohering a sculptural idiom that had become problematically disbanded.[78] To be sure, in a 1935 analysis of Hepworth's composite sculpture, the Dutch archaeologist and art reviewer Henri Frankfort declared that her compositions, though largely non-figurative, possessed a degree of interdependency more commonly found in the bodies of living things:

> One is tempted to consider these works as living organisms in which (within certain limits which to pass means death) distinct organs each with a character of its own, exist interdependently within a larger unit, which is more than the mere sum of its constituents.[79]

A comparison can be drawn here between organicist critiques of mechanism's inability to grasp the totality of an organism and Frankfort's suggestion that Hepworth's multipart sculptures signified only when comprehended as 'organic' wholes. Indeed, the phrasing of his statement closely parallels commentary in the popular press on the subject of the waning of the mechanism–vitalism controversy. For example, a 1930 *Listener* essay by A.S. Russell – on the subject of the rise of organicism in biology – significantly foreshadowed how Modernist critics spoke of the integrated, organic wholeness of composite sculpture:

> [The organicists] say that it is ludicrous to regard a living organism as a machine. The parts of a machine can be separated without alteration of their properties, and the properties do not depend on whether the machine is at motion or rest. This, of course, does not apply to the parts of a living organism. The whole organism, in all its efficiency of coordination and purposiveness, is destroyed if any part of it is removed.[80]

Julian Huxley used a similar set of metaphors to explain the integrated unity of living things in his bestselling book of 1923, *Essays of a Biologist*. Observing how evolution had brought about a gradual increase in the physiological complexity of organisms, Huxley argued that this process had not only necessitated the increasing specialisation of organs but also that the organism had become better amalgamated as a whole, each of its parts working in cooperation with its fellows:

> [There] has been an increase in the harmony [of the] parts [of the organism], and consequently in the unity of the whole. Delicate mechanisms for co-ordination have been developed, and arrangements whereby one portion becomes dominant over the rest, and so a material basis for unification is given.[81]

Of course, the idea that an artwork could merit comparison with the creatures of the natural world is ancient in the history of aesthetics.[82] Termed the theory of 'organic unity', from its earliest conception the idea denoted a significant relationship between the parts of a work that corresponded to a particular structural arrangement found only in nature. Plato's classic dictum 'that every speech should be put together like a living creature, as it were with a body of its own, so as not to lack either a head or feet, but to have both middle parts and extremities, so written as to fit both each other and the whole' is the earliest exposition of the concept and has influenced aesthetics since its inception.[83] There is a striking equivalence between Frankfort's analogy, for example, and the presumed unity between living things and exemplary building practice as proposed by the Renaissance architect Leon Battista Alberti, in his classic treatise of 1450, *On the Art of Building*:[84]

> [Just] as the head, foot, and indeed any member must correspond to each other and to all the rest of the body in an animal, so in a building, and especially a temple, the parts of the whole body must be so composed that they all correspond one to another, and any one, taken individually, may provide the dimensions of all the rest.[85]

Alberti's paralleling of living bodies and architectural elements, and stress on modular construction, certainly find resonance in Hepworth's multipart sculptures of the mid-1930s, such as *Three Forms* of 1935 (Plate 4), in which several ovoid forms, all of which appear as formal variants of each other, define the overall composition.

Frankfort was not alone in suggesting an ontological equivalence between the creatures of the natural world and an artwork. Across Europe, Modernists of a biocentric persuasion were drawn to organicist strategies as a means of demonstrating that, far from being merely superfluous embellishment, all of the parts of an abstract painting or sculpture, when perceived collectively, were crucial to the overall meaning of the artwork. An essay by Wassily Kandinsky, originally published in his *Punkt und Linie zu Fläche* (1926) and reproduced in *Axis*, epitomised this trend by analogising between the structural constitution of an artwork and a fish. In both cases, he wrote, the 'composition is the *organised* sum of the *interior* functions (expressions) of every part of the [object]'.[86] An important figure in the proselytising of a biologistic approach to artistic creativity, Kandinsky's organicist theory of composition suggested how even a Modernist artwork which lacked

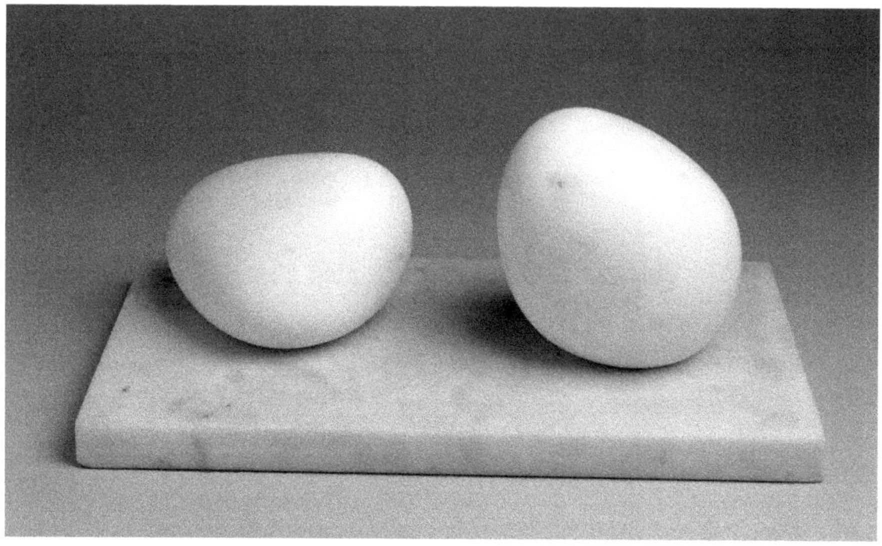

Figure 3.5

Barbara Hepworth, *Two Forms (White Marble)*, 1935. Private collection

clear-cut representational prompts could be understood to be *organic* in terms of the existential co-dependency of its parts.[87] The autonomy of life enshrined in organicist and neo-vitalist dogma was thus press-ganged by Modernists into a theory of abstraction which, in contrast to reductionist or formalist readings of modern art, preserved the autonomy of the artwork by accentuating its organic integrity.[88] Organicism, if pertinent to theories of Modernist abstraction, was particularly relevant to interpretations of multipart sculpture, which appeared to lack the physical wholeness of the 'static, simultaneous sculptural body' which had, up until the 1930s, characterised traditional forms of statuary.[89] Indeed, the physical incompleteness of a composite sculpture was, in the eyes of Modernist critics such as Frankfort, only exacerbated by the apparent motility of its components:

> [In] one or two cases [Hepworth has] omitted to fix one of the three elements upon the base. It means that she has so exhaustively realized all possible spatial interrelations between the elements of these particular carvings that each new grouping, each displacement of the mobile element within the limits of the work [...] results in fresh harmonies.[90]

Frankfort recognises here that the potential for rearrangement elicited by the seemingly movable elements of the sculpture serves to complicate its basic compositional structure, further undermining any sense of volumetric wholeness. By noting that the elements could be reorganised or even removed, he held that the composition itself was open to structural collapse and could only ever exist

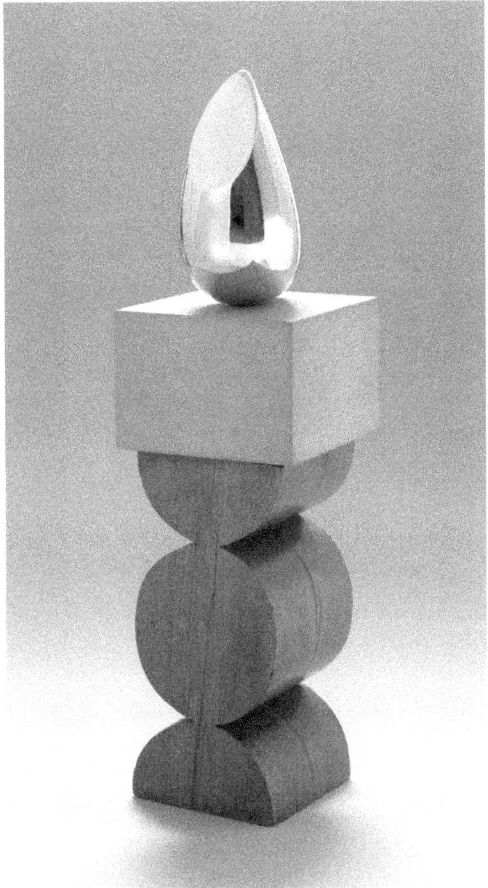

Figure 3.6
Constantin Brancusi, *Young Bird*, 1928.
Museum of Modern Art (MoMA), New York

provisionally between incomplete states. In this capacity, Hepworth's composite sculptures – such as *Two Forms (White Marble)* (1935) (Figure 3.5) – were structurally akin to those of Brancusi and Arp, which Porteus, somewhat disingenuously, cited as antecedents to her practice. He saw them as little more than 'a solitary clutch of Brancusi eggs, with a few Arp scraps'.[91]

Certainly, aside from the iconographic kinship between Hepworth's egg-like forms and Brancusi's embryological imagery, there is a structural relationship between her multipart compositions (and indeed those of Moore) and Brancusi's columnar bases, which, in consisting of various forms piled one on top of the other, indicate a susceptibility to reshuffling (Figure 3.6).[92] Similarly, the motile nature of

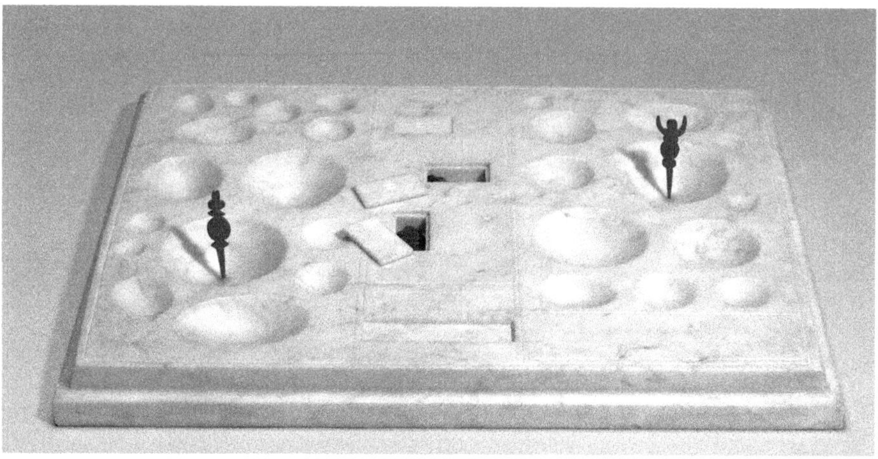

Figure 3.7

Alberto Giacometti, *No More Play*, 1931–32.
National Gallery of Art, Washington

Arp's sculptures, which openly invite the spectator to rearrange their freestanding compositional elements, acted as a powerful antecedent to Hepworth's practice, freeing her, in her own words, from 'many inhibitions' and underscoring the origins of her multipart compositions in certain types of Surrealist participatory artwork.[93] This inducement to the spectator to actively intervene in the composition of the sculpture had a precedent in Alberto Giacometti's 'board-games' of the early 1930s, which, though approximating the grounded character of Arp's part-objects, more closely encourage viewer participation (Figure 3.7). Giacometti maintained that he wanted 'one to be able to sit, walk, and lean on the sculpture'[94] and, as Krauss has remarked: 'Setting up the sculpture as though it were a game of chess is obviously an invitation to that kind of physical intimacy'.[95]

Yet whereas Brancusi, Arp and Giacometti's use of transposable bases registered as a collective attempt at frustrating contemporary conventions of display, in Hepworth's case the plinth was viewed as a kind of restraint that limited the overall contingency of the composition:[96] 'It is the function of the oblong slab to give [space] definition, to delimit with precision the spatial individuality of each work as a whole'.[97] In this respect, the pedestal united the disparate parts into a compositional whole by demarcating the spatial dimensions of the sculpture and physically containing the motile elements. In fact, *vis-à-vis* his own multipart sculptures, Moore spoke of the plinth as being akin to the painted frame of a Seurat picture – an object which, while not part of the painting proper, was nonetheless an integral factor in its presentation.[98] His views subtly echoed those of Arp, who in 1917 had already called attention to the artificiality of the pedestal and demanded its abolition in order to realise sculpture more fully in the round:

> But in the same sense that the frame of a painting impedes direct and immediate emotion, the *pedestal* of the sculpture is also a *convention*. The realities, the creations have neither limits nor feet but are complete in their unity. *The sculpture without a pedestal, the work of art should be seen from all sides.* Only in these conditions can it become reality.[99]

This impression of the pedestal aesthetically defining the spectatorial conditions under which sculpture is viewed allows us to think of it as operating in the same way as the 'frame' in Jacques Derrida's essay *The Truth in Painting* (1987), in which the frame is an entity that 'is neither the work, nor outside the work […], neither above nor below, it disconcerts any opposition but does not remain indeterminate and it *gives rise* to the work';[100] likewise, the plinth in multipart sculpture not only supports the work but it delimits and legitimises it.

Even so, the presence of the base merely curbs the spatial extension of the artwork; it does not prevent the free movement of the compositional parts within its compass. It is clear, then, that organismal metaphors assisted Modernist critics in their attempts at weakening the unpredictable character of multipart compositions by implying some prior, albeit contingent, regulation of the separate parts.[101] After all – as Donna Jeanne Haraway reminds us – organicism is at heart a structuring discourse that seeks to impose a sense of order onto systems that would otherwise be conceptually unmanageable.[102] Classical expressions of organic unity typically sought to transpose 'natural' laws into aesthetic principles precisely as a means of structuring and justifying the process of artistic design.[103] This inherent capacity of organising composite structures into integrated wholes can therefore be seen to embody the guiding principles of structuralism in philosophy.[104] In any case, the use of organicism as an interpretative strategy registers as an attempt by Modernist critics to minimise the contingency of multipart composition by positing some pre-existing conceptual framework that conditions its formal aspect and actively limits the distribution of its components. The base might delimit space, but the actual composition is carefully ordered by subtle aesthetic preconditions.

In hypothesising the existential interdependency of a composite sculpture, Modernist critics rearticulated understandings of organic unity which, since antiquity, had sought to convey the autonomous, self-sufficiency of the artwork in biologistic terms. An influential model of this strategy of artistic interpretation was developed by the German idealist philosopher Georg Hegel, who spoke of a work of poetry thus: 'a genuine work of poetry is essentially an infinite organism […] in which the whole, without any visible intention, is sphered within one rounded and essentially self-enclosed completeness'.[105] This organicist conception of creativity would have a significant impact upon romanticism in the eighteenth and nineteenth centuries, especially in Germany and Britain, as philosophers responded to the 'disenchantment' of nature by a reductionist materialism.[106] Edward Young, for example, in his influential paean to artistic originality, *Conjectures on Original Composition*, anticipated the nature-centric turn of romantic philosophy when he noted, in organicist terms, that: 'An *Original* may be said to be a *vegetable* nature; it rises spontaneously from the vital root of genius; it *grows*, it is not *made*'.[107] As Alan Powers has remarked, 'the re-enchantment of nature is a quintessentially Romantic idea', and the popularity of organicism among Modernists thus partly appears as a neo-vitalist stratagem aimed at revitalising a society that had degenerated through the reductionist imperatives of modernity.[108] Jean Hélion, an enthusiastic proponent of aesthetic organicism, underlined the neo-romantic and nature-centric priorities of Modernist abstraction by emphasising, in 'Poussin, Seurat and Double Rhythm', that as an abstract artwork *originated* in nature, like a living being, it could not be understood to be 'abstract':

> In a picture an element is real when it behaves like nature, when it coincides with its currents. The reality of the picture lies on the canvas, not at all in the origin or theme or of its elements. The field of conception belongs to nature. This is why '*abstract*' is a misplaced word.[109]

In making this statement, Hélion drew upon a strand of romantic thought that privileged the 'naturalness' of spontaneity over the synthetic limitations of 'manufactured' goods. This neo-romantic idealisation of naturalness first emerged, in part, as a response to the devastation of the First World War, which figures such as the writers Hugh I'A Fausset and Ivor Armstrong Richards blamed on the stultifying effects of science, which had distorted people's sensibilities and dampened their appreciation of living nature.[110] A creative sensibility approaching the condition of nature was, in this fashion, feted as a means of revivifying a Modernist culture that had become enamoured with the reductive, one-sided conception of reality that scientific modernity engendered.[111] Organicism, as a philosophical strategy, bestowed upon objects a property of life which, like entelechy in neo-vitalism, possessed a self-directed, animating principle. Correspondingly, as an *a priori* concept, an organicist approach postulated the existential sovereignty of the artwork through presupposing that any rearrangement of the parts would, as in a living body, destroy the vital integrity of the whole.[112]

If the revival of organicism within Modernist circles registered as a rejection of an overly reductionist, scientistic culture, as a paradigm it nonetheless drew upon organicist notions of synchronised interdependency that had materialised in the New Biology, firstly through the speculations of neo-vitalists, such as Driesch, and then, in the 1920s and 1930s, through the work of radical holists like Whitehead and John Scott Haldane. Certainly, in assuming that subtle ordering systems conditioned the overall structure of an organism, the New Biology hypothesised that the unity evident in living things embodied an overarching formative impulse. Although the agent deemed responsible for this developmental urge varied widely (from the entelechy favoured by neo-vitalists to an indeterminate substance, of biochemical origin, proposed by holistic mechanists such as Huxley), organicists found common ground in proposing that the interrelations between the separate parts were coordinated by a delicate structuring mechanism. Developing out of studies in experimental physiology, this mechanism was described, in resolutely anti-vitalist terms, by Huxley, Wells and Wells as a complex system of interrelating structures which provided a sense of morphological unity to the organism:

> No organism, and especially no higher animal, could work if it remained only a mosaic of independent parts. And so the next step is to give back to the embryo the unity it has so lately lost [through differentiation], to make its parts interlock and cooperate again. But the machinery by which the new adult unity is achieved is quite different from the underlying unity of the segmenting egg. Instead of mere gradients of activity tying the whole together, we have now coming into play the ubiquitous streams of blood and lymph and the fine branching of the nerves […]; they provide a common internal environment, a postal delivery system, a telephone system, and a controlling headquarters.[113]

As Huxley and the Wellses imply, an integrationist conception of living matter – in which binding interrelations between the parts composed the constitution of the whole – rebutted reductionist views of the organism as a static system of independent structures by adopting a morphogenetic approach in which organic material was seen as a spatio-temporal unfolding of unity.

These dynamic, structuring ideas are manifestly present in Moore's treatment of sculptural organicism. In 'The Sculptor Speaks', for example, he accentuated the manifest unity of multipart sculpture by noting the organic correspondences that existed between physically unconnected components. Rejecting the reductionist approach that characterised earlier forms of Modernism, he concluded:

> [It] may no longer be necessary to close down and restrict sculpture to the single (static) form unit. We can now begin to open out. To relate and combine together several forms of varied sizes, sections and directions into one *organic* whole.[114]

Moore's talk of an expanded sculptural unit in which a variety of different forms are aptly arranged brings to mind many of his multipart sculptures of the time, such as 1934's *Head and Ball* (Figure 3.8). Compositional arrangements of two, three or four components set atop bases of varying dimensions typify the format of this composite sculptural oeuvre. Although structurally analogous to the multipart tableaux that Hepworth was also producing during this period, Moore's compositions lack the simple regularity of shape characteristic of her experiments in composite design. Tumescent pebble forms and perforated blocks coexist within Moore's artworks, eliciting speculation as to the precise nature of their aesthetic rapport. By connoting an interdependent system of miscellaneous units (differing in terms of magnitude, positioning and temporality), Moore comes close to paraphrasing Whitehead's organicism in which dynamic systems replace static ones and the qualities of the whole are dependent upon the interactivity of

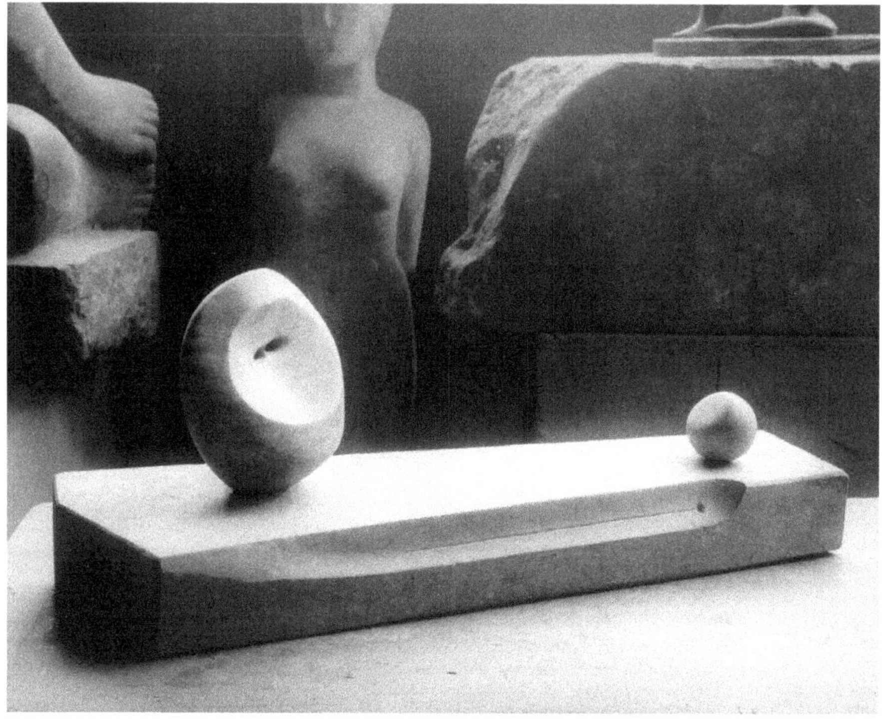

Figure 3.8

Henry Moore, *Head and Ball*, 1934. Henry Moore Foundation

the parts. One passage from Whitehead's *An Enquiry Concerning the Principles of Natural Knowledge* of 1919, in particular, suggests the template over which Moore mapped his own organicism:

> In biology the concept of an organism cannot be expressed in terms of a material distribution at an instant. The essence of an organism is that it is one thing which functions and is spread through space. Now functioning takes time. Thus a biological organism is a unity with a spatio-temporal extension which is the essence of its being. This biological conception is obviously incompatible with traditional ideas.[115]

Whitehead's dynamic, spatio-temporal reading of organicism underlay many Modernist critiques of multipart composition, not least in the case of the critic E.H. Ramsden, who quoted from Whitehead's *An Enquiry Concerning the Principles of Natural Knowledge* in a 1943 review of Hepworth's composite sculptures. Here, Ramsden's notion that Hepworth's compositions unleashed a 'dynamic force' and 'a motion or pull' between the separate components indicates a conception of existential interdependency that was indebted to Whitehead's organicist philosophy.[116] As we will see, Whitehead's emphasis upon the experiential unity of the natural world had a significant impact upon Modernist art theory and would serve to underscore the holistic and neo-romantic connotations of multipart sculpture.

Parts grown out of the whole: towards an organismal theory of sculpture

In his eristic treatise of 1940, *Annals of Innocence and Experience*, Herbert Read explained how Whitehead's work had taught him to think of art in organismal terms;[117] and, in a 1926 review of *Science and the Modern World*, he enthusiastically claimed that Whitehead's philosophy embodied 'the material of a revolution in our whole concept of life or being [by seeking] to reinterpret, not only the categories of science and philosophy, but even those of religion and art'.[118] This flowering of an organicist sensibility in art, like the emergence of organicism in the New Biology, represented a holistic revolt against positivism, materialism, formalism and what proponents of this integrative worldview broadly defined as exaggeratedly reductionist approaches to life, art, science and culture. As Christopher Lawrence and George Weisz have reasoned: 'holism has served as one style of cultural and political critique aimed at various crises of modern Western society [...]. If the machine has symbolized the inhumanity and fragmentation of modern society, metaphors of organicism and wholeness have represented the solution'.[119] In the arts, this holistic attitude found expression in biologistic critiques of artistic Modernism in which the stylistic partisanship of modern art was condemned as symptomatic of a reductionist outlook.

In an organicist critique of 1936, S. John Woods lamented that: 'It is perhaps a sign of decadence when a part is taken for the whole it helps to form [...]. Since art left service and, enthroned in a studio, invested itself with a capital A, parts have flourished at the expense of the whole'.[120] In his privileging of a holistic form of creativity, Woods came close to paraphrasing the theories of Hugh I'A Fausset, who similarly condemned modernity's encroachment on 'the creative unity of life. The part has usurped the place of the whole; analysis has eaten not only into the body, but into the spiritual nerve-centres of life'.[121] Like Grigson in 'Comment on England', Woods carried this organicist concern over into a critique of geometric abstraction, whose reductionist approach to form had 'carried art from a limpid concern with the world of natural phenomena into a completely new world, unconcerned with natural phenomena and unconcerned with abstractions from organic forms'.[122] Framed in terms of biocentrism and critical neo-vitalism, Woods' argument diverged from Modernist vindications of biomorphic abstraction in sidestepping altogether the question of similitude and, instead, focusing upon an organicist conception of artistic creation.[123] Indeed, alone among the Modernists, Jean Hélion came closest to attaining Woods' organicist ideal of a 'living' art by creating a space 'in which his elements [could] move, develop, not only forwards and backwards and from side to side, but round, spinning on their own axes or encircling each other, creating a unity in space and of space, chunks of space and chunks of non-space opposed'.[124]

Alan Powers has identified Woods' experiential, nature-centric theory of art with a neo-romantic yearning to commune with nature[125] and, in stressing that 'out of living comes a live art', Woods' organicist aesthetic closely approached the experiential, neo-romantic organicism of Whitehead.[126] While addressing the philosophy of nature in romanticism, for example, Whitehead took pains to stress that philosophical organicism recognised that *all* perception was based upon a bodily experience of oneness with the natural world:

> [We] have to admit that the body is an organism whose states regulate our cognisance of the world. The unity of the perceptual field therefore must be a unity of bodily experience. In being aware of the bodily experience, we must thereby be aware of aspects of the whole spatio-temporal world as mirrored within the bodily life.[127]

This organicist sense of experiential unity with nature, in which the inter-dependency of the compositional elements echoes the body's perception of the wholeness of lived experience, allowed Modernist critics to hypothesise an existential kinship between the artwork and life which transcended the formal limitations of biomorphic abstraction. Hélion's painting was, in this respect, seen as particularly representative of this organicist sensibility. In a review of his work, for instance, Myfanwy Evans anticipated Woods' organicist interpretation by emphasising, in terms at once neo-vitalist and organicist, the 'living', experiential nature of Hélion's 'organic' compositions (Figure 3.9):

> He does not paint limbs and trees and clouds, but he tries to express the same inward and outward growth, the same static and moving attitudes with other elements – simple unrepresentational shapes, some flat and some modelled. His 'abstract' painting is essentially organic, his massing and coordination of shapes is so living that it is not even a symbol of life – it is a piece of life.[128]

This interpretation of Hélion's paintings as 'organic' owed much to the impression that his own philosophy of organicism had on Modernist circles during the 1930s.[129] Hélion's most influential essay – 'From Reduction to Growth' (1935) – linked the decline of naturalism with the rise of geometric abstraction, which he condemned as an aesthetic 'cul-de-sac' that had reduced the creative options of the artist.[130] Over time, Modernist abstraction had gradually removed all non-essential elements from the picture-plane, replicating in art the impoverishment that reductionism had achieved in philosophy: 'Considering the totally abstract attitude in this light, it reacts against the freedom, or anarchy, of other tendencies. It offers certitude, order, clarity, but also extreme limitation […]. However perfect and beautiful the works of the master of this attitude […], the position is that of [a] reduction of possibilities'.[131] To overcome the reductionist bias of Modernist abstraction, Hélion thus recommended a type of aesthetic organicism – indebted

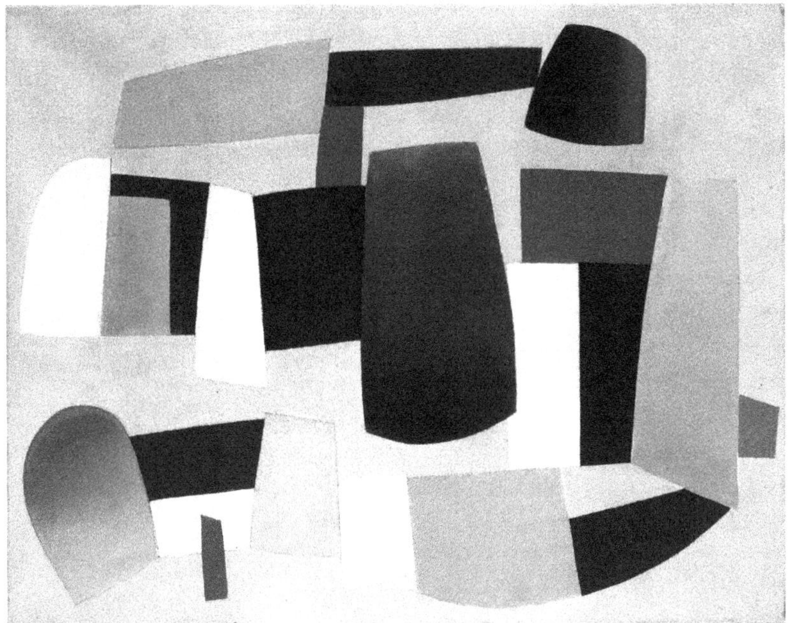

Figure 3.9

Jean Hélion, *Abstract Composition*, 1934. Tate, London

to philosophical organicism but with deep roots in its biological variant – which could revivify, by means of strengthening the experiential rapport between art and life, the aesthetic of artistic Modernism. 'The elements grow out of the whole', he wrote in an outpouring of this organicist sensibility, 'and the whole grows out of the elements. They cannot be separated'.[132] Neither, for that matter, could the underlying conception of the painting be isolated from the means of its expression: 'It is the essential point of art to identify spirit and material through sight: to create a continuity: man – fact – world'.[133]

While Hélion's organicism notably diverged from other biocentric theories of art through its emphasis upon interdependency and unity as ciphers of 'living' form, his neo-vitalist stress upon reawakening Modernist abstraction's connection to nature rearticulated, in organicist terms, the Hungarian critic Ernö Kállai's bioromantic notion that biomorphic Modernism (and related genres) symbolised humankind's biophilia and rootedness in nature.[134] In 'Poussin, Seurat and Double Rhythm' (1936), Hélion accordingly accentuated the experiential, nature-centric character of his organicism by comparing the unity of a composition in art to the dynamic interconnectivity of the human body, claiming that the smaller, self-contained elements of the image (the 'egg-structures') had to fully integrate with its overall spatial configuration (the 'tree'):

> In the field of animal life, the human body offers an obvious illustration for this: in
> a man is always evident the image of the egg-foetus but, on the other hand, he has
> arms and legs and fingers, like branches of a tree, holding space, measuring it, working
> it [...]. The egg-head has a nose going outside, a mouth going inside, hair receiving
> vibrations, ears open to sounds, letting the exterior space come into man. There is a
> continual trade between exterior and interior, without harming unity; on the contrary,
> building it.[135]

With its focus upon the experiential unity of the human body, this passage thus encapsulates the neo-romantic leanings of Hélion's organicist aesthetic by suggesting the possibility of a spiritual reawakening of art through a more profound engagement with phenomenological reality.[136]

The centrality of the body to Hélion's organicist aesthetic offers a parallel with Moore's own sculptural organicism – which typically relied upon figurative elements to emphasise the integrity of a multipart composition. For example, the inscription of partial profiles on some of Moore's sculptures (such as the eye and mouth-like line carved onto the large V-shaped element of *Four-Piece Composition: Reclining Figure*, 1934 – Plate 5) serve to underscore how the various parts interrelate by representing a disassembled though still legible body: a conceit that helps to 'pull the sculpture back into a semi-figurative reading'.[137] Moore himself was sympathetic to a partially figurative reading of this sculpture, as he specifically labelled each component as 'a head part, leg part, body and small round form which is the umbilicus and which makes the connection'.[138] It was this impression of a body that had disintegrated yet still retained some sense of corporeal unity that motivated Porteus' holistic critique of Moore's composite sculptures:

> Moore's earlier work, bunions raised in rock, lunar landscapes with female features,
> budding eyes or breasts, boiling mud-beds transfixed on the point of eruption, have
> gently disbanded into simpler elements, disparate but juxtaposed.[139]

A comparison can here be drawn with D'Arcy Wentworth Thompson's holistic appreciation of morphology. As Christa Lichtenstern has demonstrated, Moore acquired a first edition of Thompson's thesis while a student at Leeds College of Art, although it is likely that it was only through his relationship with Read that the artistic use-value of its ideas were to fully crystallise in his mind.[140] Certainly, *On Growth and Form* furnished Moore with an organicist model of morphology which powerfully shaped his creative practice and biologistic philosophy of art.[141] In one passage, for instance, Thompson spoke of the morphological bonds of 'indissoluble association' that tied an organism together into a single, composite unit:

> Muscle and bone [...] are inseparably associated and connected; they are moulded one
> with another; they come into being together, and act together and react together [...];
> there can be no change in the one but it is correlated with changes in the other; that
> through and through they are linked by indissoluble association; that they are only
> separate entities in this limited and subordinate sense; that they are *parts* of a whole
> which, when it loses its composite integrity, ceases to exist.[142]

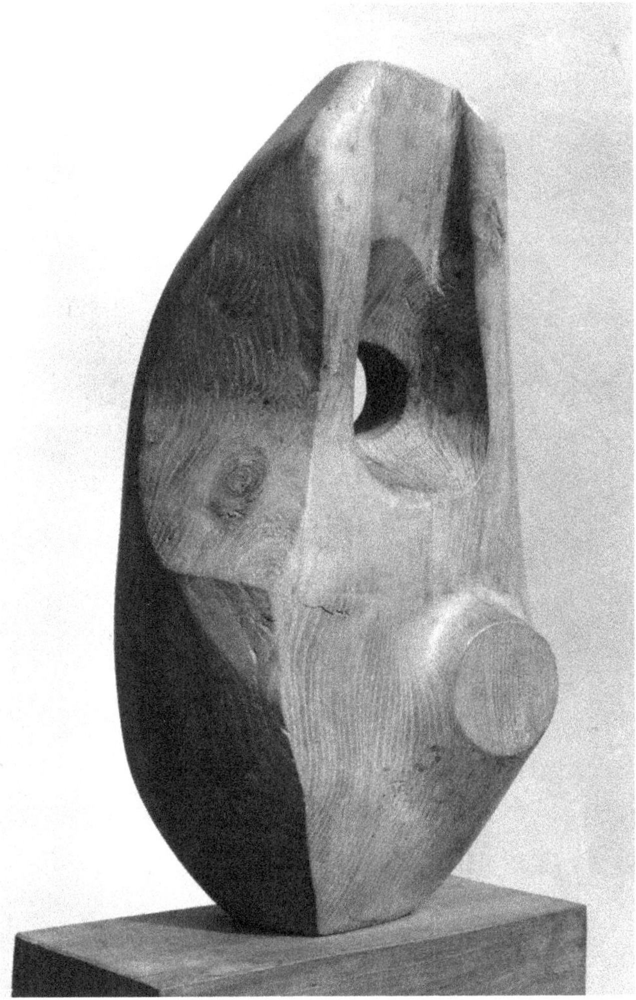

Figure 3.10

Henry Moore, *Hole and Lump*, 1934. Henry Moore Foundation

Moore's organicism possessed a deeply neo-romantic streak that was inspired by the organicist writings of the romantic poet Samuel Taylor Coleridge,[143] although his grasp of the principles of natural form was firmly indebted to his reading of Thompson.[144] His comment, in *Unit 1*, that '[b]ones have a marvellous structural strength and hard tenseness of form, subtle transition of one shape into the next', for example, appears as a paraphrasing of Thompson's discussion of the morphological forces to which bones are subject during their growth cycles.[145]

Non-figurative sculptures, such as *Hole and Lump* (1934) (Figure 3.10), highlight Moore's organicism by emphasising an existential rapport between the

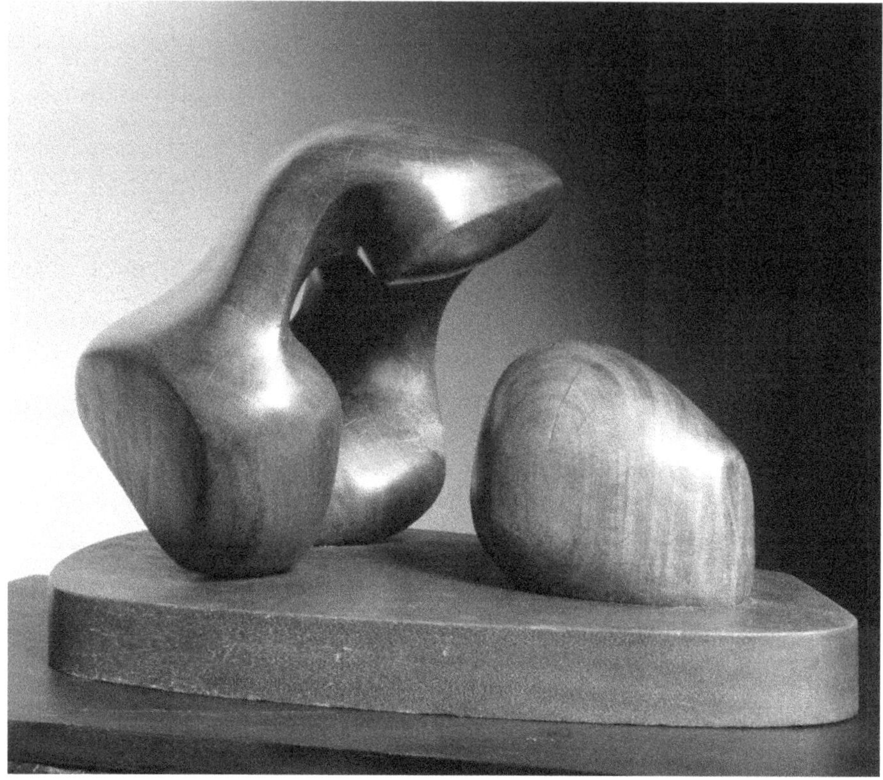

Figure 3.11

Henry Moore, *Two Forms*, 1934. Museum of Modern Art (MoMA), New York

elements of the composition. Of this artwork Moore remarked: 'I was consciously concerned with simple relationships of form: here is the hole which is *opposite* to the lump. I was putting an emphasis on this consciousness of form'.[146] Such a relationship is particularly evident in multipart compositions such as his *Two Forms* (1934) (Figure 3.11). Here, the larger form reaches out towards a smaller counterpart, which, in turn, appears perfectly calibrated to 'fit' inside the larger element's clutching contour. In this manner, Moore suggests an organic rapport, as both forms, while separate, have clearly been shaped in response to one another – one cannot be separated from the whole without disrupting the purpose of the parts and the integrity of the composition.

A comparable impression of existential interdependency is induced by F.E. McWilliam's fragmentary, surrealistic sculptures of the human figure. In *Eye, Nose and Cheek* (1939) (Plate 3), for example, a sense of connectivity is implied between the eyeball (which, like Moore's part-objects, *appears* detachable) and the larger, irregular form which comprises the 'face'. The voids which perforate, or split, the sculpture's profile, far from disrupting any sense of physiognomic continuity, encourage viewers to psychologically 'complete' the face in their mind's eye – the space thus enclosed by the sculpture emphasises the organic rapport between the different parts of the sculptural volume while simultaneously evoking, by virtue of the sculpture's weighted contours, the facial elements that McWilliam chose not to include.[147] As Michel Remy notes: '[One] is irresistibly tempted to reconstruct the human face by supplying the missing parts, so that these become essential'.[148] McWilliam himself said – in terms highly reminiscent of Moore's organicist declarations – that his discontinuous artworks concerned 'the play of solid and void, the solid element being the sculpture itself while the "missing" part inhabits the space around the sculpture'.[149] This sense of organic connectivity between the separate elements of McWilliam's sculpture was touched upon by Nash in a 1939 review in the *London Bulletin*. Here he framed McWilliam's work in neo-vitalist, organicist terms and suggested that its innate sense of vitality stemmed from the organismal inference of a complete physiognomy emerging from a collection of fragments, 'where the human attributes are implied, rather than openly confessed [...]. They are further remarkable [...] for their evidence of growth. Undoubtedly they are alive'.[150]

This sense of existential interdependency between components is, arguably, a trait that Moore's and McWilliam's composite sculptures share with Arp's multipart compositions. As Krauss notes on the relationships evident between the elements of Arp's grounded part-objects: 'each part of the still life seems to be metamorphically derived from its neighbour'.[151] Indeed, it is this impression of morphological connectivity – of each part appearing perfectly calibrated to fit the whole – that legitimised organicist critiques, allowing Modernist critics to speak, in biologistic terms, of the artwork as 'a living organism' by virtue of the dynamic interdependency expressed between the components rather than any principle of verisimilitude.

The impression of morphological organicity that emerged through the modelling and subtle positioning of the compositional elements in multipart sculpture therefore allowed biophilic Modernists to achieve the neo-romantic ideal of an artwork that appeared 'natural' but not, in any simple sense, 'naturalistic'.[152] Just as organicist critiques of interwar culture underscored the qualitative discrepancy between a society that was organised according to 'natural' as opposed to 'artificial' criteria, so artistic Modernists employed organicist metaphors – drawn largely from the New Biology – to legitimise a composite sculptural idiom as something that appeared *grown* but not *made*.[153] In this sense, sculptural organicism represented a distinctive component of the neo-romantic, biologistic discourse that developed in Modernist circles in the 1930s, one which united Driesch's neo-vitalist notion of entelechy with a holistic paradigm – rooted in Whitehead's organicism – which extolled dynamic interdependency and existential wholeness.

Notes

1 Porteus, 'New Planets', p. 22, original emphasis.
2 Curtis, *Sculpture*, pp. 126–7.
3 See C. van Eck, *Organicism in Nineteenth-Century Architecture: An Inquiry into Its Theoretical and Philosophical Background* (Amsterdam: Architectura & Natura Press, 1994), esp. pp. 41–62.
4 Hélion, 'From Reduction to Growth', p. 23.
5 Mundy, *Biomorphism*, p. 13.
6 Read, 'Books of the Quarter: Review of *Science and the Modern World*', p. 581.
7 *Ibid.*, p. 586.
8 Lawrence & Weisz, 'Medical Holism', p. 6. To describe something as 'organic' naturally relies upon the etymological ambiguity of the metaphor. For, on the one hand, 'organic' can refer to the products or processes of nature, while, on the other, it possesses a metaphorical or applied meaning that indicates particular kinds of relationship, both organisational and, more complicatedly, societal. Originally, in the thirteenth century, 'organ' denoted an instrument – its later biological association deriving from various parts of the body being understood as instruments of sensation. It is useful to note that the terms 'organise' and 'organisation' originate from this early meaning, relating to the arrangement of the parts of an instrument. 'Organic', however, pursued an alternative route – often being used in opposition to words such as 'organise' or 'mechanical'. Through the life sciences in the eighteenth century, it acquired a definite reference to things living and developing, and in terms of social organisation, it referred to a society that had been 'grown' rather than 'made'. The convoluted history of the term – its simultaneous application to both nature and artificial organisation – evinces its semantic complexity in interwar debate. See R. Williams, *Keywords: A Vocabulary of Culture and Society* (London: Fontana Press, 1983), pp. 227–8.
9 Lawrence & Weisz, 'Medical Holism', p. 6.
10 A.S. Russell, 'The Origin of Life', in *The Listener*, Vol. X, No. 245, 20 September 1933, p. 411.
11 A.S. Russell, 'Mechanism and Biology', in *The Listener*, Vol. III, No. 60, 5 March 1930, p. 411.
12 K. Wellman, 'Materialism and Vitalism', in J.L. Heilbron (ed.), *The Oxford Companion to the History of Modern Science* (Oxford: Oxford University Press, 2003) (Oxford Reference Online, www.oxfordreference.com/views/ENTRY.html?subview=Mainandentry=t124.e0443, accessed 2 July 2007).
13 See Haraway, *Crystals, Fabrics, and Fields*, p. 26; J. Pickstone, *Ways of Knowing: A New History of Science, Technology and Medicine* (Manchester: Manchester University Press, 2000), pp. 87–8.
14 Burwick & Douglass, 'Introduction', p. 1.
15 Bowler, *Reconciling Science and Religion*, pp. 122–3.
16 Bergson, *Creative Evolution*, p. 271.
17 Bowler, *Reconciling Science and Religion*, pp. 160–1.
18 Allen, *Life Science in the Twentieth Century*, p. 103.
19 Bowler, *Reconciling Science and Religion*, p. 166. See also Russell, 'Mechanism and Biology', p. 411.
20 J.S. Haldane, *The Philosophical Basis of Biology* (London: Hodder & Stoughton, 1931), p. 12.
21 Needham, *Order and Life*, pp. 109, 139.
22 Haraway, *Crystals, Fabrics, and Fields*, pp. 24–5.
23 A.N. Whitehead, *Science and the Modern World* (Cambridge: Cambridge University Press, 1926), pp. 111–12.
24 Whitworth, 'The Clothbound Universe', pp. 62–4.
25 Bowler, *Reconciling Science and Religion*, pp. 382–3.
26 Whitehead, *Science and the Modern World*, p. ix.
27 *Ibid.*, p. 112.
28 *Ibid.*, p. 146.
29 J.S. Haldane, J.H. Woodger and Joseph Needham would all come to incorporate aspects of Whitehead's concept of the 'event' and 'organism' into their work of the 1920s and 1930s. See Haldane, *The Philosophical Basis of Biology*; J.H. Woodger, *Biological Principles: A Critical Study* (London: Kegan Paul, 1929), esp. pp. 328–9; Needham, *Order and Life*, esp. pp. 109–10.

30 Botar, *Prolegomena to the Study of Biomorphic Modernism*, p. 8.

31 Allen, *Life Science in the Twentieth Century*, pp. 103–5. See also Loeb's discussion on the divisibility of matter in: J. Loeb, *The Dynamics of Living Matter* (London: Columbia University Press, 1906), pp. 29–30.

32 Haldane, *The Philosophical Basis of Biology*, p. 15.

33 Haraway, *Crystals, Fabrics, and Fields*, pp. 15, 137–8.

34 For a discussion of debates surrounding the reception of organicism by the British scientific community in the 1920s and 1930s, see Bowler, *Reconciling Science and Religion*, esp. pp. 171–8.

35 See Botar, 'Defining Biocentrism', pp. 15, 18; Botar, *Prolegomena to the Study of Biomorphic Modernism*, pp. 6–7, 47.

36 Williams, *Keywords*, p. 228.

37 F.R. Leavis & D. Thompson, *Culture and Environment: The Training of Critical Awareness* (London: Chatto & Windus, 1933), p. 87.

38 Thomas Hobbes, for example, in *Leviathan* (1651) speaks of the state as being 'but an Artificiall Man [...] in which the *Soveraignty* is an Artificiall *Soul*, as giving life and motion to the whole body', in which all the citizens perform duties analogous to *artificial* joints and nerves in the 'Body Politique'. Such a concept of organic unity relied upon seventeenth-century comprehensions of God as the supreme watchmaker, whose workmanship was evident in nature and which people dutifully copied in manufacturing artefacts. See Thomas Hobbes, *Leviathan*, C.B. Macpherson (ed.) (Harmondsworth: Penguin Books, 1968), pp. 81–2; and (on mechanism) Pickstone, *Ways of Knowing*, pp. 66, 88.

39 G. Orwell, *Coming Up for Air* (1939) (London: Secker & Warburg, 1967), p. 114.

40 See A.-K. Mayer, '"A Combative Sense of Duty": Englishness and the Scientists', in C. Lawrence & A.-K. Mayer (eds), *Regenerating England: Science, Medicine and Culture in Interwar Britain* (Amsterdam: Editions Rodopi, 2000), pp. 68–9.

41 R. Williams, *Culture and Society, 1780–1950* (London: Penguin Books, 1961), p. 256.

42 Henry Slesser quoted by E.S.P. Haynes, 'Law', in H. Kingsmill (ed.), *The English Genius* (London: Right Book Club, 1939), p. 88.

43 C. Lawrence & A.-K. Mayer, 'Regenerating England: An Introduction', in C. Lawrence & A.-K. Mayer (eds), *Regenerating England: Science, Medicine and Culture in Interwar Britain* (Amsterdam: Rodopi, 2000), p. 12.

44 C. Williams-Ellis, *England and the Octopus* (London: Geoffrey Bles, 1928), esp. pp. 11–20.

45 Hugh I'A. Fausset, 'The Insufficiency of Science', in *The Listener*, Vol. VII, No. 159, 27 January 1932, p. 129.

46 See Whitworth, 'The Clothbound Universe', pp. 62–4.

47 Read, 'Books of the Quarter: Review of *Science and the Modern World*', p. 385.

48 *Ibid.*, p. 385.

49 Fausset, 'The Insufficiency of Science', p. 130.

50 *Ibid.*, p. 130.

51 Lawrence & Weisz, 'Medical Holism', p. 7.

52 *Ibid.*, pp. 4–7.

53 P.C. Ritterbush, *The Art of Organic Forms* (Washington, DC: Smithsonian Institution, 1968), p. 85; see also Lawrence & Wiesz, 'Medical Holism', p. 6.

54 Mundy, *Biomorphism*, pp. 13–14.

55 J. Burnham, *Beyond Modern Sculpture: The Effects of Technology on the Sculpture of This Century* (New York: George Braziller, 1968), p. 94.

56 P.C. Ritterbush, 'Aesthetics and Objectivity in the Study of Form in the Life Sciences', in G. Rousseau (ed.), *Organic Form: The Life of an Idea* (London: Routledge & Kegan Paul, 1972), p. 52.

57 Porteus, 'New Planets', p. 22.

58 Curtis, *Sculpture*, pp. 171–2.

59 Curtis recognises in Arp's part-objects a kinship with Salvador Dalí's theory of involuntary sculpture, in which manipulated, though otherwise throwaway, items (such as a screwed-up bus ticket) acquired a distinctive, psychoanalytical meaning that derived from the transformative agency of the individual who 'made' them. Arp's ambiguous, though repetitively biomorphic, 'concretions' represent 'the solidification of the sculptor's thoughts' by playing upon the

unconscious agency of the human psyche in producing and interpreting objects. Curtis, *Sculpture*, pp. 158–9. Equally, Eric Robertson notes the spectator's assumed role in 'completing' Arp's sculptures through rearranging the pieces. This latent interactivity thus connects Arp's sculpture to Marcel Duchamp's creation of artworks which invited the viewer's participation. Robertson, *Arp*, pp. 116–17.

60 Robertson, *Arp*, p. 114.

61 Georges Bataille, quoted in H. Foster, R. Krauss, Y.-A. Bois & B. Buchloh (eds), *Art Since 1900* (London: Thames & Hudson, 2004), pp. 247–8. See also Robertson, *Arp*, p. 114.

62 Indeed, even though the degree of representational similitude could vary widely, accounts of Modernist sculpture published in Britain during the early 1930s – such as Herbert Read's *The Meaning of Art* (1931) – were generally in agreement that the sculptor's task was to realise a specific form 'in the particular block of stone he has before him' and that, by and large, that form was relatively unitary. See Read, *The Meaning of Art*, p. 151.

63 C. Giedion-Welcker, 'Preface to 1937 Edition', in *Contemporary Sculpture: An Evolution in Form and Space* (London: Faber & Faber, 1954), pp. xi–xii. The original English edition of this book was titled *Modern Plastic Art: Elements of Reality, Volume and Disintegration*, from the German *Moderne Plastik: Elemente der Wirklichkeit, Masse und Auflockerung*. The 1954 edition lost this emphasis on dematerialisation.

64 *Ibid.*, pp. xii–xix.

65 A. Jakovski, 'Brancusi', *Axis*, No. 3, July 1935, pp. 3–9.

66 Jakovski's text mirrored other contemporary critical accounts which chronicled the relationship between modern physics and artistic Modernism. Gavin Parkinson has demonstrated how Surrealist critics, such as Karl Einstein, readily analogised between Cubism's fragmentary aesthetic and the language of dispersal and acausality of modern physics. G. Parkinson, *Surrealism, Art and Modern Science* (New Haven: Yale University Press, 2008), pp. 146–7. Like Jakovski, Porteus and Giedion-Welcker were similarly attentive to this epistemological rapport. Recent scholarly work, most particularly in relation to the mobiles of Alexander Calder, has attempted to highlight the deep shadow that modern physics cast over the theory of Modernist sculpture. See V.V. Malloy, 'Rethinking Alexander Calder's Universes and Mobiles: The Influences of Einsteinian Physics and Modern Astronomy', in *Immediations*, Vol. III, No. 1, 2012, pp. 9–25.

67 Jakovski, 'Brancusi', p. 4.

68 *Ibid.*, p. 9.

69 *Ibid.*, p. 9.

70 The emergence of holism has been described by scholars as 'a revolt against formalism'. See Lawrence & Weisz, 'Medical Holism', p. 6.

71 Moore, 'The Sculptor Speaks', p. 338.

72 Curtis, *Sculpture*, p. 171.

73 Harrison, *English Art and Modernism*, p. 268.

74 Festing, *Barbara Hepworth*, pp. 98–100.

75 Barbara Hepworth quoted in P. Curtis, *Barbara Hepworth* (London: Tate Publishing, 1998), p. 33.

76 Robertson, *Arp*, p. 114.

77 Stokes, 'Miss Hepworth's Carving', p. 310.

78 Anne Wagner, in particular, has written extensively on Stokes' use of a critical lexicon drawn from Melanie Klein's theory of psychoanalysis to explain the interrelations evinced in Hepworth's sculptures of the early 1930s. See Wagner, 'Miss Hepworth's Stone *Is* a Mother', pp. 53–74.

79 H. Frankfort, 'New Works by Barbara Hepworth', in *Axis*, No. 3, July 1935, p. 14. It is worth noting that before he settled on an 'organic' metaphor Frankfort first likened Hepworth's multipart sculptures to music in an effort to comprehend the spatial intervals between the components.

80 A.S. Russell, 'Mechanism and Biology', p. 411. While in this 1930 article Russell still labelled organicism a form of neo-vitalism, in 1933 he confirmed that the neo-vitalist position could be more correctly termed the 'organismal view', proponents of which maintained 'that many happenings can only be explained in terms of the organism as a whole'. Russell, 'The Origin of Life', p. 411.

81 J. Huxley, *Essays of a Biologist* (1923) (Harmondsworth: Penguin Books, 1939), p. 35.

82 See G.N.G. Orsini, 'Organicism', in *The Dictionary of the History of Ideas* (Charlottesville: University of Virginia, 2003), Vol. III, pp. 422–7 (online edition http://etext.virginia.edu/cgi-local/DHI/dhi.cgi?id=dv3-52, accessed 19 September 2006); G.N.G. Orsini, 'The Organic Concepts in Aesthetics', in *Comparative Literature*, Vol. XXI, No. 1, winter 1969, pp. 1–30.

83 Plato, *Phaedrus*, C.J. Rowe (trans.) (Warminster: Aris & Phillips, 1986), p. 99.

84 Early Renaissance architectural theory put much emphasis upon the presumed unity between living things (and especially the human body) and exemplary building practice, partly resulting from the rediscovery of classical culture and the humanist desire to recreate the spirit of the antique by archaeologically 'reconstructing' its motifs and forms in painting, sculpture and architecture. See: van Eck, *Organicism in Nineteenth-Century Architecture*, pp. 42–62; F. Ames-Lewis, *The Intellectual Life of the Early Renaissance Artist* (New Haven: Yale University Press, 2000), pp. 109–40.

85 L.B. Alberti, *On the Art of Building, in Ten Books*, J. Rykwert, N. Leach & R. Tavernor (trans.) (Cambridge: MIT Press, 1988), Book VII, p. 199.

86 W. Kandinsky, 'Line and Fish', in *Axis*, No. 2, April 1935, p. 6, original emphasis.

87 On Kandinsky's biologism, see Barnett, 'Kandinsky and Science', pp. 207–23.

88 Mundy, *Biomorphism*, p. 121.

89 Rosalind Krauss identifies a critical interest in the 1930s in the apparent disparity between sculpture's inherently static nature and the theme of temporality and speculates that this concern impacted profoundly upon the development of new forms of sculptural Modernism. Krauss, *Passages in Modern Sculpture*, pp. 3–4.

90 Frankfort, 'New Works by Barbara Hepworth', p. 14. In point of fact, the parts of these sculptures were pinned, although most contemporary reviewers seemed unaware of this fact.

91 Porteus, 'New Planets', p. 22.

92 M. Gale, 'Brancusi: An Equal Among Rocks, Trees, People, Beasts and Plants', in C. Giménez and M. Gale (eds), *Constantin Brancusi: The Essence of Things* (London: Tate Publishing, 2004), p. 53. Often Brancusi would rearrange the elements of the bases to create new combinations.

93 Barbara Hepworth quoted in Festing, *Barbara Hepworth*, p. 100. On the spectator's role in 'realising' Arp's composite sculptures, see Robertson, *Arp*, pp. 116–17.

94 Giacometti quoted in Krauss, *Passages in Modern Sculpture*, p. 118.

95 Krauss, *Passages in Modern Sculpture*, p. 118.

96 On Brancusi's bases see A.C. Chave, *Constantin Brancusi: Shifting the Bases of Art* (New Haven: Yale University Press, 1993), pp. 225–6; on Arp's bases, see Robertson, *Arp*, pp. 115–16.

97 Frankfort, 'New Works by Barbara Hepworth', p. 14.

98 See Tate online catalogue: 'Henry Moore: *Four Piece Composition: Reclining Figure – 1934*', full catalogue entry T02054. First published in *The Tate Gallery 1976–8: Illustrated Catalogue of Acquisitions* (1979) (www.tate.org.uk/art/artworks/moore-four-piece-composition-reclining-figure-t02054/text-catalogue-entry, accessed 7 March 2007).

99 'Mais dans le meme sens que le cadre du tableau empêche l'émotion directe et immediate, le *socle* de la sculpture est aussi une *convention*. Les réalités, les creations n'ont pas de bouts ni de pieds mais sont completes dans leur unite. *La sculpture sans socle, l'oeuvre d'art doit être vue de tous les côtés*. Seulement dans ces conditions elle peut devenir réalité'. Hans Arp as recorded by Tristan Tzara. T. Tzara, 'Un Art nouveau: deux solutions sur le principe de l'immédiat, postulées par H. Arp et formulées par Tristan Tzara' (c.1917), reprinted in T. Tzara, *Oeuvres complètes, Vol. I: 1912–1924* (Paris: Flammarion, 1975), p. 558.

100 J. Derrida, *The Truth in Painting*, G. Bennington & I. McLeod (trans.) (Chicago: University of Chicago Press, 1987), p. 9.

101 This is an implication that underlies much of the classical thought on the subject of 'part-hood' or mereology. See V. Harte, *Plato on Parts and Wholes: The Metaphysics of Structure* (Oxford: Oxford University Press, 2005), pp. 155–7.

102 Haraway, *Crystals, Fabrics, and Fields*, pp. 61–3.

103 van Eck, *Organicism in Nineteenth-Century Architecture*, p. 62.

104 As Jean Piaget states, central to the concept of structure are 'three key ideas: the idea of wholeness, the idea of transformation, and the idea of self-regulation'. J. Piaget, *Structuralism* (London: Routledge & Kegan Paul, 1971), p. 5.

105 G.W.F. Hegel, *Philosophy of Fine Art*, Part III, quoted in Orsini, 'Organicism', p. 426.

106 See A. Harrington, *Reenchanted Science: Holism in German Culture from Wilhelm II to Hitler* (Princeton: Princeton University Press, 1996), esp. pp. xv–33.

107 E. Young, *Conjectures on Original Composition*, E.J. Morley (ed.) (Manchester & London: Manchester University Press and Longmans, Green & Co., 1918), p. 7, original emphasis.

108 Powers, 'The Reluctant Romantics', p. 265.

109 J. Hélion, 'Poussin, Seurat and Double Rhythm', in M. Evans (ed.), *The Painter's Object* (London: Gerald Howe, 1937), p. 104.

110 See Powers, 'The Reluctant Romantics', pp. 265–6.

111 On the rise of neo-romanticism in Britain in the 1920s and 1930s, see P. Fuller, 'The Visual Arts', in B. Ford (ed.), *The Cambridge Guide to the Arts in Britain: Since the Second World War* (Cambridge: Cambridge University Press, 1988), Vol. IX, pp. 100–1.

112 Orsini, 'The Organic Concepts in Aesthetics', pp. 4–5.

113 Huxley, Wells & Wells, *The Science of Life*, p. 319.

114 Moore, 'The Sculptor Speaks', p. 338, emphasis added.

115 A.N. Whitehead, *An Enquiry Concerning the Principles of Natural Knowledge* (Cambridge: Cambridge University Press, 1919), p. 3.

116 E.H. Ramsden, 'Barbara Hepworth – Sculptor', in *Horizon*, Vol. VII, No. 42, June 1943, p. 420.

117 Read, *Annals of Innocence and Experience*, pp. 227–8.

118 Read, 'Books of the Quarter: Review of *Science and the Modern World*', p. 581.

119 Lawrence & Weisz 'Medical Holism', p. 7.

120 S.J. Woods, 'Time to Forget Ourselves', in *Axis*, No. 6, summer 1936, p. 19.

121 Fausset, 'The Insufficiency of Science', p. 130.

122 *Ibid.*, pp. 19–20.

123 As Mundy has noted, while – for Grigson – 'biomorphic' did not designate a literal resemblance to natural form, it nonetheless implied that the artwork had been *derived* or *inspired* by the organic world. This is quite different to Woods' neo-vitalist theory of a 'living' art, which hypothesises an *existential* relationship between the artwork and nature. This focus upon the internal inter-relationships of a composition thus brings his theory much closer to organicism than it does the bioromantic leanings of Grigson's biomorphism. Mundy, 'Comment on England', p. 28.

124 Woods, 'Time to Forget Ourselves', p. 20.

125 Powers, 'The Reluctant Romantics', p. 261.

126 Woods, 'Time to Forget Ourselves', p. 21.

127 Whitehead, *Science and the Modern World*, p. 113.

128 M. Evans, 'Hélion Today: A Personal Comment', in *Axis*, No. 4, November 1935, p. 4.

129 For a brief discussion of Hélion's impact on Modernists in the 1930s, see Botar, *Prolegomena to the Study of Biomorphic Modernism*, p. 52.

130 Hélion, 'From Reduction to Growth', pp. 19–22.

131 *Ibid.*, p. 22.

132 *Ibid.*, p. 22.

133 *Ibid.*, p. 22.

134 See Botar, *Prolegomena to the Study of Biomorphic Modernism*, p. 52. Kállai's bioromanticism is discussed at greater length in Chapter 4.

135 Hélion, 'Poussin, Seurat and Double Rhythm', p. 104. This essay was first published in *Axis* in the summer of 1936. Of course, the idea that the parts of the body served as basic units from which everything else was composed was a long-standing theme in Renaissance aesthetics; the model of the *homo quadratus* – the body of a well built man whose outstretched hands and feet fitted into a square and circle – was considered the apogee of proportional unity and was frequently drawn upon as a paradigm of harmony by artists of the fifteenth century. See van Eck, *Organicism in Nineteenth-Century Architecture*, pp. 41–2.

136 Powers, 'The Reluctant Romantics', p. 268.

137 See Foster, Krauss, Bois & Buchloh, *Art Since 1900*, p. 269.

138 Moore quoted in R. Morpeth, *Illustrated Catalogue of Acquisitions: The Tate Gallery, 1976–8* (London: Tate Publishing, 1979), pp. 116–18, reprinted in D. Mitchinson, '1930–1940', in *Henry Moore: Early Carvings, 1920–1940* (Leeds: Leeds City Art Galleries, 1982), p. 33.

139 Porteus, 'New Planets', p. 22.

140 'Moore was pleased to tell [Christa] Lichtenstern that he had discovered this book [*On Growth*

and Form] in Leeds before he met Read'. Quoted in D. Kosinski (ed.), *Henry Moore: Sculpting the Twentieth Century* (New Haven: Yale University Press, 2001), p. 264.

141 See Lichtenstern, *Henry Moore*, pp. 57–9.

142 Thompson, *On Growth and Form*, p. 1019.

143 Lichtenstern, *Henry Moore*, p. 264.

144 Mundy, *Biomorphism*, p. 130.

145 See Mundy, *Biomorphism*, p. 130, and, for a discussion of Thompson's influence on Moore's interest in bones, Lichtenstern, *Henry Moore*, p. 59.

146 H. Moore, *Henry Moore: Wood Sculptures* (London: Sidgwick & Jackson, 1983), p. 87, emphasis added.

147 See also A. Windsor, 'Frederick Edward McWilliam', in A. Windsor (ed.), *British Sculptors of the Twentieth Century* (Aldershot: Ashgate, 2003), p. 140.

148 Remy, *Surrealism in Britain*, p. 196.

149 F.E. McWilliam in an unpaginated quote cited in Remy, *Surrealism in Britain*, p. 196.

150 P. Nash, 'F.E. McWilliam', in the *London Bulletin*, March 1939, p. 12.

151 Krauss, *Passages in Modern Sculpture*, p. 140.

152 On the opposition between 'nature' and 'naturalistic' hypothesised by Modernist artists, see Powers, 'The Reluctant Romantics', p. 258.

153 On the social connotations of 'organic', see Williams, *Keywords*, p. 228.

4. The morphology of art

Writing in his semi-autobiographical work of 1940, *Annals of Innocence and Experience*, Herbert Read claimed that, over the course of his research into the morphology of art, it had gradually become 'clear [to him] that the aesthetic experience was not a superficial phenomenon, an expression of surplus energy, a secondary feature of any kind, but rather something related to the very structure of the universe'.[1] For Read, this insight into the profundity of the aesthetic experience emerged out of a lengthy career as a critic, in which the study of artworks had gradually divulged to him an 'underlying structure' which could be 'reduced to abstract terms'. In turn, the pervasiveness of this underlying structure within the composition of artworks led him to a startling conclusion:

> It was a short and obvious step to recognise at least an analogy and possibly some more direct relation, between such a morphology of art and the morphology of nature. I began to seek for more exact correspondences, first by making myself familiar with the conclusions reached by modern physicists about the structure of matter, and then by exploring the quite extensive literature on the morphology of art.[2]

Like many within the British avant-garde, Read was heavily influenced by the biocentric currents emanating out of central Europe and his interpretation of the morphology of art was noticeably informed by the theories of the Hungarian theoretician and Bauhaus associate Ernö Kállai.[3] A close friend of the Constructivist Naum Gabo, Kállai was astonished by the visual marvels of the new scientific photography and just how closely these images resembled the products of contemporary art.[4] Recognising that technology had only lately made the 'deep structure' of the world visible to humankind, he contended that modern artists had intuitively been making this facet of physical reality perceptible to audiences through their abstract compositions. Art that sought to represent the underlying structure of nature he thus termed 'bioromantic'.[5] Yet while Read's analogy between the morphology of art and the morphology of nature owed something to Kállai's bioromantic philosophy, his reference to the primary structure of matter as it was revealed through the calculations of physicists occupied a more equivocal territory that was suspended somewhere between the representation of the organic and the mathematical.[6] Indeed, by analogising between modern art and natural or scientific imagery, 'bioromanticism' subtly diverged from other biologistic trends in Modernist discourse (such as organicism) which hypothesised a relationship between art and nature forged along abstractly structural or morphogenetic lines.[7]

In 1940, the critic Reginald Wilenski – who was pivotal in conjecturing the relationship between Modernist sculpture and the morphology of nature in Britain – employed, in his *Modern French Painters*, the term 'associationism' to describe his analogical approach of likening modern art to the deep structures of nature.[8] In so doing, he identified a major cultural shift, following the First World War, which 'enormously favoured the dissemination of the Associationist concept', involving the re-establishment of intercontinental travel, the spread of radio technology, the rise of motoring and aviation and – perhaps most importantly for the promulgation of *biologistic* associationism – the emergence of cheap print technologies which enabled the public to 'slake [its] thirst for universal knowledge and to appease the new impulses to capture contacts in all directions and compare things and find correspondences and associations across continents and ages. In the field of art there was now a vast and every-growing mass of material available for these processes'.[9]

A psychological school which had its origins in Aristotle's philosophy of sensation, associationism developed in Britain between the seventeenth and nineteenth centuries (under the auspices of such philosophical luminaries as John Locke and David Hume) and primarily attended to the *sequential logic of ideas* in memory or imagination, whereby an experience or idea 'may serve to recall another which *resembles* it or was *contiguous* with it in former experience'.[10] Although the analogical character of associationism self-evidently served to influence Kállai's bioromanticism,[11] associationism's broader emphasis on the notion that mental phenomena might be accounted for by a loose association of ideas perhaps provides a freer psychological framework – one that lacks the specificity of bioromanticism's focus on scientific imagery – in which to conceptualise the rapport hypothesised by Read and Wilenski between the morphology of art and the underlying structures of nature.[12]

This chapter will explore the role that the New Biology played in stimulating Modernist interest in morphology. Commencing with a historical survey of the 'new morphology', it will examine the fascination biophilic Modernists exhibited towards the spiral form as an abstract symbol of 'life'. The neo-romantic, idealist notion that certain fundamental structures, such as the spiral, pervaded the organic and inorganic worlds will subsequently be discussed in relation to the philosophical doctrine of monism – a holistic, pantheistic conception which postulated the unity of nature – and its appeal to synthesising, biologistic Modernists such as Read and Grigson. The final section will investigate the political imperatives propelling Modernist interest in form 'constants' and how Raoul France's science of biotechnics provided Modernists with a functionalist, biologistic philosophy to justify their neo-romantic privileging of morphology and associationist/bioromantic understanding of the relationship between Modernist sculpture and the deep structures of the natural world.

The rise of the new morphology

The analysis of form in biology stems from the nineteenth-century field of morphology, which, in a literal sense, means 'the study of form'. Originating in Goethean thought as the method through which the universal prototypes of plants might be ascertained,[13] Victorian morphology eventually succumbed to the influence of Darwinian theory and came to be associated with a specific set of ideas used in evolutionary science. Certainly, the incorporation of morphology into evolutionary studies meant that the idea of the 'type' as an abstract intellectual category was lost in favour of an assumption that the family tree of any species could be traced back to some historic archetype. Such a time-oriented outlook was in fact the very opposite of Goethe's concept of timeless forms.[14] While late-Victorian morphology took in a wide variety of disciplines, such as comparative anatomy, cytology and palaeontology, the chief aim of the subject remained unearthing the archetypal form of different species through comparing fossil records and living animals in an effort to determine the ultimate schema underlying diversity in nature.[15]

In spite of providing science with more rigorous experimental methods through the stimulus of the new evolutionary synthesis, the New Biology remained enthralled to the concerns of nineteenth-century morphology. For although history tends to stress the eclipse of a descriptive natural history by a causal analytic science,[16] in reality, the rise of new technologies in the 1920s and 1930s – such as X-ray diffraction analysis and ultraviolet microscopy – saw a return to static structural analysis, leading to the functions of structures being attributed to properties of form. Despite the holistic mind-set of the New Biology, the study of molecular and cellular structure invariably meant that views converged on the issue of whether or not structured elements persisted from generation to generation in living creatures.[17] Of course, this prioritising of form was indebted to the general theory of evolution, which conceived of time as a feature 'whose effects could be perceived directly in distinct but complementary forms: fossils, embryos and rudimentary organs'.[18] In this sense, the structure of a fossil embodied petrified time; the form of the embryo, functional time; and the shape of the rudimentary organ, retarded time.[19]

Yet at the same time, biochemistry played a crucial role in bringing form back to the table of biological discussion, as new experimental techniques had demonstrated how structure was present at all levels of organic matter. Living things, as the biochemist Joseph Needham attested, were divided into different organisational tiers, the highest level consisting of colonies of cells, then cells, followed by the parts of cells, composed of colloidal aggregates, molecules and atoms. Typically this hierarchy tallied with disciplinary boundaries, morphology dealing with higher-level anatomical structures while biochemistry dealt with the lower levels.[20] For Needham, the 'crux of the problem [lay] in determining the nature of [these] organizing relations at the coarser or higher levels' by means of combining biochemistry with morphology.[21] In point of fact, the notion that continuities existed between different grades of structure was an idea that Needham shared with other scientific reformers, such as Desmond Bernal, who likewise 'outlined a scale of form reaching from quantum-mechanical systems to the metazoa'.[22] And in this respect, the theory of transformations devised by D'Arcy Wentworth Thompson was understood by Needham to provide an effective means of charting the organising relations between these diverse layers of structure, especially when tracking the expansion of organic material in embryology.[23]

In light of the holistic prejudice of the New Biology, interwar morphology was typically framed in terms of contingency and temporality. Early-twentieth-century scientific trailblazers, such as Edward Stuart Russell and James Johnson, had proved the falsity of the mechanism–vitalism dichotomy, arguing that contemporary understandings of form and function were largely redundant, as, in effect, matter was a *contingent* entity, meaning that neither form nor function could completely take precedence. In *The Mechanism of Life* (1921), for example, Johnson quipped that: 'In a sense it is wrong to speak about an animal as a structure or even as a *thing*. It is "something happening"'.[24] These conclusions were so far reaching that, by 1936, Needham reckoned that 'such old antitheses as that of form and function need not [...] detain us', as, through the influence of organicism, it was evident that 'form [was] simply a short time-slice of a single spatio-temporal entity'.[25] Far from bankrupting morphology, indeed, the New Biology had simply reaffirmed that 'the central problem of biology [was] the form problem'.[26] Gauging the character of the relation between objects of different sizes in bodies was, for Needham, the true problem vexing the New Biology. 'What is the relation', he asked, 'between those large particles which we call elephants, trees or men, and those extremely small ones we call molecules or electrons?' This question would, to some extent, also haunt the Modernist imagination as sculptors and critics sought to understand the primary structure of matter and its relationship to the morphology of art. While a neo-vitalist enthralment to life's quintessence largely drove this philosophical enquiry, as we will see, a neo-romantic fascination with idealist morphologies also steered Modernists, first to Goethe's morphological writings and then to the idealist hypotheses of twentieth-century morphologists.

The curve of life:
morphology, the Golden Section and the spiral form

[W]hen we once begin, in these great divisions of the universe, to look for spirals, it [is] astonishing to find the enormous number and variety of such formations which can be discovered. In plants, spirals are observable from seeds and seed cases and cells, to stems and flowers, and fruit. In animals, and in man, the spiral may be said to follow the whole course of vital development from the spermatozoon to the muscular structure of the heart.[27]
Theodore Andrea Cook, *The Curves of Life*, 1914

The reappearance of morphology as a topic worthy of biological analysis in the early twentieth century was accompanied by a wave of biologistic, Modernist literature which sought to apply morphological principles to the study of art. Beholden to the same neo-vitalist impulse which saw the critic Anatole Jakovski portray Hans Arp's oeuvre, in animistic terms, as representing 'an upward spurt',[28] this critical bibliography – described by Read as elucidating 'the morphology of art' – investigated the primary architecture of nature and art in explicitly geometrical terms, assessing the morphological longevity of certain types of shape within the natural world and the history of art.[29] In particular, within the neo-vitalist, biocentric interpretative framework established by biophilic Modernists in the 1930s, the spiral was privileged as a geometric structure that symbolised, *par excellence*, the morphological rapport between natural form and the mechanism of growth.[30] Many of the Modernists who incorporated this geometric form in their art – such as Naum Gabo – were, perhaps surprisingly, only tangentially related to the 'movement' of biomorphism;[31] yet, like their biomorphic colleagues, they similarly sought to abstractly convey the hidden forces which underpinned the creation of form in nature and in human works.[32]

In Britain, the first Modernist art critic to systematically hypothesise the morphology of art in the interwar period was Read, who highlighted 'the significance which the science of form [had] for the theory of art' by scouring the annals of the New Biology for scientific evidence of 'the motives that [led] man to prefer one shape to another'.[33] These motives were, he argued, either conscious or intuitive: 'Either man makes his selection because he believes, after rational thought and observation, that one shape is "better" than another; or he does not think about it at all – he acts, as we say, intuitively'.[34] This intuitive sensibility stemmed from 'a very early stage in the history of human thought [when] man discovered that certain proportions – that is to say, certain shapes – were constant in nature'. It was the Greeks who first perceived the mathematical harmony of nature, expressing its principles in Pythagorean number theory and formulating an ideal law of proportion that came to be known as the 'Golden Section' – a distinctive ratio in which each quantity in the sequence equals the sum of the two quantities preceding it.[35] Morphologically expressed as a spiral structure, this ratio was deemed virtually omnipresent in the natural world:

> For it was early discovered [...] that this ideal proportion, so logically and rationally determined by pure thought, plays a preponderant part in the morphology of the natural world, both organic and inorganic. The forms of crystals and shells, of plants and flowers, and of the human body itself, appeared in an almost miraculous way to resolve into this equation of proportion.[36]

The very pervasiveness of the Golden Section in the natural world therefore recommended it as a universal principle of morphology. More remarkable still was the frequency in which the Golden Section found expression as a compositional standard in art. Citing the curvilinear layout of Hans Holbein's *The Virgin Child with the Burgomaster Meyer and His Family* of 1526 as an example of how the Golden Section was utilised in art, Read inferred that just as development in nature possessed a quantitative or geometric expression, so could the rules governing composition in art: 'We might be tempted to abandon any geometrical explanation of the picture's harmony did we not remember that curves in nature follow definite laws. The spiral of a snail's shell is an example, which is mathematically exact and known as the constant or logarithmic spiral'.[37]

Developed between the years 1931 and 1934, Read's morphological exegeses – though indebted to a variety of sources – were particularly inspired by the work of the English critic Theodore Andrea Cook, whose 1914 morphological tome – *The Curves of Life* – had electrified the Modernist community through its encyclopaedic analysis of the Golden Section in nature and art.[38] Preoccupied by spiral structures in art and nature, Cook studied examples that ranged from the twisting forms of plants and horns to the spiral staircase designed by Leonardo da Vinci for the Château of Blois. He thus recognised the spiral as 'an abstract conception of perfect growth' which faultlessly corresponded to the proportions of the Golden Section. The logarithmic spiral was, he concluded, 'the best formula [not only] for Perfect Growth, and a better instrument than has yet been published for kindred forms of scientific research; but also it [suggested] an underlying reason for artistic proportions'.[39]

Aside from its formative influence on Read, Cook's morphological thesis encouraged the biologistic speculations of the Canadian theoretician Jay Hambidge, and the Romanian Modernist polymath Matila Ghyka, who also examined the ubiquity of spiral formations in nature and art.[40] Whereas Hambidge's 1920 text, *Dynamic Symmetry: The Greek Vase*, aimed to revivify contemporary design by applying Golden Section principles, drawn directly from nature, to artistic practice, Ghyka's 1931 treatise, *Le nombre d'or*, was much broader in scope, attempting to explain the frequency of the Golden Section in everything from philosophy and architecture to morphology and poetry.[41] Within Ghyka's morphology, the asymmetry of form present in, for example, the corkscrewed shells of crustaceans or the spiral arrangement of petals on a flower epitomised the growth cycle of life and was a vital product of the way in which organisms developed. Whereas inorganic entities – such as crystals – expanded by a process of agglutination (the clumping of particles), living things grew through a process of imbibition (the absorption of nutrients in solution) which produced what he termed 'successive forms' – that is, a morphology that expanded spirally through uniform scaling.[42]

Inherent to these various morphological analyses of spiral structures was the assumption that the presence of spiral forms in art signified an intuitive representation of the artist's psychological oneness with the energies of the natural world. Cook reasoned that the ubiquity of spirals in artistic composition indicated a biophilic desire 'to get a look of life [...]; of those principles of growth which have brought [the shell] so near to the logarithmic spiral without ever exactly reproducing the mathematical curve'.[43] Elsewhere in the text he observed that 'creativity' was the only appropriate expression for the way in which an artist uniquely reproduced the primary forms of life:

> His artistic work is in a strict sense 'creation', for he does not merely copy the natural beauty which he sees around him, he 'creates' a fresh stock of beauty for us all. His picture, therefore, (or his statue, his building, or his music), may be fairly compared with the living things in the world of which he is at once a part and an interpreter. These organisms have reached certain forms through the essential processes of life. The creations of his art are [...] equally essential (if more subtle) manifestations of those same processes.[44]

Developing these ideas in his *Art and Industry* (1934), Read likewise emphasised the psychologically impulsive character of the morphology of art – how spiral formations were instinctively apprehended by artists due to their presence, not just in nature, but in the very fabric of the human body:

> Perhaps we derive from our own physical constitution, from the apparatus of breathing, of focusing, from the symmetry of our bodies and the symmetry of nature, a generalised sense of harmonic form which we intuitively apply when we 'apprehend' a work of art. Artists, undoubtedly, sometimes work to the Golden Section by instinct, but not deliberately.[45]

This belief in humankind's ability to subconsciously replicate the morphology of nature through art was, in effect, a strain of psychobiology. Originating in the work of the nineteenth-century French physiologist Hippolyte Bernheim, 'psychobiology' was a biologistic school of psychology which attempted to incorporate biological approaches into the study of the mind and its processes.[46] Holistic in its outlook, and thus sympathetic to the organicist imperatives of interwar society, this school of psychology held a psychobiological approach to be appropriate only when applied to the dynamics of whole organisms, apprehended not reductively but as exemplars of organic complexity.[47] In aesthetic terms, a psychobiological perspective assumed that, because humankind was part of nature, if it simply followed its intuition it would reach forms corresponding to those in nature without necessarily ever having seen these forms in the flesh.[48] Due to its power to explain the presence of biological forms within modern art, a psychobiological approach therefore became an accepted explanatory model through which Modernists – such as Ghyka, Gabo and Read – hypothesised the relationship between the morphology of art and nature.[49] Henry Moore readily adopted a psychobiological standpoint, for example, to explain the aesthetic appeal of certain types of natural form to his practice: 'There are universal shapes to which everybody is subconsciously conditioned and to which they can respond if their conscious control does not shut them off'.[50]

The impression that spiral forms were fundamental to living nature appealed to Modernists whose neo-romantic fascination with the question of life had been kindled by neo-vitalist dogma, in particular, Driesch's notion of entelechy and the teleological implications of Bergson's biologistic metaphysics.[51] Read, for one, was philosophically dissatisfied with mainstream evolutionary theory and drew upon the Bergsonian concept of a creative impetus to substantiate that his aesthetics, like art itself, was rooted in physical sensation and hence inherently biological.[52] Above all, the passage of living time delineated by the spiral form was yoked to the neo-vitalist concept of *duration*, a Bergsonian idea which postulated the continuity of lived experience and the philosophical indivisibility of the *élan vital*.[53] Thus drawing upon neo-vitalist philosophy, Ghyka attributed the manifestation of the equi-angular or logarithmic spiral to a vital impulse present in living matter:

> These geometric forms and 'pulsating' rhythms, that one *never* finds in unorganized matter, which appear like a signature, the sign of the reactions of life (knowledge and growth) [...] are [...] the pentagonal symmetries and rhythms, the growth spirals which resemble this same pentagonal symmetry.[54]

Here, Ghyka's notion that entelechy inscribes itself upon organic matter as part of the agency of life is extremely close to Bergson's comprehension of duration as an indivisible movement that impresses itself upon living matter:[55] 'Wherever anything lives, there is, open somewhere, a register in which time is being inscribed'.[56]

For Modernists – especially those who identified with the neo-vitalist sympathies of biomorphism – the spiral's capacity to express shape and kinesis made it the ideal vehicle through which to evoke the quintessence of life's processes.[57] Finding inspiration 'in the study of plant form',[58] the zigzagging forms of sculptures such as *Spring Tree* (c.1932) (Plate 6) by Richard Bedford powerfully express the centrality of the spiral to Modernist conceptions of vital energies, echoing the view of the sculptor Etienne Béothy that the spiral was a morphological index of life: 'The gigantic developmental spiral of organic life shows a certain tendency towards blooming in full'.[59] Bedford's watercolour titled *Spiral* (c.1938) (Figure 4.1), in particular, appears as a stylised composite of the various illustrations of spiriform growth that Cook reproduced in *The Curves of Life*, ranging from the squiggling flagella of bacteriophages to the torqued structures of umbilical cords and sponges (Figure 4.2). It was this type of motif – highly geometrical yet characteristic of the natural world – that led Wilenski to state in 1932 that contemporary sculptors had begun 'to seek and find analogies between the characteristic forms in life and the characteristic forms in art, between formal principles in natural structures and formal order in geometry, architecture and sculpture'.[60]

The association of helical patterns with living matter had an antecedent in romantic nature philosophy, where the search for an ideal unity of plan within the organic kingdom led poets and philosophers to investigate the underlying structure of plants. For example, in his studies, the German naturalist Lorenz Oken was entranced by the spiral formations present in the vessels of plants and hypothesised an idealist morphology in which spiral configurations appeared more frequently in higher-level plants than in fungi or mosses, while ferns possessed only a solitary bundle of spiral vessels.[61] Similarly, the nineteenth-century poet-scientist Johann Wolfgang von Goethe speculated that a 'spiral tendency' might

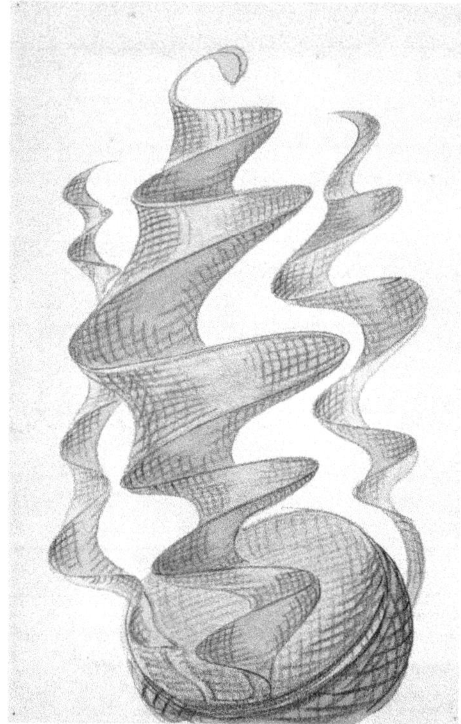

Figure 4.1

Richard Bedford, *Spiral*, c.1938. Victor Batte-Lay Foundation Collection

be responsible for growth in plants, their structure deriving from the competing developmental tendencies of verticality and spirality: 'In a thousandfold twistings around its center, it performs the miracle that enables a single plant to derive infinite reproductions from within itself'.[62] Like Oken, Goethe conjectured the possibility of an idealist science of form, although his morphology posited that all plants were variations of one rudimentary type.[63] In this way, he envisaged the smallest, spiriform vessel of a plant as a primary unit of life.[64]

Goethe's morphology ultimately derived from an eighteenth-century scientific practice which sought to find, conceptually if not necessarily in practice, the underlying archetype from which all morphological phenomena originated.[65] While individual organisms rarely embodied 'typical' characteristics, the perceptive morphologist was deemed able to intuit the archetype from summative experience. For Goethe, the particular served as a mutable template for the plan of the ideal, which was thus 'true to nature' in the loose sense that it was based on direct observation.[66]

The influence of Goethe's idealist morphology of spiriform archetypes upon the neo-romantic sensibility of the Modernist imagination can be gauged, not

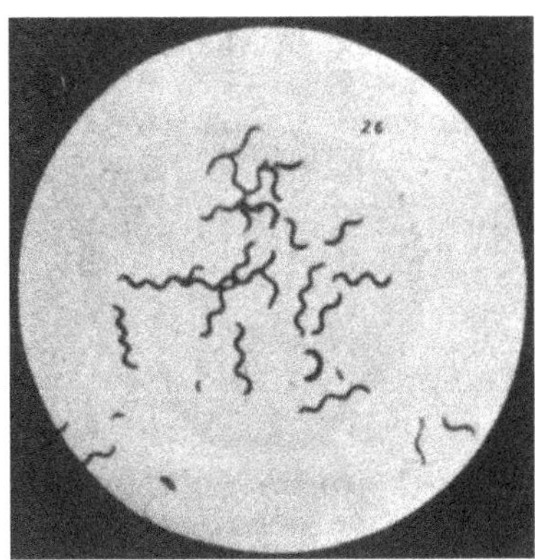

Figure 4.2

**Theodore Andrea Cook, 'Spirillum Rebrum', Fig. 6 in *The Curves of Life*
(London: Constable & Co., 1914)**

least, by the prominence his work was given by Cook in *The Curves of Life*.[67] For his part, Moore revealed a lifelong fondness for Goethe's writings in a 1977 interview with Christa Lichtenstern in which he singled out *Conversations with Eckermann* (1836) as meritorious due to its impression of 'universal understanding'.[68] Presented as a dialogue between Goethe and his disciple Johann Peter Eckermann, this book awakened Moore's curiosity in those sections where Goethe explained the immanence of organic form as a prerequisite of beauty in nature.[69] The impact of Goethe's thought on Moore's practice is particularly evident in a passage in 'The Sculptor Speaks' where Moore deliberates on the role of morphology in teaching the artist the quintessence of shape:

> [The] sensitive observer of sculpture must also learn to feel shape simply as shape, not as a description or reminiscence. He must, for example, perceive an egg as a simple single solid shape, quite apart from its significance as food, or the literary idea that it will become a bird. And so with solids such as a shell, a nut, a plum, a pear, a tadpole, a mushroom, a mountain peak, a kidney, a tree-trunk, a bird, a bud [...], a bone.[70]

Read, in his 1934 study of Moore, correspondingly drew upon Goethean morphology to propose that artists apprehended in nature the archetypal shapes from which all forms ultimately derived:

> Mr Moore, in common with artists of his type throughout the ages, believes that behind the appearance of things there is some kind of spiritual essence, a force of immanent being which is only partially revealed in actual living forms. Those actual forms are, as it were, clumsy expedients determined by the haphazard circumstances of time and place. The end of organic evolution is functional or utilitarian and, spiritually speaking, a blind end. It is the business of art, therefore, to strip forms of their casual excrescences, to reveal the forms which the spirit would evolve if its aims were disinterested.[71]

The pantheistic, neo-vitalist undertones in this passage exemplify Read's cognisance of romantic nature philosophy and the larger, intellectual debt Modernists owed to neo-romanticism in the 1930s.[72] As Lynn Gamwell has demonstrated, pantheism arose in the early nineteenth century as a response to the mind–matter dichotomy proposed by Cartesian science. In an effort to close the perceptual gap between the subjective world of the mind and the material world, romantic artists and philosophers hypothesised a doctrine of cosmic harmony in which spirit (mind) and nature (matter) formed an indissoluble unity that flowed from the same, amalgamating source: the godhead. Accordingly, romantic nature philosophers, like Lorenz Oken and Hans Christian Ørsted, investigated phenomena such as electromagnetism and the ether as a means of finding a unifying substance within the physical universe.[73] Goethe's idealist quest to unearth the morphological essence of plant form thus operated at the axes of nature deification, aesthetic sensibility and objectivity, which comprised the pantheistic basis of romantic-era morphology.[74]

The role that an essentialist, neo-romantic morphology played in Moore's practice is best illustrated by his 1935 sculpture *Carving* (Plate 7). Hewn from a single block of walnut wood, the sculpture's upward arc and self-similarity of form loosely describe the developmental curve of a logarithmic spiral as it was stereotypically illustrated in the work of Cook, Ghyka *et al.* (Figure 4.3). In *Unit*

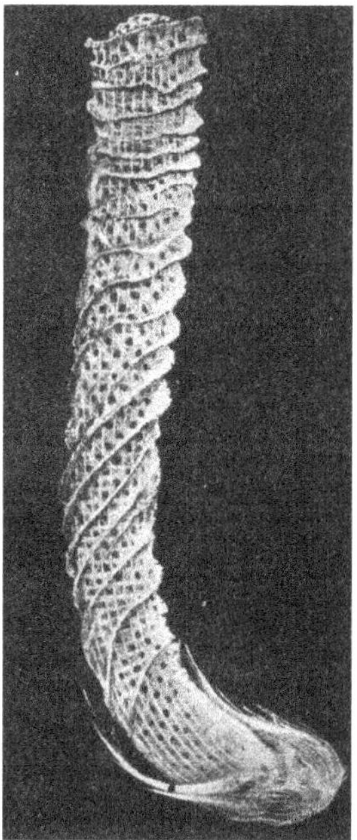

Figure 4.3

Theodore Andrea Cook, 'The Glass-Sponge', Fig. 9 in *The Curves of Life* (London: Constable & Co., 1914)

1 Moore spoke of natural objects using a terminology that unambiguously evoked Modernist disquisitions on spiriform structures. Of trees, he claimed they gave 'the ideal for wood sculpture, upward twisting movement',[75] recalling Cook's assertion that 'as the winds due to the earth's motion blow fairly steadily just when the trees are growing fast, the young tree may take a permanent twist, from this cause, which it never loses'.[76] By the same token, while extolling how shells disclosed 'nature's hard but hollow form',[77] Moore claimed they possessed 'a wonderful completeness of single shape' – an idealist conception of shell form which originated in morphological tracts which praised how shells manifested 'the most beautiful and the most easily examined instances of spirality in organic life'.[78]

As the nineteenth century passed into the twentieth – despite moving away from the pantheistic speculations of the romantics – scientific enthralment to spiral formations in nature continued unabated, equipping artistic Modernists with up-to-date, morphological information about the significance of spirality in natural form.[79] For example, while largely materialist in philosophical orientation, D'Arcy Wentworth Thompson acknowledged the beauty of the spiral form and explored its conformations exhaustively in his 1917 treatise on morphology.[80] He corroborated the idealist viewpoint that the spiral patterning found in shells, the florets of a sunflower, locks of hair and the torqueing shape of horns was representative of 'the remarkable curve known as the equiangular or logarithmic spiral', though not without a caveat. What united these manifestations of spiral form was their status as a 'by-product' of organic development which – though a consequence of biological processes – was in no meaningful sense 'alive'. Rather than simply *growing*, the forms which presented the logarithmic spiral *accumulated* matter over time: '[All] alike consist of stuff secreted or deposited by living cells; all grow, as an edifice grows, by accretion of accumulated material; and in all alike the parts once formed remain in being and are thenceforward incapable of change'. As a result, Thompson saw the logarithmic spiral as a three-dimensional representation of development, each part differing in age and 'in magnitude in the strict ratio of their age'.[81]

Despite his opposition to a neo-vitalist or pantheistic understanding of development, Thompson's tome served to galvanise Modernist interest in morphology, most especially the spiral form, by demonstrating that ideal structures existed in nature due to the efficiency of physical energies.[82] In *Art and Industry* Read described *On Growth and Form* as a 'fundamental work, investigating the laws under which organic life [evolved]'[83] and Thompson's work hence served as a mouthpiece through which an idealist morphology, abstractly conceived, could be presented to Modernists as an alternative to the phylogenetic speculations of Darwinism or the more teleological assumptions of the neo-vitalists.[84] Indeed, far from invalidating the neo-romantic, pantheistic suppositions of artistic Modernists, the New Biology served to enhance the status of the spiral as a 'sign of life' which perfectly conveyed the form and movement, equilibrium and epigenesis, characteristic of organic material and, ultimately, of life itself.[85]

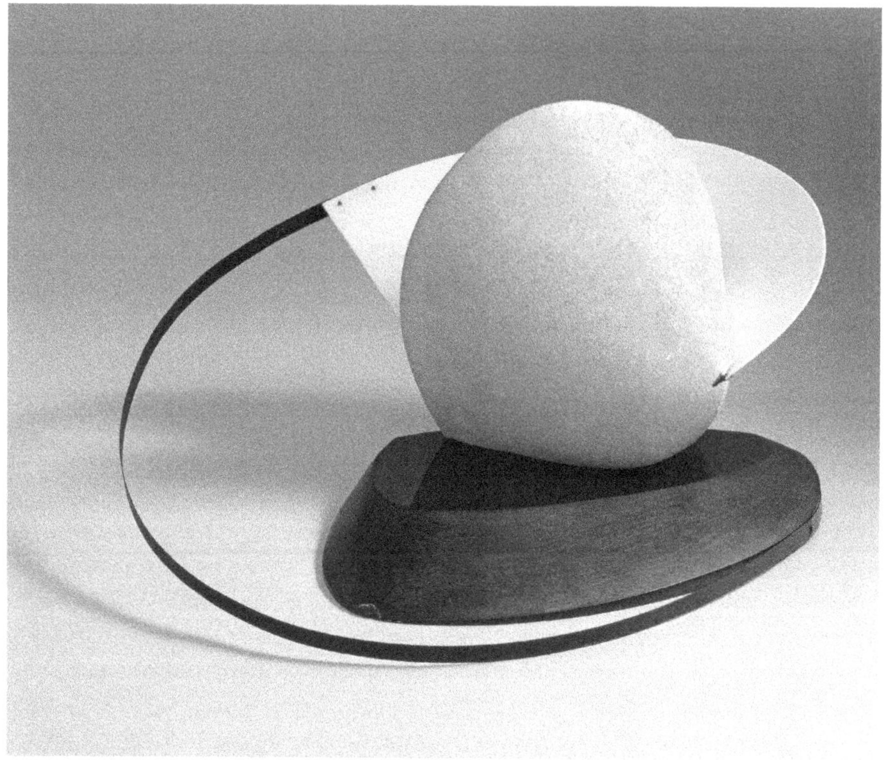

Figure 4.4

Naum Gabo, *Stone with Collar*, 1933. Tate, London

For the popular science writer Raoul France, the spiral was one of seven 'fundamental technical forms' which collectively underpinned the morphology of nature. It was the morphological product of 'everything which has the function of boring and squeezing through', ranging from spirochaete bacteria to the 'light, screw form of the wings of the maple-seed [which] propel it through the air'.[86] Just as Cook described the spiral as 'representing [the] equal distribution of energies' and the curve as symbolising 'a distribution of energy in some way comparable to equipotential',[87] so France spoke of the spiral as the acme of mechanical efficiency: 'spiral movement occurs with less expenditure of energy than movement in a straight line [...]. If something is moving forward, the least inclination towards the spiral makes its passage easier; and the other resistance which it encounters models it mechanically'.[88] Thus the meandering, spiral form that Naum Gabo incorporated into his construction *Stone with Collar* (1933) (Figure 4.4) appears as a Modernist attempt to produce 'a profound and universal realism' by abstractly representing the biomechanical forces and vital energies which underlie the natural world.[89]

Recognition of the similarity of form between Modernist sculpture and the morphology of nature encouraged Wilenski in the development of his 'association-ist' aesthetic. In a speculative *Apollo* article of 1930 – titled 'Ruminations on Sculpture and the Work of Henry Moore' – he charted the shift from the pared-down orthogonality of Vorticist sculpture to the 'next stage' in contemporary art, which, he argued, involved 'the creation of objects imbued with rhythmic relations which symbolise rhythmic relations divined or discovered by artists in the organic world'.[90] While the analogical nature of Wilenski's argument – in spite of its neo-classical and Primitivist inclinations – was underpinned by the neo-romantic, morphological idealism which pervaded artistic Modernism during the early 1930s,[91] as we will see, the impression that physical forces influenced the creation of certain 'form constants' in nature (and thus informed the organic, biologistic character of the new sculpture) fed into a monistic belief in the fundamental unity of the organic and inorganic worlds.

Nature as a unity: morphology, monism and crystallography

At the end of an essay on 'The Golden Section' – published in the *Listener* in 1931 – Read concluded that 'whether the radii vectors of [the spiral of a shell] have any constant relation to the Golden Section is a mathematical question [for which] the science of aesthetics may bring us exactly the same conclusion as the science of physics, the mathematical universe of James Jeans'.[92] While he claimed 'not [to] know whether there is any scientific hypothesis to explain this strange coincidence', he saw fit to ask if the scientific correspondent (A.S. Russell) 'might be persuaded to deal with this aspect of the question in one of his fortnightly notes in *The Listener*'.[93]

In his response, Russell was highly sceptical that 'ratios such as those derived from the Golden Section are of importance in nature', although he acknowledged that modern science was enthralled to the notion of universal constants, albeit in terms of the relative weights of atomic structures.[94] 'Curiously enough', he wrote, 'there has been a return to Greek simplicity in recent attempts to establish relations between the fundamental constants of Nature, and between the masses of the electron and the proton'.[95] Indeed, the newfound atomic makeup of matter had breathed new life into the philosophical notion that fundamental structures underlay the outward appearance of the physical world – the discovery of the electron and the proton, in particular, adding extra credence to this idea. In an earlier article in 1931, Russell confirmed that the task of the new science was to find the 'relations between the fundamental constants of nature[,] [between] the massive unit of positive electricity, the proton, and the mass-less unit of negative energy, the electron'.[96] One outcome of this interest had been the observation that these units behaved in accordance with the laws of symmetry: the electric charge of the electron was found to be 137, a number 'arrived at from a consideration of the number of symmetrical terms in 16 rows and 16 columns (which is 136) and by adding 1'. In this respect, Russell was clear that the nature of matter had, in effect, been reduced to the numerical character of subatomic relationships.[97]

Read's exchange with Russell emphasises the degree to which Read's biologistic aesthetic was framed by morphological debate within physics – albeit of an organicist variety. As George Woodcock has contended, throughout his career Read claimed to base his aesthetics on biology even though his philosophy was notably heterodox, cherrypicking those elements from Jungian psychoanalysis, Platonic cosmology and Bergsonian metaphysics that best fitted his biologistic standpoint, even if they ran counter to mainstream opinion.[98] While *On Growth and Form* had shown Read – in his own words – that 'certain fundamental laws determined even the apparently irregular forms assumed by organic growth', he left no doubt that his understanding of morphology stemmed from neo-vitalist and organicist tendencies in the philosophy of science, singling out Alfred North Whitehead's 'organismal' philosophy for particular praise.[99]

If artistic Modernists had increasingly come to see biology as the prototypical science of the twentieth century, this was significantly due to the influence of Whitehead, who had long hypothesised a uniquely biologistic conception of a science which was based upon organic 'events' rather than simple units or atoms.[100] Aiming to create a more holistic scientific culture, Whitehead observed that science was 'taking on a new aspect that [was] neither purely physical, nor purely biological'.[101] The basic unit of any scientific system was envisaged as the 'event', a dynamic entity which had 'to do with all there is, and in particular with all other events'.[102] Conceptualised in organicist terms, the event therefore acted as a panacea to the mechanism–vitalism controversy by promoting a holistic view of reality in which interactivity supplanted the study of static systems or immaterial life forces:

> Accordingly, a non-materialistic philosophy of nature will identify a primary organism as being the emergence of some particular pattern as grasped in the unity of a real event. Such a pattern will also include the aspects of the event in question as grasped in other events, whereby those other events receive a modification, or partial determination [...]. The concept of an organism includes, therefore, the concept of the interaction of organisms.[103]

While Whitehead rejected claims that his work simply represented a repackaging of neo-vitalism, his research into particle flux within electromagnetic fields had caused him to regard molecules as social entities, with living organisms as merely larger, self-conscious convergences of particulate matter.[104] This quasi-vitalist belief in communities of matter – whose quality of existence was determined by hierarchies of interactivity – though beholden to organicism, had deep roots in Bergson's biologistic theories, which Whitehead praised for introducing 'into philosophy the organic conceptions of physiological science'.[105] Indeed, Whitehead's biocentric desire to perceive everything as simply a component of an integrated 'nature' articulated a viewpoint that was central to neo-romanticism and which profoundly affected the holistic approach of the New Biology, especially in the field of morphology.[106] Joseph Needham's studies of embryological morphogenesis, for example, discounted Driesch's entelechy while adopting Whitehead's organismal model to hypothesise the qualities of relationship evident between different grades of biochemical matter.[107]

This biologistic sea-change in scientific attitudes was supported by the quantum field pioneer Niels Bohr, who maintained – in a *Discovery* article of 1933 – that changes in classical mechanics had produced an environment in which it was possible to rethink the relationship between physics and biology. Atomic theory, he remarked, had advanced to such a degree that it was now possible to explain the behaviour of an organism in purely atomic terms. Everything from the respiration of plants to the stability of atomic structures in organic compounds such as chlorophyll and haemoglobin was deemed to be rooted in the performance of photo- and physico-chemical processes.[108] So intricately woven together were the particles of living matter, Bohr averred, that it was impossible 'even [to] tell which atoms really belong to a living organism, since any vital function is accompanied by an exchange of material, whereby atoms are constantly taken up into and expelled from the organization which constitutes the living being'. From these findings he extrapolated that were we 'able to push the analysis of living organisms as far as that of atomic phenomena, we should scarcely expect to find any features differing from the properties of inorganic matter'.[109] In other words, Whitehead's conception of the 'event' – as an organismal activity apprehended at any point in the scale of relations between the organic and the inorganic – found explicit confirmation in the findings of the new physics.

Read's appreciation of Whitehead stemmed from Whitehead's ability to discern a unity of pattern in the fabric of nature, which, in turn, laid the foundation for a possible synthesis between physics and aesthetics.[110] Indeed, the syncretic vision of a unified nature postulated by modern science enjoyed credibility among biophilic Modernists who, inspired by the neo-romantic and neo-vitalist currents of the day, were concerned with expressing a more holistic understanding of life and art. Bohr's notion that, at an atomic level, living organisms were not fundamentally different from inorganic material, for example, was echoed by the Constructivist architect Leslie Martin in an organicist declaration in *Circle*:

> It is a matter of common knowledge that in science, the world of 'appearances', of parallel systems in which the dead world and the world of living organisms are clearly marked off from one another, has been abandoned. The world of appearances has given place to a world in which things unrelated to each other in appearance are united in the completeness of a single system.[111]

Such uncompromising conclusions regarding the synchrony of matter bore witness to the influence of monism, a 'theory or system of thought or belief, that [assumes] a single ultimate principle, being [or] force',[112] which – to paraphrase the Victorian philosopher Edmund Gurney – was an attempt to transcend the Cartesian dualism that had traditionally characterised the metaphysical relationship between mind and matter.[113] Deriving from early-nineteenth-century German nature philosophy, monism represented an effort at reconciling pantheism with a Darwinian conception of the theory of evolution, inspiring, in turn, a cogent neo-romantic doctrine of physical and spiritual evolution. The scientists who fronted this philosophical movement, such as the experimental psychologist Gustav Theodor Fechner, argued that natural selection supplied the mechanism through which primary units of matter (monads) progressively developed levels of complexity, 'first becoming molecules, then cells, and so on to the brain and ultimately the universe'.[114]

In the mid-nineteenth century, monism took on a new philosophical hue through the hypotheses of the German biologist Ernst Haeckel. A zealous Darwinist, Haeckel rejected the metaphysical speculations of early-nineteenth-century monism by positioning it within the domain of natural science, convinced as he was of the fundamental unity of organic and inorganic nature.[115] In this respect, his writings – most particularly *The Riddle of the Universe at the Close of the Century* (1900) – had a transformative impact on British scientific consciousnesses by rebuffing the idea of a separate spiritual realm and arguing that matter and mind were simply differing aspects of the same primary substance.[116] While Haeckel's monism was avowedly materialist, his belief that even the lowliest particles of matter experienced a rudimentary consciousness and pantheistic identification of God with nature meant that his philosophy possessed strongly neo-vitalist overtones.[117] Intellectually indebted to the revival of neo-romanticism at the *fin de siècle* and neo-Lamarckian notions of progressive evolution, Haeckel's monism thus paved the way for the more teleological suppositions of Bergson's neo-vitalist apologia of 1907, *Creative Evolution*.[118]

The importance of scientific monism to Modernist thinking on art was made apparent in a 1935 essay by Read, 'Ben Nicholson and the Future of Painting'. Offering a biologistic interpretation of Nicholson's gesso-covered, hardboard reliefs, Read compared his hard-edged abstractions to Moore's fluid biomorphism, suggesting – in terms that recalled Haeckel's monism – that, despite appearances, both artists represented the same reality, effecting a synthesis of the organic and the inorganic:

> While I consider that the most general [...] of these intimations of reality are of the organic type, and intimately linked to the essential forms of life, there are other aspects of reality of a more mathematical and crystalline nature which may equally form the basis of artists' creations. Perhaps Ben Nicholson's tend in this direction; while Henry Moore's, for example, are more obviously organic. But it would be a mistake to make any hard and fast distinction, because reality is a unity, of which organic and inorganic forms are but polar aspects.[119]

As David Thistlewood has demonstrated, throughout his career Read hypothesised the possibility of a synthesis of artistic Modernism in which the antithetical styles of Surrealism and abstraction would find resolution in the work of certain artists, such as Moore, Nicholson and Hepworth, who possessed the ability to resolve the polarity of modern art as part of a dialectical process.[120] This conception of Modernism explicitly sought to supersede aesthetic dualism with a monistic paradigm, in which matter (represented by Surrealism) and mind (represented by Modernist abstraction) found a creative synthesis. In the 1936 edition of *Art Now*, Read staged this synthesis as an aesthetic continuum: 'Though at their extremes – a Mondrian against a Dalí – these two movements have nothing in common, yet the space between them is occupied by an unbroken series, in the middle of which we find artists like Picasso and Henry Moore whom we cannot assign confidently to either school'.[121] By 1948, he had developed this syncretic model into a monistic hypothesis whereby the metaphysical (or inorganic) could find resolution with the vital (or organic) as part of a monistic unity. Within this syncretic paradigm, the various manifestations of contemporary art were thus 'arranged along a polar axis, with transcendental metaphysics at one end and an intense self-awareness of physical vitality at the other end. It is along the same axis that we can place abstraction and realism in art [...]. [Yet] the axis exists *within* the individual artist only if he can become aware of it'.[122]

This syncretic model was indebted to the theories of the German art historian Wilhelm Worringer, who in 1908 produced a biologistic theory of art in which he argued that 'the work of art [is] an autonomous organism' in terms of its self-governing systems of development and responsiveness to environmental conditions.[123] Worringer conjectured that throughout history humankind had equivocated between an urge to empathy and an urge to abstraction, which aesthetically manifested themselves in the organic and the inorganic respectively. The urge to empathy represented a psychological state of profound contentedness with the natural world that was artistically symbolised by a reified type of naturalism which engendered the 'sensation of happiness that is released in us by the reproduction of organically beautiful vitality'.[124] The urge to abstraction, by contrast, embodied a sense of psychological estrangement that stemmed from a 'physical dread of open places' and a fear of 'the extended, disconnected, bewildering world of phenomena'.[125] This tendency was displayed most clearly in artworks of a geometrical character, which, in their semblance to inorganic form, exemplified an aesthetic of lifelessness.[126]

Worringer acknowledged that his binary paradigm was problematical in that it disregarded the naturalistic work of, say, the Palaeolithic cave painters and had little power to explain organic or biomorphic modes of Modernist abstraction. To overcome this thorny issue he utilized a psychobiological explanation, arguing that 'organic' abstraction was intuitively created in accordance with the laws of nature and was thus a type of *empathetic* naturalism rather than a form of pure abstraction.[127]

Read had been introduced to Worringer through the work of the English philosopher T.E. Hulme, and always regarded himself to be the older scholar's disciple.[128] Bound to Worringer's problematic duality, Read expended considerable energy attempting to reconcile the divergent sensibilities of contemporary art before hypothesising a holistic, 'organic' aesthetic.[129] Although he was ambivalent about the nature of a biologically inflected form of Modernism, his recognition of a nature-centric style, poised halfway between Surrealism and Constructivist abstraction, was part of the self-same syncretic, bioromantic impulse which had led to Grigson's coinage of biomorphism.[130] Indeed, Grigson's 1935 definition of biomorphic abstraction clearly provided the syncretic model upon which Read based his own monistic philosophy of art:

> Abstractions are of two kinds, geometric, the abstractions which lead to inevitable death; and the biomorphic. The biomorphic abstractions are the beginning of the next central phase in the progress of art. They exist between Mondrian and Dali, between idea and emotion, between matter and mind, matter and life.[131]

The monistic concept that the inorganic and organic worlds constituted a unity had, since the end of the nineteenth century, been a viewpoint promulgated by crystallography. Indeed, Haeckel's own monism had been inspired by the study of liquid crystals in the early 1900s, in which certain inorganic structures when mixed with organic plasma appeared to create flowing, crystalline forms that responded dynamically to changes in temperature by spinning filigrees of 'worm-like' feelers. Otto Lehmann, who first discovered these 'living' crystals in 1904, believed that their very motility and sensibility indicated that they partook of an animal-like state which, unlike inorganic matter, possessed a soul.[132] In 1917 Haeckel published *Crystal Souls: Studies of Inorganic Life*, which drew upon Lehmann's research while introducing a neo-vitalist conception of the liquid crystal to his pantheistic philosophy of monism. Claiming that natural science had validated the monistic worldview, Haeckel proposed that all things, organic and inorganic alike, thus possessed life:

> [The] year 1904 was an outstanding milestone in the history of natural philosophy [...] the most important achievement in all this may be [...] the definitive conviction of the fundamental unity of all natural phenomena, which finds, in the concept of 'monism', its simplest and clearest expression. At one blow the artificial boundaries, which up to now have been erected between life and death, between natural science and moral science, have fallen down. All substances, inorganic as well as organic, possess life; all things have psyches, crystals as well as organisms. There arises, anew and unbreakable, the old conviction of the inner unitary interdependence of all happenings, of the unbounded rule of generally valid laws of nature [...].[133]

Early in his career, Haeckel had studied radiolarians – marine micro-organisms which possessed a striking crystalline geometry – and Lehmann's sensible crystals appeared to provide him with further evidence of the unity of the inorganic and organic worlds. As Spyros Papapetros has put it: 'If the radiolarians were crystal animals, then the rheocrystals were animal-like crystals – the one was the inverted image of the other'.[134]

Monistic faith in the synchrony of the organic and the inorganic realms found apparent confirmation, in the years before the Great War, in the experimental research of the French biochemist Stéphane Leduc. Leduc's research into 'osmotic growths' seemed to verify Haeckel's monistic speculations as, like Lehmann's liquid crystals, they appeared to imbibe nutrients and develop into the burgeoning 'forms of madrepore, fungus, alga and shell'.[135] Like Lehmann's, Leduc's research chiefly focused on the properties of colloidal solutions, whose peculiar ability to aggregate into cell-like clusters under certain electro-thermal conditions encouraged Leduc to believe that he had synthesised a type of living matter. Like many conversant with romantic philosophies, Leduc was fascinated by the idea that the morphology of life was defined by irregular curvature, whereas inorganic matter tended towards the angular geometry of crystals. As a result, the snaking, membranous products of his research suggested to him a 'striking analogy' with living beings, which he attributed to the 'morphogenetic power of osmosis', which created form through imbibition as opposed to agglutination.[136] First published in France in 1906 as *Les bases physiques de la vie et la biogenèse* (and in Britain in 1911 as *The Mechanism of Life*) with artful photographs of 'gardens' of osmotic growths that he had cultivated (Figure 4.5), Leduc's experiments to synthesise life with colloidal solutions caused something of a sensation and highlighted a sense of confusion within the scientific community over the distinction between living and inorganic matter.[137]

While Leduc's research had, by the 1930s, been largely disclaimed (not least because, as Evelyn Fox Keller notes, of the 'conspicuous artificiality of his osmotic growths'), his belief in a materialist explanation for the mysteries of life granted him common ground with morphologists, such as D'Arcy Wentworth Thompson, who similarly sought to comprehend natural morphology through physical law.[138] Like Haeckel, Leduc had an essentially monistic understanding of matter, seeing his attempts to synthesise life from inorganic substances as compelling evidence that, at a fundamental level, the kingdoms of the animal, vegetable and mineral were contiguous:

> [We] are totally unable to define the exact boundary which separates life from the physical phenomena of nature, we may fairly conclude that no such separation exists. This is in conformity with the 'law of continuity' – the principle asserts that all the phenomena of nature are continuous in time and space [...]. All the forms and phenomena of nature are united by insensible transition; it is impossible to separate them, and in the distinction between living and non-living things we must content ourselves with relative definitions, which are far from being precise.[139]

Although Leduc's attempts to synthesise life with inorganic compounds was ultimately unsuccessful, his monistic hypotheses nonetheless anticipated the work of crystallographers in the interwar years who postulated synchrony, at a basic level, between the organic and inorganic realms. J. Killian, in his populist *Crystals:*

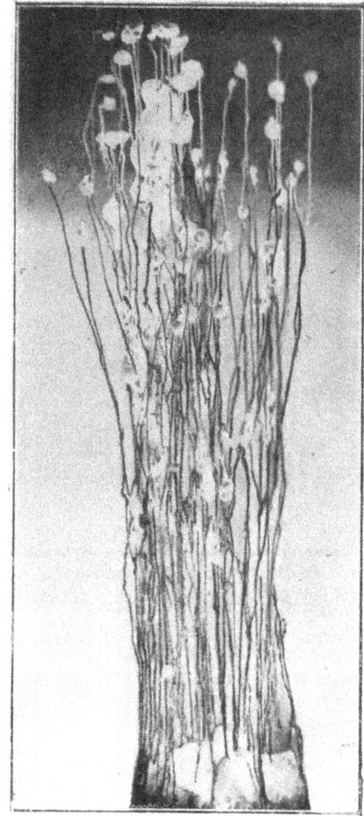

Figure 4.5

Stéphane Leduc, 'Osmotic growth', Fig. 38 in *The Mechanism of Life* (London: Rebman, 1911)

Secrets of the Inorganic (1941), for example, though adamant in his belief that the crystalline possessed a more coherent geometry that colloidal substances, nonetheless suspected that certain inorganic compounds formed the basis of living matter and, in particular cases, still abstractly reflected the morphology of life:

> As the crystal finds its image even in the bodies of plants, animals and men, in the shape of mathematical form, foreign to life (symmetry, axes and sagittal planes), it is itself illuminated in its highest members by the ray of distant life (e.g. curved faces on the diamond crystal; determination of habit according to the changing surroundings, etc.). The winding shapes of iron ore accommodate themselves towards the organic world and the fibrous serpentine appears to reflect plant-like forms (such as flax or hemp).[140]

Killian's research primarily focused on precipitated crystals and largely avoided the animistic speculations that accompanied Lehmann's work on liquid crystals,

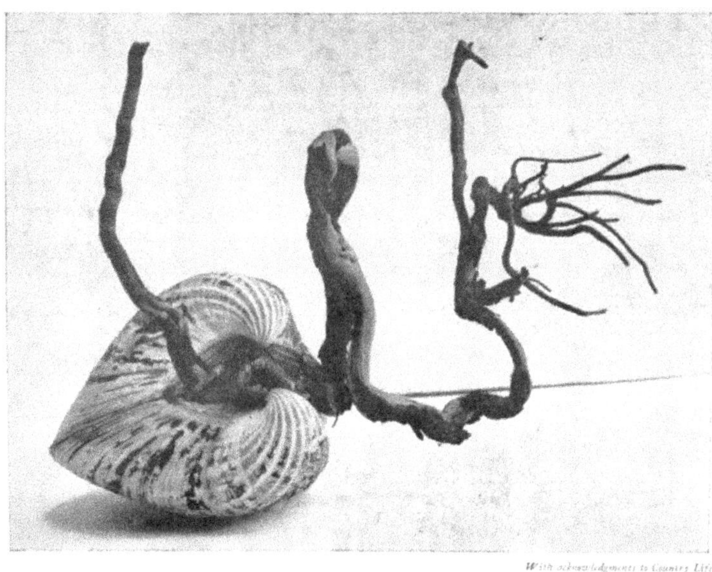

Figure 4.6

Paul Nash, image of *The Bark is Worse than the Bite*, c.1937. Tate, London

aside perhaps from a neo-vitalist conviction in the capability of a crystal to achieve completion 'as in the growth or healing of a living organism'.[141] As Killian's investigations emphasised, it was X-ray crystallography that had transformed scientific knowledge of crystalline structure in the 1920s and 1930s and, in so doing, had narrowed the perceived borders between the inorganic and organic. For instance, in a wide-ranging discussion of the impact of X-ray crystallographic analysis on biology in *Order and Life*, Joseph Needham spoke of how the new crystallography had muddied the waters between the living and the non-living by demonstrating how closely certain cellular structures approximated crystalline states:

> It has been said that the crystalline state is the natural state of solid matter. The significance of this for biology is that there may be phases within the cell where such arrangement exists, for we certainly cannot suppose that all the solids of living cells are in a state of true solution within them. Up to the present time the main advance in this direction has been the mapping of the crystalline arrangement in the coarser animal and plant fibres. In the case of natural silk, wool, etc., the crystals are of protein, in the case of cotton they are of cellulose.[142]

Up until the 1920s, it had been largely assumed that materials such as silk and cotton possessed an amorphous, glutinous structure but then X-ray crystallographic analysis had disclosed patterns of repeating crystalline sections entrenched along the fibre's principal axis. Insoluble fibrous proteins were shown to possess

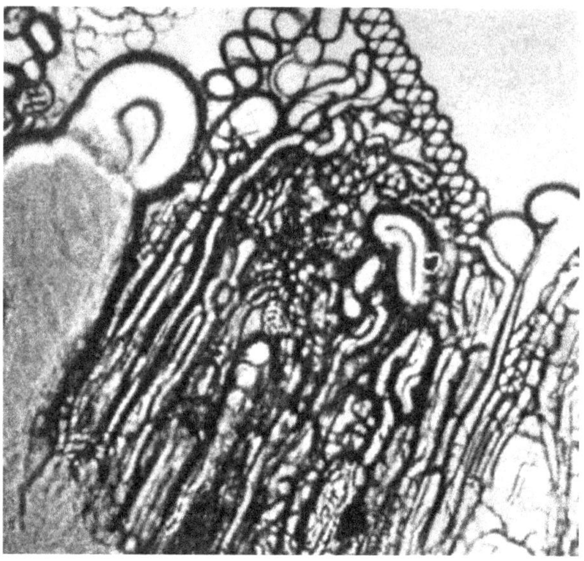

Figure 4.7

Joseph Needham, 'Morphological Forms Produced by Wetted Lecithin',
Fig. 43 in *Order and Life* (New Haven: Yale University Press, 1936)

anything but a nebulous structure as they were seen to enjoy a composition that was strikingly akin to the crystalline.[143] In providing a synthesis of biology, chemistry and physics, the impact of such crystallographic research on the New Biology was highly significant. As Needham acknowledged: 'The importance of this work on the crystal structure of animal fibres can hardly be overestimated. Is not biology the exploration of fibre properties?'[144] Everything from the molecular contractibility of muscle tissue to the polarity and symmetry of egg cells could now be explained in purely crystalline terms.

Artistic interest in the monistic potentiality of crystallographic imagery was anticipated in some of the rayograms produced by Surrealist and Constructivist photographers, like Man Ray and László Moholy-Nagy, during the interwar years, which appeared to mimic the entwining, fungal forms captured by Leduc in his emulsions.[145] The branch-like filigree of living crystals is similarly evident in some of the Surrealist objects fashioned by Nash, whose neo-vitalist notion 'of giving life to inanimate objects' came close to Leduc's monistic quest to find a link between animate and inorganic nature.[146] The lithe, snaking form which emerges from a shell in *The Bark is Worse than the Bite* (c.1937) (Figure 4.6), for example, emulates the liquescent, gelatinous morphologies of liquid crystals as they were reproduced not only in Leduc's publications but also in textbooks such as Needham's *Order and Life* (Figure 4.7).[147]

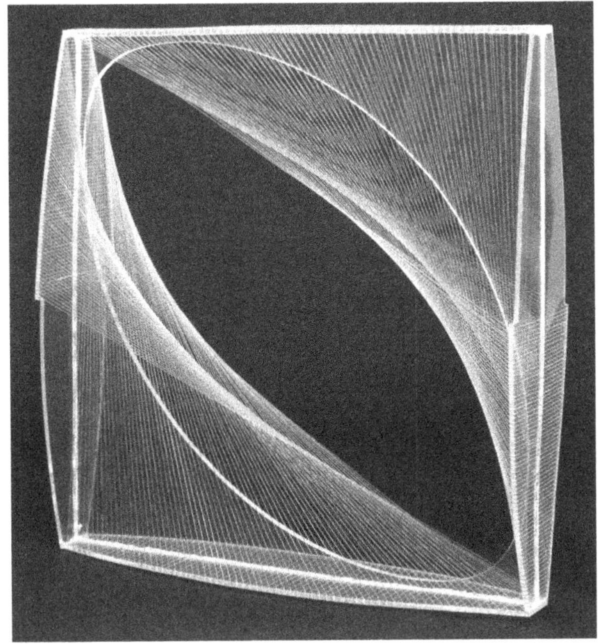

Figure 4.8

Naum Gabo, *Linear Construction No. 1*, 1942–49. Tate, London

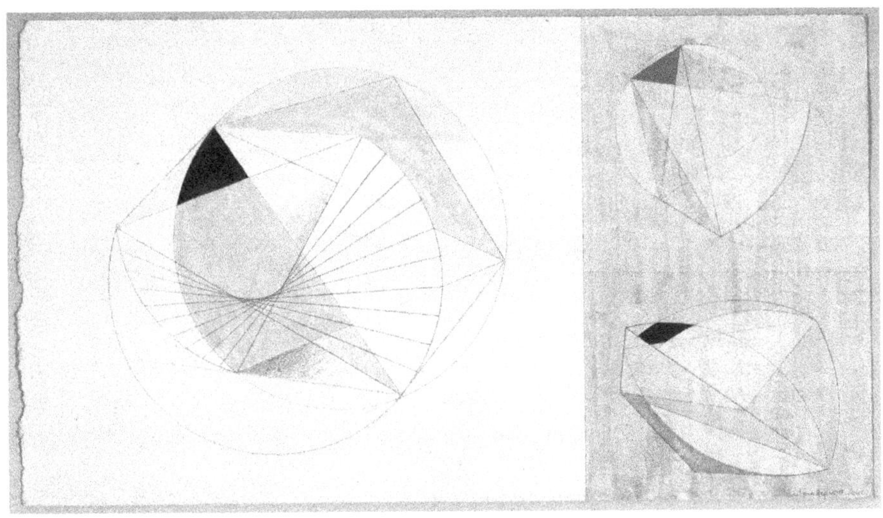

Figure 4.9

Barbara Hepworth, *Drawing for 'Sculpture with Colour' (Forms with Colour)*, 1941. Tate, London

For his part, Naum Gabo disregarded the crystalline–organic dichotomy, owning that geometric form signalled a deeper fidelity to nature's internal rhythms than biomorphic Modernism (which represented only the *superficial* appearance of organisms). In a letter of 1950, to the abstract painter Alfred Jensen, he summarised the monist viewpoint that had defined his artistic philosophy for over twenty years:

> You can make a work of art which may have geometrical elements in it [...], but in its entirety may convey to you some structure so much akin to the structure we find in Nature, that it might reasonably be called more organic than another work of art supposedly organic which only represents the surface of an organism, its appearance. That is why I believe so much in the Constructive approach and am striving to show in my work that structure is as much an organic phenomenon as appearance [...]. The actual terms, 'geometric' and 'organic' are therefore, to me, rather misleading.[148]

Constructions by Gabo, such as the otherworldly *Linear Construction No. 1* (1942–49) (Figure 4.8) (which, though geometrical, present a fragile, tissue-like space which evokes – in the words of Paul Nash – images of 'the womb, seedpods or cocoons'), provide visual expression to this idea, negotiating the boundaries between the representations of the crystalline and of the organic.[149] These monistic themes are similarly current in Hepworth's linear drawings of the early 1940s, which – like Gabo's constructions – depict crystalline nettings of form that suggest the subtle geometry of webs or certain types of nest. Indeed, for Hepworth especially, friendship with the physicist and avant-garde collaborator Desmond Bernal had brought an understanding of X-ray crystallographic representation; and it is possible, if hypothetical, that images such as her *Drawing for 'Sculpture with Colour' (Forms with Colour)* (1941) (Figure 4.9) present schematic illustrations of magnified crystalline or organic structures.[150]

Underscoring these Modernist attempts to amalgamate the realms of the living and inorganic, an anonymous review of Richard Bedford's drawings in the *Glasgow Herald* in 1938 remarked that his sketches of flowers combined a vocabulary of graceful, interlacing line with a heightened sensitivity towards the crystalline (Plate 8):

> R.P. Bedford's drawings of flowers show how forms that are essentially frail and delicate like the petals of the cyclamen or a rose can be translated into bulky masses that would achieve their final meaning only in marble. These drawings of his are a halfway house in the process of transformation from the vegetable to the mineral world.[151]

This notion of morphogenetic crystallisation, in which the vegetable is slowly transformed into the inorganic, was a concept that had currency within the field of crystallography itself. Killian, for example, maintained that as space was the domain of the crystalline, to exist in space the organic necessarily had to partake of the inorganic:

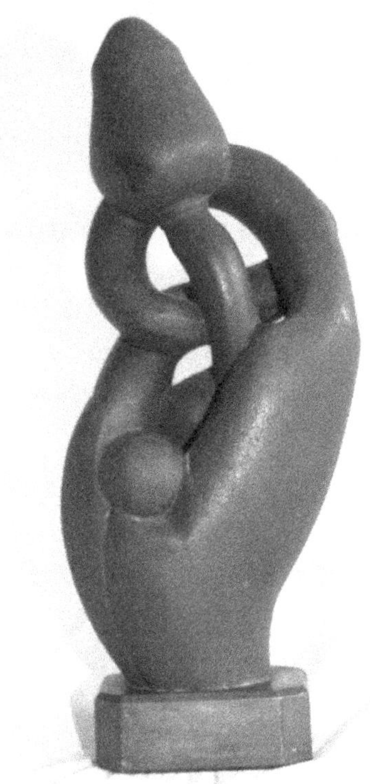

Figure 4.10

Richard Bedford, *Small Tulip*, date unknown (c.1930s). Private collection

As soon as the organic makes an appearance in space, i.e.: in the inorganic world, it must somehow take on aspects of the inorganic type. That the plant world exhibits more or less striking examples of crystalline proportions is not to be marvelled at – it is included in the structure of being. Square stems, the symmetrical positions of twigs, leaves and flowers, the forms of leaves [...] – shortly, every mathematical aspect of the plant organism – are scarcely to be differentiated from an appearance of the inorganic (crystalline type).[152]

Thus the impression that, in crystallographic terms, the organic world bordered on the inorganic did much to fuel monistic speculation within Modernist circles. Read's monistic hypothesis – that while 'the most general and most accessible of [...] intimations of reality are of the organic type, and intimately linked to the essential forms of life', there were 'other aspects of reality' that comprised the crystalline – underlines how far crystallographic theory – from Leduc's 'osmotic growths' to Killian's 'plant-like forms' – had entered the philosophical consciousness of the avant-garde.[153] Indeed, Wilenski's associationist supposition that Modernists had begun to see the geometric (i.e. the nominally inorganic) as

149

representative of the biological provides an aesthetic counterpoint to crystal-lographic attempts to narrow the gap between the realms of the living and non-living:

> [The] time came when the modern sculptors knew enough about the matter to refute these critics, when they were able to remind them that man himself is a geometric composition, that geometric form abounds in the animal and vegetable world, and that there is said to be reason for presuming that the earth which we inhabit is a sphere; the time came, in other words, when they had abandoned the presumed antithesis between organic and geometric form and had begun to presume instead that geometric form is *symbolic* of the organic.[154]

Aptly, in light of the comments by the art critic at the *Glasgow Herald*, it was partly Bedford's sculptures of plants and flowers which inspired Wilenski's monistic, associationist interpretation of contemporary sculpture.[155] The simple, undulant forms of a sculpture such as *Small Tulip* (Figure 4.10), for example, appeared to him both geometric and – by virtue of its bulbous, swollen aspect – organic:[156] a monistic vision of the unity of nature that was strikingly reflected in Killian's crystallographic assertion that: 'As soon as the organic makes an appearance in space, i.e.: in the inorganic world, it must somehow take on aspects of the inorganic type'.[157]

The form constants of nature: Marxism, the Bauhaus and biotechnics

The form–function debate that raged within the New Biology throughout the 1920s and 1930s, and which would lead to the establishment of an organicist conception of morphogenesis, was mirrored by a political debate within Modernist artistic circles over the usefulness of abstraction and whether artistic form ought to obey a social *function* rather than simply exist as a means to an end in itself. Ambivalence towards the perceived role of artistic Modernism stemmed not only from the newfound political awareness which the self-consciously radical movements of Surrealism and Constructivism had brought to Britain during the 1930s but also from a growing dissatisfaction among Modernist critics with formalism as an explanatory model – an interpretative framework which was increasingly seen as tautological due to its inability to shed any light on meaning in art save that pertaining to 'significant form'. For example, in a scientistic *Listener* article of 1930 on 'Art and Decoration', Read rejected the 'obfuscations of puritanism' that passed for contemporary art criticism, lamenting that the English were 'continually trying to find an excuse for art. Nowhere else in the world does the phrase "art for art's sake" give rise to such melancholy associations. Our empirical frame of mind will not allow of such a meaningless abstraction. We want art to be of some use'.[158]

Andrew Causey has observed that Read's dismissal of formalism stemmed from his increasing political engagement with Marxism and a wider perception that Bloomsbury-era criticism – epitomised by the formalist dogma of Roger Fry – had too elitist a conception of culture to meaningfully engage with the complexities of modern life.[159] This political attitude emerged within British Modernist circles after exposure to the politicised artistic currents flowing out of Europe. Constructivism's origins in Soviet Russia, for example, had imparted to the movement a deep-seated ambivalence about Modernist individualism alongside a Marxist belief in art's obligation to participate in social reconstruction due to its firm entrenchment in the capitalist system.[160]

During the 1930s, Marxism became something of a *cause célèbre* for British intellectuals as the severity of the Great Depression had heightened expectations that capitalism was experiencing its death throes. For many political freethinkers, the out-and-out bankruptcy of *laissez-faire* economics justified, if not the wholesale assumption of a Marxist philosophy, then certainly the incorporation of socialist principles into the management of society.[161] In terms of the arts, the most celebrated product of this leftward political shift was the establishment of the Artists International Association (AIA) in 1933, whose chief purpose was to mobilise visual artists so as to demonstrate a commitment to social responsibility in art and encourage political radicalism in artistic practice.[162] Amid mounting concern within the AIA leadership over which art styles were most appropriate to the socialist cause, a series of lectures and debates was organised to resolve this issue, the most significant by-product of which was the polemical anthology *5 on Revolutionary Art*, published in 1935.[163] In her foreword, Betty Rea adopted the rhetoric of crisis that had become a pervasive feature of what Richard Overy has christened 'the morbid age' to pronounce: 'The future of art hangs on the future of civilization. It is time artists began to think about what sort of future they want, and what they can do to get it'.[164]

One contention held by left-wing intellectuals was that, far from being culturally apathetic, Modernist sculpture embodied functionalist qualities that would prove invaluable to society as a whole. This hypothesis essentially operated under the psychobiological notion that contemporary art subconsciously replicated the forms of the natural world, which, once reworked into original aesthetic motifs, would enable better industrial and architectural design. Bernal observed that through Constructivism and functionalist architecture, 'the artist began to see new and more rigid harmonies which the necessities of engineering and economics had imposed on the technician'. One result of this contact had been the reduction of abstract art into 'the simplest geometrical forms, rectangles, spheres, cylinders and their positive and negative aspect, of which the bas-reliefs of Ben Nicholson [...] stand as a fitting example'.[165] The elements of this 'pure' art of formal elements led Herbert Read to conclude that abstract art would 'occupy, in the future, a relationship to industrial design very similar to the relationship pure mathematics bears to the practical sciences', due to the elemental laws of design being examined by such artworks.[166] By thus preserving and scrutinising the fundamental constants of art, Read believed that artistic Modernism could contribute to the establishment of a 'new classless society', by providing the aesthetic building blocks with which more efficacious techniques of industrial design and urban planning could be realised.[167] In his opinion, this practice of 'social action' was already underway 'in architecture and to some extent in the industrial arts', thus claiming a link between the work of the Bauhaus architects and the new generation of sculptors.[168]

The ubiquitous influence of Constructivist dogma and functionalist theory on British aesthetics in the 1930s is indicated by the spate of English-language books published in conjunction with these movements at this time – an event that dramatically coincided with the arrival in London of Bauhaus émigrés fleeing from persecution in Nazi Germany.[169] For example, the architectural theorist and Bauhaus director Walter Gropius (who was resident in England during the period 1934–37) maintained in his manifesto-like *The New Architecture and the Bauhaus* (1935) that the conventional relationship between art and manufacture had become distorted due to the unfamiliarity of the artist with the technical requirements of construction and the technician's inability to combine form with economic design.[170] One route through which the Bauhaus group attempted to realise 'a modern architectonic art' was by creating a 'federative union of all the different "arts" [so that] every branch of design, every form of technique – could be coordinated and find its appointed place'.[171] Consequently, Gropius envisaged the workshops of the Bauhaus as laboratories in which practical new designs for functional artefacts could be worked out. In this manner, 'type-forms' could be created 'that would meet all technical, aesthetic and commercial demands'. Although the prototypes of these models would be reliant on handicraft, Gropius asserted that the designer would still need to be 'intimately acquainted with factory methods of production and assembly'.[172]

These considerations embodied an ideology that felt that the difference maintained between industry and handicraft was less to do with divergent methodologies than to do with 'the subdivision of labour in the one and undivided control by a single workman in the other'.[173] As a result, the Bauhaus group sought to disassemble these antithetical positions and reintegrate art into the mainstream of life. This aesthetic philosophy opposed formalism by stressing that all kinds of creative work were interdependent and so form was contingent upon the functionalist demands of culture: 'Thus our informing conception of the basic unity of all design in relation to life was in diametrical opposition to that of "art for art's sake", and the even more dangerous philosophy it sprang from: business as an end in itself'.[174] These ideas acted as a stimulus to the functionalist notions of Read, who noted in *Art and Industry* that 'I have no other desire in this book than to support the ideas [...] expressed by Dr Gropius' – quoting a statement made by Gropius in the *Journal of the Royal Institute of British Architects* in 1934 on the topic of reintegrating art into the wider fabric of culture.[175] Within this context he claimed that aesthetics needed to acknowledge the importance of the artist to everyday design:

> In every practical activity the artist is necessary, to give form to material. An artist must plan the distribution of cities [and] of buildings [...]; an artist must plan the interiors of such buildings [...], down to the smallest detail, the knives and forks, the cups and saucers and the door-handles. And at every stage we need the abstract artist, the artist who orders materials till they combine the highest degree of practical economy with the greatest measure of spiritual freedom.[176]

Artistic functionalism worked in tandem with the New Biology in two ways. In the first place, just as the New Biology understood organic matter to be contingent, so, too, Modernist art theory posited that artistic form was *not* inseparable from social context: form and function were reliant upon each another. In fact, holism was an important part of this artistic philosophy – Gropius stressing that the aim of the Bauhaus was to demonstrate that design was 'one great cognate whole – the mirror of the indivisibility and immensity and underlying unit of life itself, of which it is an integral part'.[177] Secondly, economy of form and functionalism in design depended upon an acute awareness of the properties of materials – an issue that enabled adherents of the Bauhaus to utilise the findings of modern science in the course of design development. The German artist Joseph Albers made the point that 'Economy of form depends on function and material. The study of material must [...] precede the investigation of function. Therefore our studies of form begin with studies of materials'.[178] In this fashion, the educational strategies of the Bauhaus – popularised in Britain by Moholy-Nagy's *The New Vision* (1934) and Bayer *et al.*'s anthology *Bauhaus: 1919–1928* (1939) – laid stress upon a pseudo-scientific analysis of materials that would facilitate an enhanced knowledge of 'formal qualities'. Albers listed these qualities as 'harmony or balance, free or measured rhythm, geometric or arithmetic proportion, symmetry or asymmetry, central or peripheral emphasis'.[179] Indeed, for several Bauhaus theorists, only a systematic introduction to the methodologies of science could benefit student awareness of the physical laws of materials. *A propos* this educational scheme, Paul Klee noted in *Bauhaus: 1919–1928* that studies 'in algebra, in geometry, in mechanics characterise teaching directed toward the essential and the functional in contrast to the apparent'.[180] Only by these means did Klee believe that an artist could learn the basics of form and how to create inventive designs.

One offshoot of this novel attentiveness to scientific method was the subject of biotechnics – an area of study that aimed to reconcile the observation of material in nature with functional application.[181] The father of this discipline was the Hungarian biologist and design theorist Raoul Francé, who devised the subject on the basis that '[every] process in nature has its necessary form' – observing that the action of physical forces impelled organisms to adopt more efficacious designs.[182] Emerging out of the form–function debate in biology, Francé's hypothesis echoed the principles of the New Biology in its claim that 'every form is only the frozen momentary picture of a process' and that, as such, every function in nature possessed its necessary form.[183] His most influential text, *Plants as Inventors*, developed this idea by examining a wide range of examples in the world of botany, looking at how everything from turbines to pumps had a perfectly engineered counterpart in nature. On this basis, Francé entreated engineers, designers and architects to look to the forms of the natural world for practical solutions to problems of design, as, in terms of ingenuity, the 'laws of nature' were deemed to be 'always true and consequently practical'.[184]

Francé's ideas drew upon a long tradition in Germanic culture that valued nature not solely as a sourcebook of formal motifs but as a model for the designer,[185] a strand of thought that complemented the broader, neo-romantic rejection of machinist culture that swept through German society in the early twentieth century.[186] In this virulently anti-mechanistic environment Francé tapped into a *Zeitgeist* that had peaked during the Art Nouveau movement and been sustained by German and Dutch Expressionism until postwar theorists, such as Hugo Häring, profited from a contemporary enthusiasm for ecologically sound planning by promoting 'organic' methods of design in the 1920s.[187] Certainly, Häring shared with Francé a contempt for manufactured design, explaining in his 'Approaches to Form' (1925–26) that:

> Forms which result from functional criteria are created by life. Forms, in this respect, that derived from function were praised as fundamentally eternal and universal because they [were] constantly regenerated by life [whereas] forms created for the sake of expression [were by contrast] ephemeral and exposed to changes in human cognition.[188]

Just as the experimental morphology of D'Arcy Wentworth Thompson announced that the 'permanence or equilibrium' of organic form could be 'explained by the interaction or balance of forces',[189] biotechnics assumed that '[a] whole series of [organic] phenomena [corroborate] the assumption that the interplay of natural forces with various kinds of matter finds its equipoise in constantly recurring forms'.[190]

One figure who was particularly influential in promoting biotechnical ideas to a British audience was the Czech functionalist architect Karel Honzík, whose essay 'A Note on Biotechnics' was printed in *Circle* in 1937.[191] Arguing that engineers and architects generally ignored natural design, Honzík vouchsafed that natural objects expressed an economical compromise between form and function in ways that *anticipated* the products of human design. By way of an example, he pointed to the system of joists and ribbing that underpinned the leaves of a *Victoria regia* lily, noting that its composition might provide the engineer with a model for successfully spanning the roofs of modern buildings with reinforced concrete.[192] Building on Francé's notion that each form possessed an 'optimal' character that, through the operation of accretive and erosive forces, represented its 'essential form of rest',[193] Honzík explained how sand always slides down a slope at 34–37 degrees, so as to create a pyramid, which was 'its natural form at rest, its expression of equilibrium between conflicting forces of gravitation and friction among its component grains':

> The cone, the pyramid, parallel lines, the plane and the globe are all of them 'constants' of Nature's technique and organization. Living and dead matter seems to follow a single impulse, displacing and reorganizing itself, or growing, with the object of attaining equilibrium, under the direct influence of internal and external forces. Nature seeks that ideal state of equilibrium, or rest, in the balance of these forces, and the moment she succeeds in this she ceases to be formless and assumes a characteristic shape in flowers, crystals and other organisms that are the expression of the forces that mould them.[194]

Perceiving the action of a 'universal law' that sought a state of harmony in physical matter, Honzík, like Francé, conceived of this principle in vaguely neo-vitalist terms – even though his Platonic belief that form preceded function was somewhat at odds with Francé's doggedly functionalist philosophy. An echo of Francé's dogma was likewise present within Read's functionalist assertion in *5 on Revolutionary Art* (1935) that: 'abstract art has a positive function. It keeps inviolate, until such time as society will once more be ready to use them, the universal qualities of art – these elements which survive all changes and revolutions'.[195] Indeed, by merging psychobiology, Marxist philosophy and idealist morphology, Read set forth a powerful defence against the conservative, left-wing opinion that Modernist abstraction – in the words of the Marxist aesthetician F.D. Klingender – represented a 'frantic flight from content [and] social reality',[196] suggesting that its very abstractness was itself symbolic of a materialistic attitude and functionalist potential:

> It is at least arguable that the purely formal element in art does not change, that some canons of harmony and proportion are present in primitive art, in Greek art, in Gothic art, in Renaissance art and in the art of the present day. Such forms [...] are archetypal, due to the physical structure of the world and the psychological structure of man [...]. The recognition of such universal formal qualities in art is consistently materialistic. It no more contradicts the materialistic interpretation of the history of art than does a recognition of the relative permanency of the human form, or the forms of crystals in geology.[197]

By 1937, Read had developed this pseudo-Marxist argument into a powerful functionalist doctrine which justified Modernist abstraction on idealist, morphological grounds:

> The claim, therefore, of the abstract artist is that the forms he creates are of more than decorative significance in that they repeat in their appropriate materials and on their appropriate scale certain proportions and rhythms which are inherent in the structure of the universe, and which govern organic growth, including the growth of the human body. Attuned to these rhythms and proportions, the abstract artist can create microcosms which reflect the macrocosm [...]. He has no need of natural appearances – of the accidental forms created in the stress of the world's evolution – because he has access to the archetypal forms which underlie all the causal variations presented by the natural world.[198]

In effect, these psychobiological precepts recapitulated the neo-romantic understanding of idealist morphologies which had preoccupied Read in the early

1930s – especially in relation to the sculpture of Moore – with one important caveat. Whereas his earlier analyses privileged the unique role of the artist as creator, the political imperatives of aesthetic discourse in the mid-1930s saw Read shift his position to one in which abstract artists, by virtue of their form experiments, were seen to possess a *social* value by equipping architecture and the industrial arts with blueprints for design:[199]

> [It] is freely admitted that many of the great advances in civilization (for example, wireless telegraphy and aeronautics) are dependent on the services of pure mathematics. The function of abstract art must be envisaged in a similar way, with architecture and the industrial arts as the field of its practical application. These arts need for their aesthetic justification a heightened sensibility to the purity of form, the economy of means and the relevance of colour which can best be induced and refined by the creation and appreciation of non-representational works of art.[200]

While it is debateable whether or not Francé's biotechnical theories directly influenced Read's increasingly functionalist aesthetic (as Francé is never mentioned by Read in his writings, unlike morphologists such as Thompson and Ghyka), his collaboration with biocentric Constructivists – Honzík and Gabo among them – no doubt clarified in Read's mind the biotechnical utility of Modernist abstraction, especially when combined with an idealist understanding of morphology.[201] Although Read was imprecise in his explanation of the *exact* nature of Modernist abstraction's social function, his neo-romantic belief in universal type-forms tacitly echoed Francé's idealist conception of the 'universal' and the holistic unity of natural form.[202] In 'Ben Nicholson and the Future of Painting' (1935), for example, Read argued – in psychobiological terms – that Nicholson's art had a kinship with architecture which derived from a monistic faith in the unity of reality and the artist's ability to intuit the underlying forms of nature:[203]

> Ben Nicholson [...] believes there is a reality underlying appearances, and that it is his business, by giving material form to his intuition of it, to express the essential nature of this reality. He does not draw that intuition of reality out of a vacuum, but out of a mind attuned to the specific forms of nature – a mind which has stored within it a full awareness of the proportions and harmonies inherent in all natural phenomena, in the universe itself.[204]

Although Read's monism ultimately stemmed from Whitehead's organicist philosophy, the 'mathematical and crystalline'[205] nature of Nicholson's hardboard reliefs and painted plaster constructions are cast in essentialist, morphological terms that explicitly recall Francé's idealist, neo-romantic conception of the archetypal forms of nature.[206] In *Plants as Inventors*, for instance, Francé noted that: 'The fundamental technical forms of the world are contained in the crystalline-form, the globe, the plane, the pole, the ribbon, the screw and the cone'. These forms were thus 'the basis of architecture, of the parts of an engine, of crystallography and chemistry [...], of art, of industry – of the whole world. And the world teeming with life has produced no other possible forms'.[207]

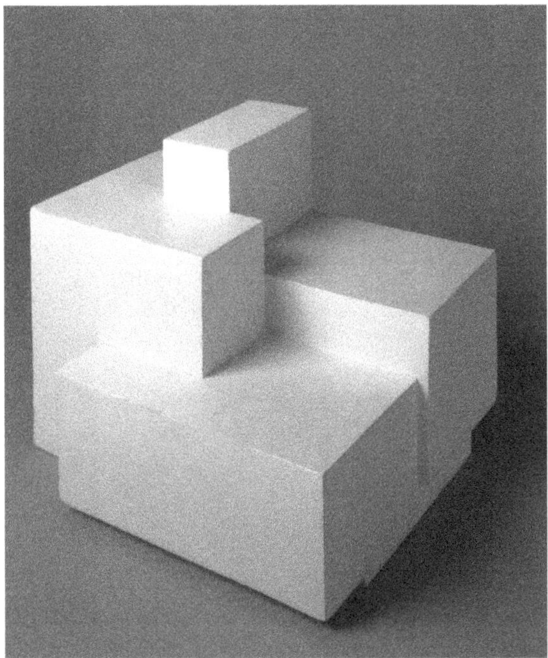

Figure 4.11
Ben Nicholson, *circa 1936 (sculpture)*, c.1936. Tate, London

Cast in the same light as the neo-romantic, morphological suppositions of Cook, Hambidge and Ghyka, Francé's and Honzík's treatises served as biologistic mouthpieces through which Modernists could obtain a functionalist, monistic understanding of natural form.[208] Nicholson's multiplanar exercises in orthogonal form, such as *circa 1936 (sculpture)* (Figure 4.11) and *1936 (sculpture – version 1)* (1936) (Figure 4.12) – which only slightly postdate Read's monistic speculations on the 'crystalline' nature of his practice – tender sculptural equivalents of the type-forms (the 'parallel lines, the plane') that Honzík claimed emerged in nature through an 'ideal state of equilibrium, or rest, in the balance of [physical] forces'.[209]

Many Modernists of a biocentric persuasion adopted a monistic worldview which corresponded – in substance if not in detail – with the idealistic tenets of Francé's science of biotechnics. Hepworth, for instance, verbalised a unitary vision of matter in which everything was simply a variant of a single, underlying structure.[210] Writing in *Circle*, she opined that: 'In the physical world we can discover in the endless variations of the same form, the one particular form which demonstrates the power and robustness of the simplified structure – the form is clear and every part of it is in precise unity with the whole'.[211] While this pronouncement lacks the specific reference to the organic–inorganic dichotomy

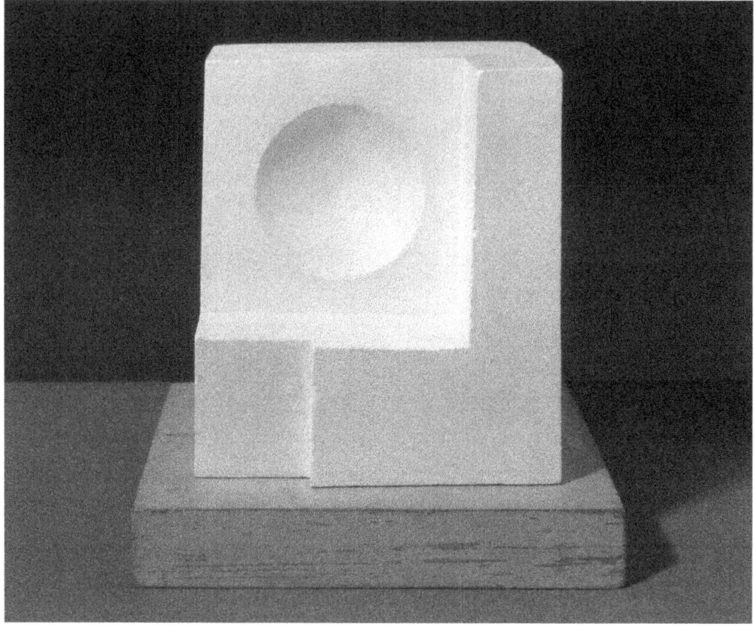

Figure 4.12

Ben Nicholson, *1936 (sculpture – version 1)*, 1936. Tate, London

which characterised the monistic hypotheses of Read, Martin *et al.*, its patent bio-centrism, allusion to organicism and emphasis upon repeating forms underscore its debt to the idealistic suppositions of biotechnics. Above all, the notion that every form was simply a remodelling of an *archetypal* form emulated the idealist principles of biotechnical theory, whereby the fundamental forms of nature represented the morphological apogee of biomechanical efficiency.[212] Summarising this position, Honzík remarked:

> A surface subject to pressure must be supported in one direction, and eventually, if the minimum quantity of matter is to be used, it must be supported in the other direction as well. This is inevitable, for there is no way of avoiding it. The resultant form is one of the most constant in nature. It is a shaping by natural forces at which man arrives by intuition, experience and calculation, and a plant by its own tropism.[213]

Echoes of Francé's and Honzík's biotechnical faith in the ability of nature to divulge – through patient observation – its elementary structure reverberate in Moore's neo-romantic speculations on morphology. For example, in *Unit 1* he spoke of the biologistic temperament of his artistic ideology, merging biotechnics with morphology to explain – in Goethean terms – how the close scrutiny of natural form had revealed to him the structural principles underpinning nature: 'I have

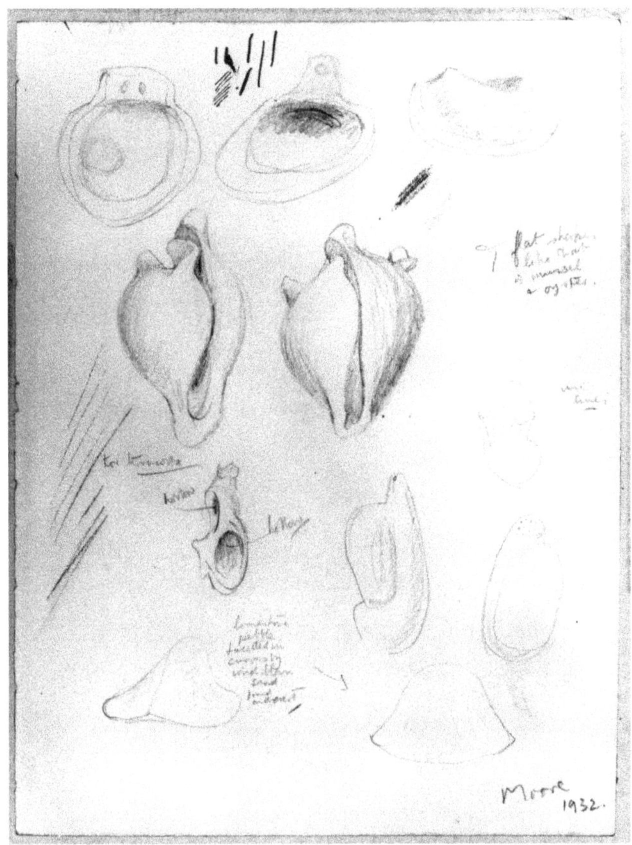

Figure 4.13

Henry Moore, 'Studies of Shells and Pebbles', 1932, HMF946.
Arts Council Collection, South Bank Centre, London

found the principles of form and rhythm from the study of natural objects such as pebbles, rocks, bones, trees, plants, etc.'[214] He creatively extemporised these bio-technical interests in a series of preparatory drawings of the 1930s (Figure 4.13). Representing a diverse array of natural objects – such as shells, roots and bones – seen from a variety of perspectives, these sketchbook images are outstanding for their manner of documenting Moore's curiosity about the elemental forms and shaping forces of nature via a suite of pencilled annotations. One study of shells and pebbles, for example, carries the legend: 'flat shapes/like that/of mussel/& oyster/hollow [...] pebble/facetted in/curves by/wind-blown/sand/found/on desert'.[215]

Whereas Moore's knowledge of the principles of natural form was beholden to his reading of D'Arcy Wentworth Thompson and Goethe,[216] the conceptual

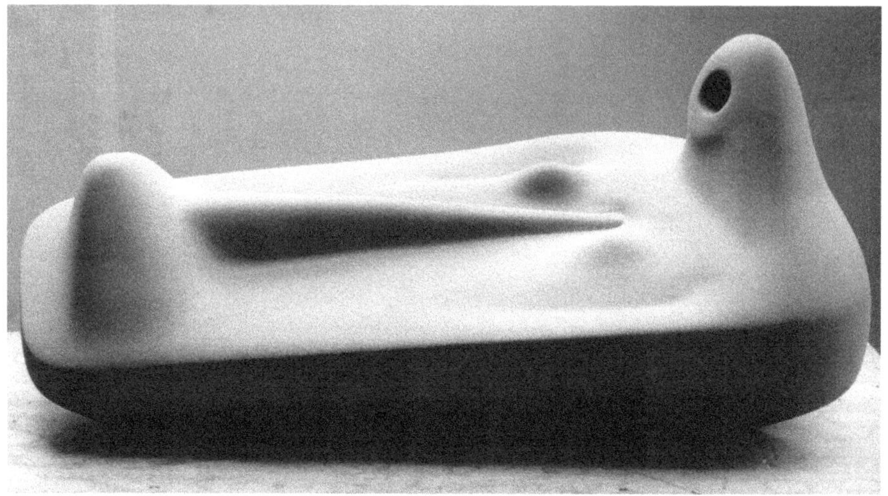

Figure 4.14

Henry Moore, *Reclining Figure*, 1933–34. Moore Danowski Trust

overlay between his morphological writings and the neo-romantic, idealist tenets of biotechnical theory underscores the degree to which biotechnics had influenced Modernists across Europe, most especially those – like Gabo – who were directly connected to emergent trends in Constructivist abstraction.[217] In this respect, the gentle, sand-smoothed dips and swells of Moore's *Reclining Figure* (1933–34) (Figure 4.14) articulate a deeply monistic conception of human anatomy in which bodily architecture has been whittled down to an essentialist, biotechnical vocabulary of cones, planes and globes. Although veering, suggestively, towards biomorphic abstraction, the composition emphasises the orthogonal, grounded character of the sculpture, which, in turn, highlights the geometricity of the funnel-like protrusions of the 'head' and 'feet'. Under the monistic terms of Read's analysis of 'Ben Nicholson and the Future of Painting', an artwork such as this avoids any straightforward distinction between 'organic and inorganic forms'; softening hard-edged orthogonality while simultaneously retaining a sense of undulant, streamlined geometry, it powerfully conveys the syncretic notion that 'reality is a unity'.[218] Indeed, it was precisely this understated impression of geometry in Modernist sculpture – of a semi-abstract naturalism crystallised, as it were, by the demands of geometric form – that led Wilenski to the monistic realisation that modern artists had distinguished in nature the common morphological principles which comprised the basis of *all* natural form:

It is by such studies that the modern sculptors have arrived at the concept of the universal analogy of form, the concept of all human, animal and vegetable forms as different manifestations of common principles of architecture, of which the geometric forms in their infinity of relations are all symbols; and at the concept of the meaning of geometric relation as the symbolisation of this universal analogy of form.[219]

Even if Wilenski's views were kindled by the revelation of Karl Blossfeldt's macro-photographs (which will be addressed in Chapter 5), his monistic aesthetic – while aligned to a neo-classical type of formalism – was clearly indebted to the syncretic paradigm which shaped the biologistic parameters of Modernist sculptural theory.[220] Consequently, the monistic unity of the 'dead world' of the crystalline and 'the world of living organisms' acknowledged by biocentric Constructivists and Modernists, such as Martin and Read, underscored the extent to which adherence to a particular style did not imply loyalty to a specific ideology.[221] Allegiance to a biologistic philosophy – as borne out by the Neo-Constructivist fabrications of Nicholson – did not lead to an intelligibly 'biological' style (such as the biomorphic abstractions of Hans Arp); rather – as Oliver Botar and Isabelle Wünsche have suggested – the lessons learnt from the New Biology, experimental crystallography and biotechnics found expression in a heterogeneous variety of styles, ranging from the hard-edged orthogonality of biocentric Constructivism to the supple contours of biomorphic Modernism.[222] Correspondingly informed by the tenets of the new morphology, psychobiology, bioromanticism, associationism and biotechnics, British aestheticians developed a radically *syncretic* theory of Modernist sculpture in which all abstract form, be it 'mathematical' or 'organic', could be reconciled within a single biologistic model: an idealist, biophilic paradigm of artistic form that had deep roots in neo-romantic nature philosophy and scientific monism.

Notes

1 Read, *Annals of Innocence and Experience*, p. 226.
2 *Ibid.*, p. 226.
3 Botar, *Prolegomena to the Study of Biomorphic Modernism*, p. 47.
4 See Hammer & Lodder, *Constructing Modernity*, p. 383.
5 O. Botar, 'Ernö Kállai and the Hidden Face of Nature', in *Structurist*, Nos 23, 24, 1983/84, p. 77.
6 Read's morphological theories were indebted, in this respect, to those of Gabo. See Hammer & Lodder, *Constructing Modernity*, p. 383.
7 Botar, *Prolegomena to the Study of Biomorphic Modernism*, pp. 14–15.
8 'There were, however, also those, myself for example, who sought a system of associations that would *explain* not only [...] works [of art] accepted as "aesthetic" and therefore "apt" by [...] critics, but also, as far as might be, the works rejected by them as "not aesthetic" and accordingly "not art"'. R.H. Wilenski, *Modern French Painters* (London: Faber & Faber, 1940), p. 272.
9 *Ibid.*, p. 272. Wilenski also identifies the contemporary museum as a symptom of the nascent associationist mind-set, in which artefacts from diverse cultures, both ancient and modern, were typically exhibited together in creative and unexpected juxtaposition.
10 H.G. Warren, *A History of Association Psychology* (New York: Charles Scribner's Sons, 1921), p. 6, original emphasis.
11 Guitemie Maldonado notes that, as early as 1929, the critic Jan Brzekowski mentioned associationism in his article 'Kilométrange', suggesting that associationist ideas were current during the

period when Kállai was developing his bioromantic aesthetic. J. Brzekowski, 'Kilométrange' (2), in *L'Art comtemporain*, No. 2, 1929, p. 50. See Maldonado, *Le cercle et l'amibe*, p. 53.

12 Maldonado, *Le cercle et l'amibe*, p. 53.

13 See Ritterbush, *The Art of Organic Forms*, pp. 1–4.

14 A. Arber, *The Natural Philosophy of Plant Form* (Cambridge: Cambridge University Press, 1950), p. 63.

15 Allen, *Life Science in the Twentieth Century*, pp. 2–3.

16 See M. Ridley, 'Embryology and Classical Zoology in Great Britain', in T.J. Horder (ed.), *A History of Embryology* (Cambridge: Cambridge University Press, 1986), esp. pp. 62–3; Allen, *Life Science in the Twentieth Century*, pp. 18–19.

17 R. Olby, 'Structural and Dynamical Explanations in the World of Neglected Dimensions', in T.J. Horder (ed.), *A History of Embryology* (Cambridge: Cambridge University Press, 1986), pp. 279–80.

18 G. Canguilhem, 'On the History of the Life-Sciences Since Darwin', in *Ideology and Rationality in the History of the Life-Sciences* (Cambridge: MIT Press, 1988), p. 108.

19 *Ibid.*, p. 108.

20 Needham, *Order and Life*, p. 110.

21 *Ibid.*, p. 113.

22 Haraway, *Crystals, Fabrics, and Fields*, p. 133; see also Olby, 'Structural and Dynamical Explanations in the World of Neglected Dimensions', p. 281.

23 See Thompson, *On Growth and Form*, pp. 1036–7; Needham, *Order and Life*, pp. 38–9.

24 J. Johnson, *The Mechanism of Life* (London: Edward Arnold, 1921), p. 4.

25 Needham, *Order and Life*, p. 6.

26 *Ibid*, p. 23.

27 T.A. Cook, *The Curves of Life* (1914) (New York: Dover, 1979), pp. 7–8.

28 A. Jakovski, 'Inscriptions under Pictures', in *Axis*, No. 1, January 1935, p. 16.

29 In a footnote to *Art and Industry*, Read notes: 'The literature of what may be called the morphology of art is considerable'. He goes on to list books by Matila Ghyka, Jay Hambidge, Theodore Cook, F.M. Lund and D'Arcy Wentworth Thompson as essential reading on this subject. H. Read, *Art and Industry* (London: Faber & Faber, 1934), footnote to p. 17.

30 See Maldonado, *Le cercle et l'amibe*, pp. 195–7.

31 Botar & Wünsche, 'Introduction', pp. 3–4.

32 Hammer & Lodder, *Constructing Modernity*, p. 384.

33 Read, *Art and Industry*, p. 17.

34 *Ibid.*, p. 15.

35 In his 1931 article 'The Golden Section', Read described the ratio thus: '[To] cut a finite line so that the shorter part is to the longer part as the longer part is to the whole. The resulting section is roughly in the proportion of 5 to 8 (or 8 to 13, 13 to 21, and so on), but never exactly so: it is always what is known in mathematics as an irrational, and this has added not a little to its mystical reputation'. H. Read, 'The Golden Section', in *The Listener*, Vol. V, No. 107, 28 January 1931, p. 142.

36 Read, *Art and Industry*, p. 17.

37 Read, 'The Golden Section', p. 142.

38 See Maldonado, *Le cercle et l'amibe*, p. 197.

39 Cook, *The Curves of Life*, p. 427.

40 Importantly, Read cited both of these writers as influences on his research. See Read, *Art and Industry*, footnote to p. 17.

41 J. Hambidge, *Dynamic Symmetry: The Greek Vase* (New Haven: Yale University Press, 1920); M. Ghyka, *Le nombre d'or: rites et rythmes pythagoriciens dans le développement de la civilisation occidentale* (Paris: Gallimard, 1931).

42 Ghyka, *Le nombre d'or*, p. 47.

43 Cook, *The Curves of Life*, p. 321.

44 *Ibid.*, p. 424.

45 Read, *Art and Industry*, p. 29.

46 D.A. Dewsbury, 'Psychobiology', in *American Psychologist*, Vol. XLVI, No. 3, March 1991, p. 198.

47 *Ibid.*, p. 198.

48 Botar, *Prolegomena to the Study of Biomorphic Modernism*, p. 500.

49 *Ibid.*, pp. 500–1.

50 Moore, 'The Sculptor Speaks', p. 339. Moore's admission also recalled the Jungian concept of collective archetypes that posited 'the existence of definite forms in the psyche', which formed the basis of humankind's mythologies. Jung's books on psychology had achieved popular acclaim in Britain; and the idea that an awareness of proportion in nature came about instinctively tallied well with the Jungian theory that instincts were 'impersonal, universally distributed, hereditary factors' – archetypes simply codifying these 'patterns of instinctive behaviour'. See for instance C.G. Jung, 'The Concept of the Collective Unconscious', in the *Journal of St Bartholomew's Hospital*, 1936/37, reprinted in H. Read, M. Fordham & G. Adler (eds), *The Archetypes and the Collective Unconscious: The Collected Works of C.G. Jung* (London: Routledge, 1980), Vol. IX, Part 1, pp. 42–3. Published at a time of mounting public anxiety over international and economic crises, Jung's texts were accepted as opportune rejoinders to the sense of psychosocial malaise that was affecting European civilisation – a melancholic disaffection that found expression in such potently pessimistic titles as Oswald Spengler's bestselling historical study *The Decline of the West* (1918). See Overy, *The Morbid Age*, pp. 9–49, 136–74.

51 See Rousseau, 'The Perpetual Crises of Modernism and the Traditions of Enlightenment Vitalism', p. 57.

52 Woodcock, *Herbert Read*, p. 196.

53 See Maldonado, *Le cercle et l'amibe*, p. 205; Bergson, *Creative Evolution*, pp. 4–7, 89–97.

54 'Ces types géométriques et ces rythmes "pulsants" qui ne se trouvent *jamais* dans la matière non organisée, qui paraissent comme la griffe, le signe des réactions de la Vie (croissance et génération), [...] sont [...] la symétrie pentagonale et les pulsations, les spirales de croissance apparentées a cette même symétrie pentagonale'. Ghyka, *Le nombre d'or*, p. 128.

55 See Maldonado, *Le cercle et l'amibe*, p. 205.

56 Bergson, *Creative Evolution*, p. 16.

57 Maldonado, *Le cercle et l'amibe*, p. 197.

58 This was how Wilenski described the biocentric character of Richard Bedford's artistic practice. See R. Wilenski, 'The Place of Sculpture Today and Tomorrow', in *The Studio*, Vol. CX, July–December 1935, p. 223.

59 E. Béothy, 'L'abstraction est la qualité spécifique de l'homme', *abstraction-création: art non-figuratif*, No. 4, 1935, p. 4.

60 R.H. Wilenski, *The Meaning of Modern Sculpture* (London: Faber & Faber, 1932), p. 157.

61 See Ritterbush, *The Art of Organic Forms*, pp. 11–12.

62 Johann Wolfgang von Goethe, 'The Spiral Tendency' (1831), in *Goethe's Botanical Writings* (1952), pp. 127–30, quoted in Ritterbush, *The Art of Organic Forms*, p. 34.

63 See L. Gamwell, *Exploring the Invisible: Art, Science and the Spiritual* (Princeton: Princeton University Press, 2002), p. 16.

64 Ritterbush, *The Art of Organic Forms*, pp. 34–5.

65 Allen, *Life Science in the Twentieth Century*, p. 3.

66 Daston & Galison, *Objectivity*, pp. 69–70.

67 On this topic see Maldonado, *Le cercle et l'amibe*, p. 197. On the influence of Goethe on biomorphic Modernism more specifically, see Mundy, *Biomorphism*, pp. 140–1.

68 The interview was conducted by Christa Lichtenstern on 12 October 1977. See Lichtenstern, 'Henry Moore and Surrealism', p. 651. It should be noted that *Conversations with Eckermann* is also (and was originally) known as *Conversations with Goethe*.

69 See, for example, Goethe and Eckermann's discussion of a creature's beauty being dependent upon its attainment of the 'summit of its natural development'. J.P. Eckermann, *Conversations with Eckermann* (1836) (London: Dent & Sons, 1930), pp. 193–4.

70 Moore, 'The Sculptor Speaks', p. 338.

71 H. Read, *Henry Moore* (London: A. Zwemmer, 1934), pp. 11–12.

72 On this intellectual debt, see Mundy, *Biomorphism*, p. 150.

73 Gamwell, *Exploring the Invisible*, pp. 15–16.

74 See Daston & Galison, *Objectivity*, p. 69.

75 Moore, Untitled statement, p. 30.

76 Cook, *The Curves of Life*, p. 31.

77 Moore, Untitled statement, p. 30.

78 Cook, *The Curves of Life*, p. 13.

79 See Maldonado, *Le cercle et l'amibe*, p. 197.

80 Notably, Thompson drew particular attention in *On Growth and Form* to Cook's work on spirals. Thompson, *On Growth and Form*, footnote to p. 748.

81 *Ibid.*, pp. 751–2.

82 Lichtenstern, *Henry Moore*, p. 57.

83 Read, *Art and Industry*, footnote to p. 17.

84 On the scientific legacy of Thompson, see M. Ridley, 'Embryology and Classical Zoology in Great Britain', in T.J. Horder (ed.), *A History of Embryology* (Cambridge: Cambridge University Press, 1986), p. 48.

85 Maldonado, *Le cercle et l'amibe*, p. 197.

86 R. Francé, *Plants as Inventors* (London: Simpkins, Marshall & Co., 1926), p. 16.

87 Cook, *The Curves of Life*, p. 91.

88 Francé, *Plants as Inventors*, pp. 16–17.

89 Hammer & Lodder, *Constructing Modernity*, p. 384.

90 R.H. Wilenski, 'Ruminations on Sculpture and the Work of Henry Moore', in *Apollo*, Vol. XII, December 1930, p. 413.

91 Andrew Causey has demonstrated that Wilenski sought, in his work, a formalist reconciliation of classicism and Primitivism. See A. Causey, 'R.H. Wilenski and *The Meaning of Modern Sculpture*', in D.J. Getsy (ed.), *Sculpture and the Pursuit of a Modern Ideal in Britain, c.1880–1930* (Aldershot: Ashgate, 2004), p. 268. The geometrical idealism hypothesised by neo-romantic nature philosophy meant that it was quite possible to harbour, as Wilenski did, distaste for the 'wild', 'romantic' representation of nature in art while possessing a fundamentally neo-romantic worldview.

92 Read, 'The Golden Section', p. 142.

93 *Ibid.*, p. 142.

94 A.S. Russell, 'Science Notes: The New Zealand Earthquake, The Golden Section', in *The Listener*, Vol. V, No. 110, 18 February 1931, p. 274. In his reply to Read, Russell noted that: 'In the spherical shape of the moon and stars, in the exactness of the dependence of musical intervals on numerical proportions, in the regularity of the hexagons of the honeycomb, and of other shapes in crystals, there is what the Greeks would have called perfection or harmony, but there are always good reasons for these in terms of matter and energy'.

95 *Ibid.*, p. 274.

96 A.S. Russell, 'Science Notes: The Proton, the Electron and the Penetrating Radiation', in *The Listener*, Vol. V, No. 106, 21 January 1931, p. 95.

97 *Ibid.*, p. 95.

98 Woodcock, *Herbert Read*, p. 196.

99 Read, *Annals of Innocence and Experience*, p. 202.

100 See Haraway, *Crystals, Fabrics, and Fields*, p. 25.

101 Whitehead, *Science and the Modern World*, p. 129.

102 *Ibid.*, p. 129.

103 *Ibid.*, p. 130.

104 See Bowler, *Reconciling Science and Religion*, p. 383.

105 Whitehead, *Science and the Modern World*, p. 183. Like Whitehead, Bergson's application of biological categories to philosophy was a formative influence on Read's incipient biologistic aesthetic and no doubt helped condition his positive response to Whitehead's organismal theory of the physical world. See Woodcock, *Herbert Read*, pp. 195–6, 222.

106 On Whitehead's place in British neo-romanticism, see Botar, *Prolegomena to the Study of Biomorphic Modernism*, p. 8.

107 Haraway, *Crystals, Fabrics, and Fields*, pp. 112–13.

108 N. Bohr, 'Light and Life', in *Discovery*, No.161, May 1933, pp. 151, 153.

109 *Ibid.*, p. 154.

110 Read, *Annals of Innocence and Experience*, p. 203.

111 J.L. Martin, 'The State of Transition', in J.L. Martin, B. Nicholson & N. Gabo (eds), *Circle: International Survey of Constructive Art* (London: Faber, 1937), p. 217.

112 *Oxford English Dictionary* online, 'monism, *n.*', www.oed.com/view/Entry/121244 (accessed 5 December 2012).

113 E. Gurney, 'Monism', in *Mind*, Vol. VI, No. 22, April 1881, p. 155.

114 Gamwell, *Exploring the Invisible*, p. 98.

115 Botar, 'Defining Biocentrism', p. 20.

116 Bowler, *Reconciling Science and Religion*, p. 135.

117 See N.R. Holt, 'Ernst Haeckel's Monistic Religion', in *Journal of the History of Ideas*, Vol. XXXII, No. 2, April–June 1971, p. 272.

118 Bowler, *Reconciling Science and Religion*, pp. 135–6.

119 H. Read, 'Ben Nicholson and the Future of Painting', in *The Listener*, Vol. XIV, No. 352, 9 October 1935, pp. 604–5.

120 D. Thistlewood, 'Herbert Read's Organic Aesthetic [1]', in D. Goodway (ed.), *Herbert Read Reassessed* (Liverpool: Liverpool University Press, 1998), pp. 225–6.

121 Read, *Art Now*, p. 146.

122 H. Read, 'Realism and Abstraction in Modern Art' (1948), in H. Read, *The Philosophy of Modern Art* (London: Faber & Faber, 1952), p. 97.

123 W. Worringer, *Abstraction and Empathy* (1908), M. Bullock (trans.) (London, 1953), reprinted in C. Harrison & P. Wood (eds), *Art in Theory: 1900–1990* (Oxford: Blackwell, 1992), p. 68.

124 *Ibid.*, p. 68.

125 *Ibid.*, p. 70.

126 *Ibid.*, p. 69.

127 Botar, *Prolegomena to the Study of Biomorphic Modernism*, p. 46.

128 Woodcock, *Herbert Read*, p. 168.

129 See Thistlewood, 'Herbert Read's Organic Aesthetic [1]', pp. 215–32.

130 Botar, *Prolegomena to the Study of Biomorphic Modernism*, p. 50.

131 Grigson, 'Comment on England', p. 8.

132 See S. Papapetros, 'On the Biology of the Inorganic: Crystallography and Discourses of Latent Life in the Art and Architectural Historiography of the Early Twentieth Century', in O. Botar & I. Wünsche (eds), *Biocentrism and Modernism* (Farnham: Ashgate, 2011), pp. 82–5.

133 E. Haeckel, 'Foreword of "Crystal Souls – Studies of Inorganic Life"' (1917) (translated by A. L. Mackay), reprinted in *Forma* (special issue: 'Crystal Souls by Ernst Haeckel'), Vol. XIV, Nos 1–2, 1999, p. 33.

134 Papapetros, 'On the Biology of the Inorganic', p. 87.

135 S. Leduc, *The Mechanism of Life* (London: Rebman, 1911), p. 149.

136 *Ibid.*, pp. 149–50.

137 Mundy, *Biomorphism*, p. 136.

138 E.F. Keller, *Making Sense of Life: Explaining Biological Development with Models, Metaphors and Machines* (Cambridge: Harvard University Press, 2002), p. 51.

139 Leduc, *The Mechanism of Life*, p. 3.

140 J. Killian, *Crystals: Secrets of the Inorganic* (London: John Gifford, 1941), p. 52.

141 *Ibid.*, p. 118.

142 Needham, *Order and Life*, p. 144.

143 Brown, *J.D. Bernal*, pp. 80–1.

144 Needham, *Order and Life*, p. 146.

145 Ritterbush, *The Art of Organic Forms*, p. 74.

146 Nash, 'The Life of the Inanimate Object', pp. 137–9.

147 As Ritterbush has observed, Leduc's popularity among artists was compounded by his neo-romantic, idealist yearning to grasp nature according to aesthetic, as opposed to solely scientific, principles. Ritterbush, *The Art of Organic Forms*, p. 74.

148 N. Gabo, carbon of letter to Alfred Jensen, 4 July 1950, Yale University, quoted in Hammer & Lodder, *Constructing Modernity*, pp. 384–5.

149 P. Nash quoted in Hammer & Lodder, *Constructing Modernity*, p. 384.

150 Anne J. Barlow discusses this possibility, although she is reluctant to claim that Hepworth had a 'literal' scientific understanding of crystallographic form. See Barlow, 'Barbara Hepworth and Science', p. 103. Bernal, was something of a pioneer in the field of X-ray crystallography, undertaking X-ray examinations of protein molecules and amino acids which helped to lay the

foundations for a physical understanding of the 'biological specificity of cell function'. Brown, *J.D. Bernal*, p. 96. (On Hepworth's relationship with Bernal, see also Brown, *J.D. Bernal*, pp. 153–4.)

151 Anon., 'Features of London: Exhibition – Drawings by Sculptors', in the *Glasgow Herald*, 13 January 1938.

152 Killian, *Crystals*, p. 22. While Killian's opinions had legitimacy in the crystallographic analyses of plant fibres they also reprised a much older way of conceptualising plant growth that originated in the writings of the seventeenth-century natural historian Nehemiah Grew. Grew believed that all plants comprised four differently shaped salt crystals, which entered the plant through the roots and coalesced to create circles or lines which expanded through the incremental accretion of microscopic crystals in geometric patterns. The perception of regularities in natural forms, he believed, was evidence that the stages of development entailed the replication of simple steps which offered the possibility of a materialistic explanation of life. See Ritterbush, *The Art of Organic Forms*, p. 8.

153 Read, 'Ben Nicholson and the Future of Painting', pp. 604–5.

154 Wilenski, *The Meaning of Modern Sculpture*, p. 157.

155 Indeed, in *Modern French Painters* (1940), Wilenski makes clear that his concept of association-ism is, in effect, a type of technophilic monism: 'At the same time, more and more books with photographs and coloured reproductions of pictures of all periods and places became available [...]. In these conditions artists and critics, captured by Associationist ideas, sought a common denominator in all these heterogeneous art activities now brought within their ken; [and] they assumed that a common *aesthetic* denominator might be found to replace existing academic standards [...]. Thus it came that in the first postwar decade the leading art critics told us *ad nauseum* that all Art is and always has been One, that its oneness is and always has been its Aesthetic Character'. Wilenski, *Modern French Painters*, p. 272.

156 Bedford's *Flower* is one of the few Modernist sculptures that Wilenski chose to illustrate his argument in *The Meaning of Modern Sculpture*.

157 Killian, *Crystals*, p. 22.

158 H. Read, 'Art and Decoration', in *The Listener*, Vol. III, No. 69, 7 May 1930, p. 805.

159 Causey, 'Formalism and the Figurative Tradition in British Painting', p. 22.

160 Harrison, *English Art and Modernism*, pp. 284–5.

161 Overy, *The Morbid Age*, pp. 50–92.

162 R. Radford, *Art for a Purpose: The Artists' International Association, 1933–1953* (Winchester: Winchester School of Art Press, 1987), p. 6.

163 *Ibid.*, pp. 64–5.

164 B. Rea, 'Foreword', in B. Rea (ed.), *5 on Revolutionary Art* (London: Wishart, 1935), p. 7.

165 Bernal, 'Art and the Scientist', p. 122.

166 Read, *Art and Industry*, p. 61.

167 H. Read, 'What Is Revolutionary Art?', in B. Rea (ed.), *5 on Revolutionary Art* (London: Wishart, 1935), p. 21.

168 *Ibid.*, p. 21.

169 See Anker, 'The Bauhaus of Nature', pp. 229–51.

170 W. Gropius, *The New Architecture and the Bauhaus* (London: Faber, 1935), pp. 42–3.

171 *Ibid.*, pp. 43–4.

172 *Ibid.*, p. 37.

173 *Ibid.*, p. 37.

174 *Ibid.*, p. 60.

175 Read, *Art and Industry*, p. 63. In fact, the passage by Gropius that Read quoted followed closely the text of *The New Architecture and the Bauhaus* on pp. 59–60.

176 Read, *Art and Industry*, pp. 63–4.

177 Gropius, *The New Architecture and the Bauhaus*, p. 67.

178 J. Albers, 'Concerning Fundamental Design', in H. Bayer, W. Gropius & I. Gropius (eds), *Bauhaus: 1919–1928* (London: George Allen & Unwin, 1939), p. 116.

179 *Ibid.*, p. 120.

180 P. Klee, 'Statement', in H. Bayer, W. Gropius & I. Gropius (eds), *Bauhaus: 1919–1928* (London: George Allen & Unwin, 1939), p. 172.

181 On this subject's particular appeal to British aestheticians see especially Anker, 'The Bauhaus of Nature', pp. 229–51, and (in relation to Moholy-Nagy) O. Botar, 'The Origins of László Moholy-Nagy's Biocentric Constructivism', in E. Kac (ed.), *Signs of Life: Bio Art and Beyond* (Cambridge: MIT Press, 2007), pp. 315–37.

182 R. Francé, *Die Pflanze als Erfinder* (Stuttgart, 1920), quoted in L. Moholy-Nagy, *The New Vision: Fundamentals of Design, Painting, Sculpture, Architecture* (London: Faber & Faber, 1934), p. 60.

183 Francé, *Plants as Inventors*, p. 60.

184 *Ibid.*, p. 8.

185 See T. Benton, 'Modernism and Nature', in C. Wilk (ed.), *Modernism, 1914–1939: Designing a New World* (London: V&A Publications, 2006), p. 323.

186 See Harrington, *Reenchanted Science*, pp. xv–xix.

187 Benton, 'Modernism and Nature', pp. 323–4.

188 H. Häring, 'Approaches to Form' (1925–26), reprinted in T. Benton, C. Benton & D. Sharp (eds), *Form and Function: A Sourcebook for the History of Architecture and Design, 1890–1939* (London: Open University, 1975), p. 103.

189 Thompson, *On Growth and Form*, p. 16.

190 K. Honzík, 'A Note on Biotechnics', in J.L. Martin, B. Nicholson & N. Gabo (eds), *Circle: International Survey of Constructive Art* (London: Faber & Faber, 1937), p. 257.

191 *Ibid.*, pp. 256–62. On this subject see also Harrison, *English Art and Modernism*, p. 282.

192 Honzík, 'A Note on Biotechnics', pp. 256–7.

193 Francé, *Plants as Inventors*, p. 11.

194 Honzík, 'A Note on Biotechnics', p. 258.

195 Read, 'What Is Revolutionary Art?', p. 21.

196 F.D. Klingender, 'Content and Form in Art', in B. Rea (ed.), *5 On Revolutionary Art* (London: Wishart, 1935), p. 42.

197 Read, 'What Is Revolutionary Art?', p. 15.

198 Read, *Art and Society*, p. 125.

199 See Harrison, *English Art and Modernism*, pp. 281–2.

200 Read, *Art and Society*, p. 125.

201 On Gabo's influence on British art, see Hammer & Lodder, *Constructing Modernity*, pp. 231–67.

202 See, for example, Francé's pantheistic, nature-centric description of environmental change and its effects on the forms of life: Francé, *Plants as Inventors*, esp. pp. 8–11.

203 Naturally, by implying that a singular reality underlies appearances, Read's analysis also bore some relation to theosophical reasoning – a state of affairs that was perhaps hardly surprising given the popularity of theosophy among the European avant-garde in the early years of the twentieth century. See C. Harrison, F. Francina & G. Perry (eds), *Primitivism, Cubism, Abstraction: The Early Twentieth Century* (New Haven: Yale University Press, 1993), pp. 198–9.

204 Read, 'Ben Nicholson and the Future of Painting', p. 605.

205 H. Read, 'Ben Nicholson and the Future of Painting', p. 605.

206 On Francé's influence on theories of abstraction (including Read's) in interwar Europe, see Hammer & Lodder, *Constructing Modernity*, pp. 382–3.

207 Francé, *Plants as Inventors*, pp. 17, 19.

208 See Botar, *Prolegomena to the Study of Biomorphic Modernism*, pp. 406–12.

209 Honzík, 'A Note on Biotechnics', p. 258.

210 In *English Art and Modernism, 1900–1939*, Charles Harrison briefly discusses the emergence of a monistic and idealist mentality in British Modernist discourse as part of a wider analysis of the politicisation of the avant-garde in the 1930s. He identifies this interest in the 'fundamental constructive principles in life' as gaining cultural traction through 'recent work in the natural sciences, particularly botany', although his evaluation stops somewhat short of linking Honzík's 'A Note on Biotechnics' to Francé's theory of biotechnics or, indeed, Modernist biologistic tendencies to neo-romantic morphologies and the New Biology more generally. Harrison, *English Art and Modernism*, p. 282.

211 Hepworth, 'Sculpture', p. 114.

212 It should also be noted that Hepworth's statement echoes monistic statements made by crystallographers about the underlying nature of matter. Sir William Bragg, for example, penned a 1937 article for *The Listener* on 'The Nature of Things', whose portrayal of molecular structures

paralleled Hepworth's nature-centric monism. '[All] things', he wrote, 'are composed of atoms of each kind being very nearly alike. Then we find these atoms exert forces on one another, and form companies called molecules [...]. In the processes of Nature the molecules are formed, broken-up and reformed, and in this way the business of the world is carried on'. W. Bragg, 'The Nature of Things', in *The Listener*, Vol. XVII, No. 436, 19 May 1937, p. 985.

213 Honzík, 'A Note on Biotechnics', pp. 257–8.
214 Moore, Untitled statement, p. 29.
215 H. Moore, annotation on *Study of Shells and Pebbles* (AG 32.57 HMF946), reproduced in Garrould (ed.), *Henry Moore, Vol. II*, p. 71.
216 See Mundy, *Biomorphism*, p. 130; Lichtenstern, *Henry Moore*, pp. 57, 59, 275.
217 Botar, *Prolegomena to the Study of Biomorphic Modernism*, p. 411.
218 Read, 'Ben Nicholson and the Future of Painting', pp. 604–5.
219 Wilenski, *The Meaning of Modern Sculpture*, p. 159.
220 For a detailed examination of the wider artistic context which informed Wilenski's *The Meaning of Modern Sculpture* – especially regarding formalism, see Causey, 'R.H. Wilenski and *The Meaning of Modern Sculpture*', pp. 267–85.
221 Martin, 'The State of Transition', p. 217; see also Hammer & Lodder, *Constructing Modernity*, p. 382.
222 Botar & Wünsche, 'Introduction', pp. 3–4.

5. Worlds beneath the microscope

<blockquote>

Saturn, trembling in the crystal lens of the telescope, makes us ponder as a glow-worm does a child: so too, the Milky Way generating universes, or a sleeping seed that a touch of moisture brings to life under the microscope. Roses of Jericho, Brownian movements of constellated atoms, no matter what, must be seen under the microscope: and our gaze must be turned within, upon our own abyss, in search of that ineffable axis which is the abscissa of the vast whole.[1]

Amedée Ozenfant, *The Foundations of Modern Art*, 1931

</blockquote>

So wrote the painter Amedée Ozenfant in his 1931 Purist apologia, *The Foundations of Modern Art*. Disclaiming the view that 'Nature' was simply the sum of perceptible appearances, Ozenfant implored the modern artist to freshly contemplate the natural world so that any painterly or sculptural composition would not simply be an act of dreary imitation. So as to evoke the universe beyond everyday sensation (and thereby point up the fallacy of artistic naturalism), he signalled to the evidence garnered by the new visualising technologies of the day: to the images of planetary motion obtained by the telescope and the pictures of microscopic life acquired by the microscope. Ozenfant was not alone in pointing to the altered character of perception following the technological breakthroughs of the early twentieth century. In the opening lines of *Surrealism and Painting* (1928), André Breton conveyed a dizzying assault upon the faculty of vision, in which nature was radically de-familiarised through a sequence of dramatic reorientations of scale: 'The eye exists in its savage state. The Marvels of the earth a hundred feet high, the Marvels of the sea a hundred feet deep, have as their sole witness the wild eye that traces all its colours back to the rainbow'.[2] In this way, Breton extolled the new dimensions of vision of the modern age, ostensibly by means of a psycho-visual process of 'zooming in', but one that implicitly referred to the new magnitudes of sight enabled by lens-based technology. The 'savage' human eye, reduced to a primal state by witnessing nature in a state of raw creation, was in fact a common trope of a technological discourse that sought to evoke the more disquieting aspects of nature when perceived at close range through a lens. For example, in his introduction to W. Watson-Baker's photomicrographic album *World Beneath the Microscope* (1935), W. Gaunt stressed the mental discomfiture experienced when observing very small things made large and the perceptual parity achieved between minuteness and vastness of scale:

> The mind is stretched, uncomfortably sometimes, but with a new fascination, to speed and profundity, to the thought of worlds that lie a million light years away from us, to the worlds that recede in evolutionary time beneath the lens, to the thought that they even merge or that by some extraordinary trick of relativity the smaller may contain the large. There is an affinity between the telescope and the microscope, between the discovery of stellar space and the discovery of the atom.[3]

Despite the fact that microscopical imagery had been first brought to the public eye in 1665 by Robert Hooke's tome *Micrographia*, the development of new technologies in the first few decades of the twentieth century – such as X-ray crystallography, ultraviolet microscopy and electron microscopy – had revealed a hitherto unseen world of grossly enlarged organic structures and micro-organisms.[4] Indeed, as the nineteenth century passed into the twentieth, microscopical observation and its photomicrographic kin – by permitting a more systematic analysis of nature's fundamental elements – had become integral to the development of what would become the New Biology.[5] As Lorraine Daston and Peter Galison have shown, the establishment of ever more accurate methods of photomicrographic registration coincided with the emergence of 'a new configuration of epistemological convictions, image-making practices, and moral comportment that aimed to quiet the observer so nature could be heard: *mechanical objectivity*'.[6] Representative of a wider tendency towards objectivity in the sciences, the development of photomicrography as a biological tool in the late nineteenth and early twentieth centuries emphasised scientific dissatisfaction with the idealising susceptibilities of romantic-era science and an attempt to constrain the propensity of scientists to aestheticise the natural world. In lieu of the fallibility of human perception, a range of mechanical technologies emerged which could systematically reproduce all manner of natural objects (fossils, bacteriophages, crystals, seeds, etc.) without the subjective interference of the human hand.[7]

Overlapping with the development of photomicrography as a technology was an upsurge in popular-science literature – especially during the 1920s and 1930s – which sought to augment biology's social role, thereby cementing its position as technological helpmeet to the paradigmatic science of the age. Yet, in the course of prosecuting this agenda, popularisers of science constructed a biologistic discourse which emphasised the spectacular nature of photomicrographic imagery, complicatedly merging together the values of the pedagogic and the entertaining.[8] At the dawn of the twentieth century, the evolutionary zoologist Ernst Haeckel had laid the foundations of this dichotomy by producing a collection of exquisite lithographs of marine protozoa and plant life – published as *Kunstformen der Natur* (*Art Forms in Nature*) in 1904 – that made public in photorealistic detail a previously undiscovered universe of minuscule beings (Figure 5.1).[9] In a 1913 book he co-authored with Haeckel, the zoologist W. Breitenbach explicitly conflated the educational with the aesthetic, stressing – in idealistic, neo-romantic undertones – how Haeckel's illustrations revealed to lay audiences the ornate splendour of marine organisms:

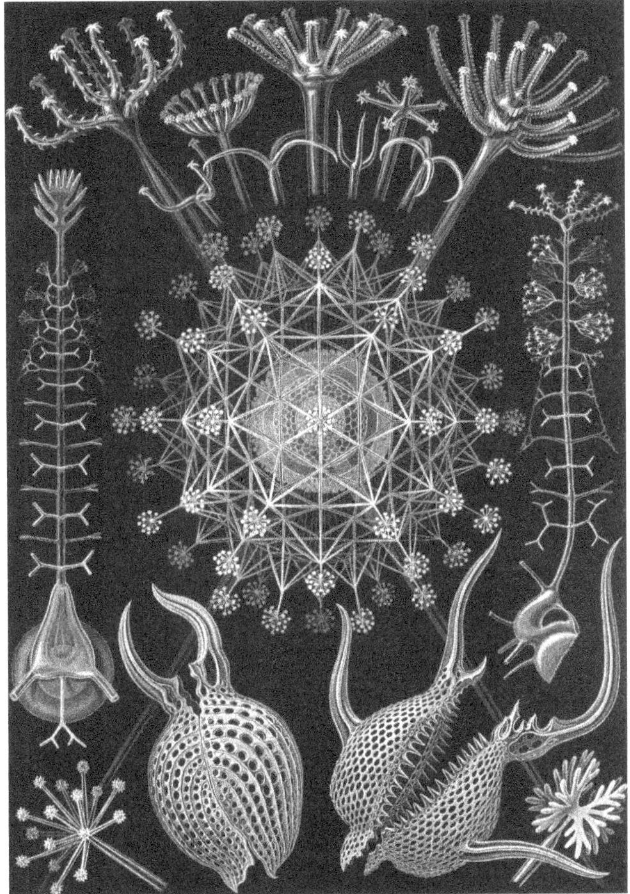

Figure 5.1

Ernst Haeckel, 'Phaeodaria', Plate 61 in *Kunstformen der Natur*
(Leipzig: Verlag des Bibliographischen Institut, 1904)

When a jellyfish such as this [...] floats on the surface of the calm ocean of warmer
climes, it resembles a swimming flowering plant whose leaves, tendrils, flowers and
fruits look as though they were made of vivid, brightly coloured, transparent crystal
[...]. Nature has scarcely produced anything so fragile and splendid in colour as these
wonderful creatures [...]. To laymen and inlanders, these daughters of the ocean seem
mysterious and fantastic, and they do not really know what to make of them.[10]

By wedding a meticulous classificatory system to a highly wrought form of visual
design, Haeckel's beguiling lithographs established the standard for what would
become the publishing industry's later attempt at marrying the educational with
the aesthetic in books of close-up nature photography.[11] Indeed, his hand-drawn
images of microscopic forms appeared to many of his colleagues – such as the

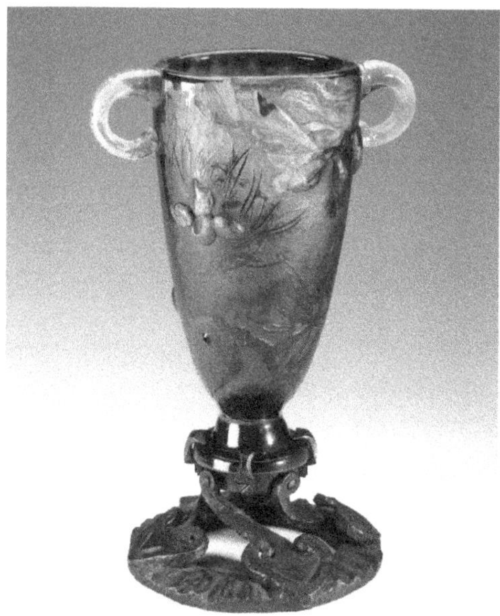

Figure 5.2

Emile Gallé, vase, 'Canthare Prouvé', 1896, Musée de l'Ecole de Nancy

Leipzig embryologist Wilhelm His – to eccentrically recall an earlier, pre-objective period of idealistic zoology in which the subjective intuitions of the scientist found expression in self-consciously artistic representations of nature.[12] Haeckel himself always maintained that his images were 'true-to-nature' and purposefully *selective* rather than mechanically *objective*, although, in an age of mechanical objectivity, his aesthetics – though flagrantly neo-romantic – were irrevocably transformed by the revelation of photomicrography, even if this was, at times, staged as something akin to a rivalry.[13] Like Goethe, Haeckel possessed a powerful appreciation of morphology and it was arguably his fidelity to the ideal of a resolution between truth and beauty, determined by the epistemological axes of subjectivity and objectivity, that safeguarded his later reputation among artists – not least because the character of his practice (framed by objective science and subjective artistry) presented him in the guise of an artist-cum-scientist.[14] The artistic enthusiasm that marked the reception of his publications can be gauged by the degree to which his images provided the designers, artists and architects of the Art Nouveau movement with a design vocabulary replete with aquatic, protozoan and unicellular forms. *Fin-de-siècle* interest in Haeckel's images was evinced in the lustrous, rainbow-hued glassware produced by Emile Gallé, which featured decorative embellishments of marine organisms (Figure 5.2) – that mirrored the decorative polychromy of Haeckel's illustrations – and the 'biomorphic' motifs of the architect René Binet, whose design for the entrance portal to the 1900 Universal

Exhibition in Paris was directly inspired by Haeckel's ornate representations of radiolarians (subaquatic organisms with eye-catching, crystalline exoskeletons).[15] This excitement, felt by late-nineteenth-century and early-twentieth-century artists for the minuscule forms of nature, was one cultural expression among many of the deeply flowing biologistic current that emerged in Europe in the years immediately preceding and following 1900 in response to the nature-centric, metaphysical undertones of the late-nineteenth-century romantic revival.[16]

By the 1920s and 1930s, improvements in photographic technology – largely driven by the objective imperatives of turn-of-the-century scientific research – had facilitated the scientific appreciation of microscopic forms and tissue samples while better-quality lighting and improved staining techniques expedited public veneration of photomicrography through augmenting the print quality of micrographic images reproduced in the pages of the popular press.[17] Equally, the development of microcinematography, through such early film pioneers as Jean Painlevé and Jean Comandon, brought to the public gaze speeded-up or slow-motion films of marine life, primitive organisms and germinating plants.[18] While the European scientific community's attitude towards such films occasionally bordered on the contemptuous – cynics viewing them as mere entertainment or, worse, unscientific reportage – microcinematography boomed as a populist genre, particularly delighting avant-garde audiences, who saw – especially in the work of Painlevé – evidence of an aesthetic sensibility choreographing the wonders of the natural world.[19] The critic Elie Faure, for example, remarked that: 'The science films of Jean Painlevé [...] in showing the dancing and glittering life of a mosquito, bring to mind the enchantment of Shakespeare and allow one to glimpse the exhilaration of the mathematician lost in the silent music of infinitesimal calculations'.[20] These revolutionary technologies were heralded by nature-centric Modernist photographers and filmmakers, like László Moholy-Nagy, as signalling a new age of mechanical objectivity that rejected the aesthetic pretensions of pictorialism by offering up a vision of nature undistorted by human gaze.[21]

The aesthetic cachet acquired by photomicrographic imagery through its popularisation in the years between the wars appealed to biophilic Modernists and, as Jennifer Mundy has explained, led to the development of biomorphism by lending itself to the particularly modern implications of biomorphic Modernism's nature-inflected vocabulary of form.[22] To be sure, the understanding that photomicrography had impacted upon the psyche of the modern artist was addressed by the English inventor of the style-label 'biomorphism', Geoffrey Grigson, in his 1943 monograph *Henry Moore*. Noting the ubiquity of micrographic imagery in everyday life, Grigson championed its acknowledgement as a source for contemporary art:

> Biology must be acknowledged [...]. The rounded limbs of a human foetus, or even the pneumococcus that chokes and ruins the lungs with pneumonia, would not, when realized with the bigness of life, be less worthy than a lounge suit in white marble or an Alsatian dog a million times smoothly reproduced in coloured China.[23]

The recognition of a similarity of form between the new scientific imagery and biomorphic Modernism has been described by art historians, such as Oliver Botar, as 'naturamorphic': an analogical paradigm which originated in the bioromantic hypotheses of the Hungarian art theorist Ernö Kállai. Certainly, Kállai's writings on bioromanticism – which began to emerge in the late 1920s and early 1930s, after a transformative encounter with the macrophotographs of Karl Blossfeldt – were the first systematic attempt made by a critic to hypothesise the relationship between Modernist art and scientific photography.[24] Astounded by the marvels of microscopic and X-ray photography, Kállai thus identified a visual relationship between the new scientific imagery and modern art, perceiving that technology could reveal the fundamental structure of the world in a way that corresponded with the psychobiological intuitions of artists.[25]

While the pan-European stimulus of photomicrography on avant-garde activity during the interwar period has been well documented by art historians, less work has been carried out on the social reception of photomicrography or the specific cultural conditions under which close-up scientific imagery was consumed by Modernists in Britain.[26] Accordingly, this chapter will examine the ways in which scientific photographic imagery influenced Modernist sculpture, firstly by assessing the wider channels through which photomicrography and microcinematography were disseminated in Britain in the period 1930–39 and then by focusing upon the phenomenal success of Blossfeldt's photo-album of enlarged plant-forms, *Art Forms in Nature*. Nothing short of a popular sensation, Blossfeldt's book, in particular, will be determined as having been a major conduit through which a bioromantic discourse was first established in Britain. Just as Kállai's biocentric philosophy was informed by exposure to Blossfeldt's photographs, so too Modernist critics, such as Anthony Blunt, found in Blossfeldt evidence of a bioromantic tradition of looking to nature for inspiration that stretched back to the Renaissance.[27]

'New light on the infinitesimal': journalistic coverage of photomicrographic technology

A 1929 review in the *Times Literary Supplement* (*TLS*) of Karl Blossfeldt's photo-book of plant forms – published in Britain as *Art Forms in Nature* – noted the 'extraordinary appeal of the 120 plates', which featured various close-up photographs of flowers, seeds, leaves and tendrils (Figure 5.3).[28] Yet the *TLS* review neglected to mention any tangible reasons for the strength of Blossfeldt's appeal (the publication would go through three editions between 1929 and 1935 in Britain) beyond a vague allusion to a contemporary interest in observing the action of physical forces on plants.[29] In actual fact, the wide-reaching impact of *Art Forms in Nature* owed much to the phenomenon of the interwar boom in popular-science literature books, which included not only texts related to

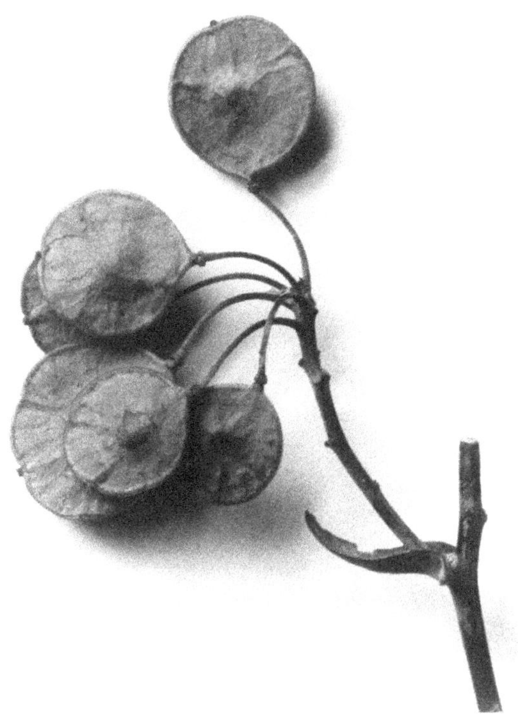

Figure 5.3

Karl Blossfeldt, *'Ptelea trifoliate* (Hoptree)', 1905–25

the explosion of interest in physics sparked off by new developments in the theory of the atom,[30] but also publications on the subject of biology that were sustained by fresh discoveries in the nascent fields of embryology, genetics and biochemistry. New findings in biology were commonly reported on by newspapers and weekly journals, and as novel methods of recording the inner workings of the natural world permitted the visualisation of such findings (and were often directly responsible for them), journalistic reports typically stressed the role of X-ray or microscopical technology in bringing forth innovative and exciting concepts in biology. From its inception, photomicrography had been closely connected to biology by allowing for the spread of scientific knowledge without recourse to either expensive illustrations or the cumbersome mounting and preservation of specimens.[31] That photomicrography provided both an image of nature otherwise imperceptible to the human eye as well as an apparently 'perfect and unerring record' made it particularly apposite to the scientific study of the physical world.[32] Arguably, such notions were integral to Modernist responses to photomicrographic technology.

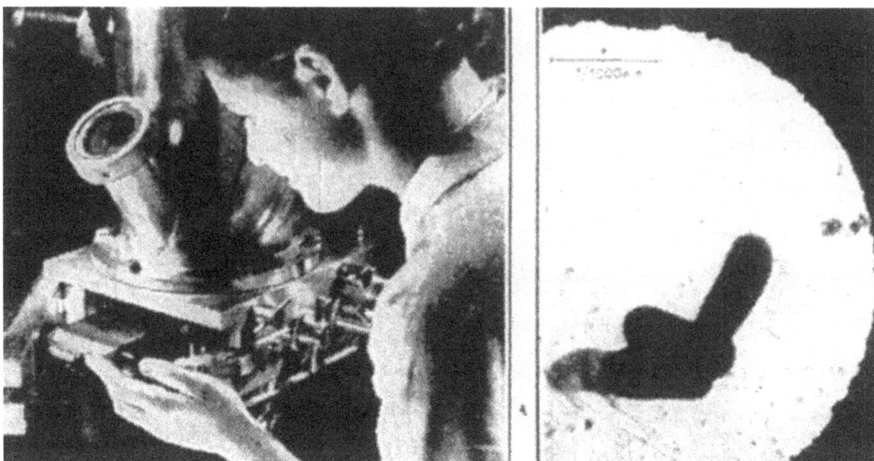

Figure 5.4

Illustration of the 'Super-microscope' and a photomicrograph of bacilli obtained
by this apparatus, printed in *The Times*, 26 November 1938

A 1931 article in *The Times* spoke praisingly of a 'technique for taking cinematographic films of living cells' so that the 'effects of irradiation on cancerous tissues' could be better investigated. As cell growth was normally too slow for practicable observation, the text stressed how continuous 'cinematography films opened new possibilities. The time factor can be manipulated so that rapid movements can be slowed down, making it possible to analyse every phase, or slow movements can be compressed so as to transform changes imperceptible to the eye into a continuous progress'.[33] The unveiling of the electron microscope in 1938 occasioned the eye-grabbing headline of 'Key to a Hidden World: New Light on the Infinitesimal' and was described as giving 'an infinitely deeper insight into the vast world of micro-organisms and micro-inorganic matter than has hitherto been considered possible'.[34] By providing much greater magnification, the electron microscope moved far beyond conventional technology, 'revealing organisms, such as viruses and bacterial individualities, which no human eye has ever beheld in the past'. Indeed, the wizardry of the microscope was seen to lie precisely in its ability to successfully bring together various schools of optics, most notably 'ultraviolet rays, photography and microscopy into collaboration' so as to be able to view even 'the minute particles of colloid chemistry'.[35]

Alongside such thrilling reviews of the insights gained from these technologies were a variety of illustrations and photographs: the *Times* report on the electron microscope included pictures of microscopical apparatus as well as a photograph of a bacillus (Figure 5.4).[36] That said, coverage of photomicrography typically veered between the edifying potential of the technology and its beguiling aesthetic qualities. On the one hand, the pedagogic capability of microscopy led educationalists to promote its usage as nothing short of a teaching revolution.[37] Yet on the other, photography journals striving to move beyond antiquated pictorialist values promoted photomicrography as an aesthetic marvel of the scientific age, lauding its clarity of technique and attention to detail.

Interwar photo-journals and the dissemination of the scientific 'close-up'

In the first edition of *Modern Photography* in 1931, G.H. Saxon Mills spoke of the 'development, scope and possibilities' of modern photography, arguing that in 'scientific observation, by means of x-rays or astronomical research, and in its amazing development in the cinematic form, its form is as yet hardly explored'.[38] Indeed, in stark contrast to the Spartan picture-editing policies of newspapers and literary reviews, photography journals rapidly became conduits through which photomicrographic images reached a wider audience, as editors sought to capitalise upon public curiosity in pictures of the physical world newly seen by science. Perceived as a symptom of the new photographic age, photomicrography featured prominently in the pages of photo-journals, and was often singled out by editors as representative of 'the various types of photography now practiced'.[39] Publications such as *Photography Yearbook* increasingly had sections devoted to 'scientific' photography, in which blown-up photographs of insects, crystals and plant life were printed alongside images obtained through radiography, infrared imaging, telephotography and aerial photography.[40]

Something of the popularity of micro- and X-ray photography among amateurs can be gleaned from the regular discussions of cutting-edge technical innovations – ranging from higher-powered lenses to new emulsion formulas for radiographic material – that took place in the yearly publication *Modern Photography*.[41] At the same time, journalistic enthralment with the workings of the physical world as revealed by science was mirrored in photo-journals singling out exhibitions featuring close-up images of nature as being worthy of special consideration. For example, Trude Weiss, reviewing the International Exhibition of Photography in 1932 for *Close-Up*, drew attention to the 'beautiful close ups of flowers and animals [...]. Very big enlargements of tissues, the human skin, flowers, never failed to have an amazing effect. [...] There were some excellent microphotographs of medical objects, as well as records of operations on the eye'.[42] Such enthusiasm was generated and sustained by newspaper accounts of the new visualising technologies, in which photography was deemed responsible for revealing the hitherto invisible makeup of nature: knowledge of the atomic composition of matter typically being communicated through reference to photographs of particle tracks produced in cloud chambers.[43]

The documentary cinema movement and the popularisation of microcinematography

Another agent responsible for disseminating images of photomicrography to the public (via microcinematography) was the cinema. The emergence of the documentary cinema movement in the 1930s had reinvigorated an industry that, in Britain at least, had been typically limited by a lack of investment and a commitment to 'theatrical' productions.[44] Short 'factual films' began largely as poor-quality 'fillers' for main cinematic features, though by the early 1930s this position had begun to change, as filmmakers such as John Grierson and Paul Rotha saw documentary as a potential source of educational outreach and genuine socio-economic change.[45] Although scientists such as John Russell Reynolds had begun making films to aid clinical diagnosis and by 1921 Kodak had begun collecting a library of medical films, the Commission on Educational and Cultural Films commented in a 1932 report, *The Film in National Life*, that Britain had largely failed to make much use of film for scientific recording and research.[46] An early exception to this rule had been the 'Secrets of Nature' series, which had been produced by the company British Instructional: an organisation that made a number of high-quality documentaries at this time for general release in cinemas. By the end of 1929 almost a hundred such films had been shown, to popular acclaim. An influential contributor was Percy Smith, who had experimented successfully with making time-lapse films of plant life, which could show the process of growth accelerated by up to 96,000 times.[47] The series continued into the 1930s and Gaumont–British Instructional (G-BI) was successful in encouraging the participation of leading biologists in the production of its nature films: the biologist Julian Huxley became biological films advisor to G-BI and helped to produce films such as *The Private Life of Gannets* (1934) alongside a variety of films about heredity.[48]

On the one hand, microcinematographic films of nature were seen to possess an entertainment value, which was reflected in their popular success at the box office. Yet on the other, the films were acknowledged as being valuable to science through faithfully recording the processes of nature.[49] Natural history features were seen as effective educational aids and used interchangeably (albeit somewhat inconsistently) as both classroom tools and as crowd pleasers in the cinemas.[50] Part of this belief in the pedagogic value of documentary and natural history film was due to a firm conviction, on the part of scientists, that science had a social value.[51] Yet pedagogues naively felt that, as an 'impartial' record, film provided an engaging and informative supplement to the routine of classroom teaching by animating the subject and allowing it to be viewed 'naturally' within its own environment.[52] As early as 1929, a review in *The Times* remarked upon the growth in the number and quality of educational films in Britain and noted that the catalogue of British Instructional Films made 'interesting reading' by detailing documentary films 'showing the habits of birds, the lives of insects [and] the

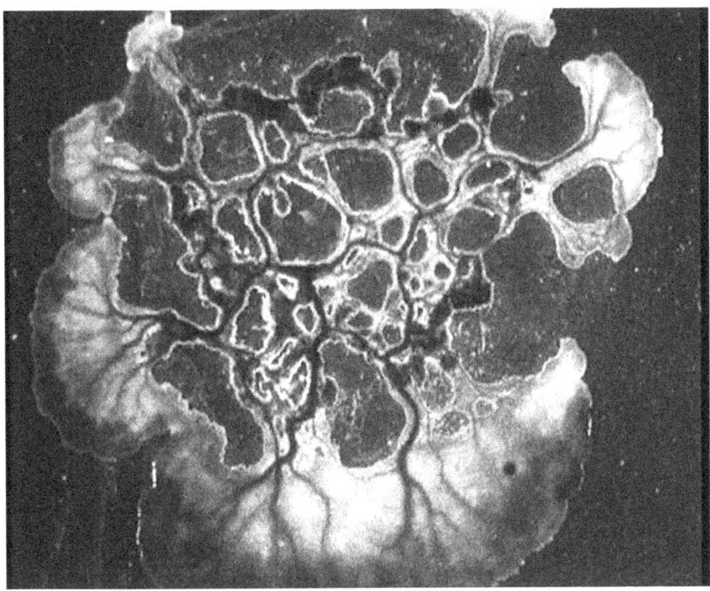

Figure 5.5

Film still of slime fungus from the series 'Secrets of Nature', *Magic Myxies*, 1931

growth of flowers'. The botanical films were deemed especially praiseworthy in their detail and focus: '*The Romance of Flowers* shows the way in which flowers distribute their pollen and in *Plant Magic* the audience can see how the elements gathered through the roots and cells of the plants are turned into grape sugar'.[53]

The long-running and commercially successful 'Secrets of Nature' series was one of the major conduits through which microcinematography was broadcast to a wider audience. Although the series encompassed a wide variety of natural history subjects, including films about animals at the zoo, bird habits and meteorology (to name but a few), a large minority of 'shorts' were microcinematographic films dealing with subjects such as pond life, germination and insect life. *An Aquarium in a Wineglass* (1926), for example, showed in microscopic detail how a countless and diverse number of infusoria spawned from the placing of a piece of hay in a wineglass. Likewise, *Magic Myxies* of 1931 showed the life cycle of a slime fungus in close-up, speeded up to take place within a ten-minute time span (Figure 5.5). While there was occasionally some criticism of the veracity of the commentary,[54] in general 'Secrets of Nature' was well received; the critical success of these films may partly be explained by the fact that they were made to high production standards, with the ultimate aim of showing them in cinemas.[55] A *Times* article of 1931 spoke of the popularity of these 'nature' pictures in the course of explaining the significance of making films of 'living cells' for medical research.[56]

Something of the ground-breaking and 'avant-garde' character of the cine-matography employed by the 'Secrets of Nature' production crew can be gauged from the fact that the radical film journal *Close-Up*[57] published a 1931 article entitled 'Personally About Percy Smith' in which the techniques and philosophy of the 'Secrets of Nature' team were expounded (Smith had been an early experimenter in microcinematography and, in collaboration with the filmmaker Mary Field, had been responsible for many of the triumphs of 'Secrets of Nature').[58] *Close-Up* viewed itself as a channel for experimental Modernist criticism and its proactive engagement with microcinematography can be regarded as a litmus test for the positive way in which the avant-garde responded to photomicrography at the cinema.[59] Microcinematography was regularly promoted within its pages, with films such as the silent documentary film *Germinating Life* – which provided a microcinematographic account of human prenatal growth – being praised for qualities like providing 'very fine shots on blood-circulation' and being 'absolutely serious and, unlike so many [other] popular scientific films, free from romance and mysticism'.[60]

Beyond the undisputed merits of the cinematography, the politicisation of biology during the 1930s – through the agency of the increasingly prominent British Eugenics Society – served only to impress the importance of microcinematography, as a scientific technology with social implications, on to the mind of the public.[61] A 1938 propaganda film produced by G-BI for the Eugenics Society, for example, entitled *From Generation to Generation*, incorporated microcinematography to demonstrate the physiology of reproduction. The film explained the workings of Mendelian genetics by interspersing images of 'pedigree' human beings with sequences of 'defectives', with Julian Huxley – a prominent eugenicist – providing the commentary. Huxley concluded his narrative with a paean to eugenic hygiene: 'If we are to maintain the race at a high level mentally and physically, everybody sound in body and mind should marry and have enough children to perpetuate their stock and carry on the race'.[62] The interplay between microcinematography's popular appeal and its political uses, framed through the proselytising activities of scientists, the popular-science industry and documentary cinema, ensured that photomicrographic imagery continued to exercise the public's imagination throughout the 1930s.

Art Forms in Nature and the photomicrographic photo-album

Upon its publication, Karl Blossfeldt's photo-album *Art Forms in Nature* (origin-ally published in Germany by Wasmuth of Berlin under the title *Urformen der Kunst* in 1928) was something of a publishing phenomenon, striking a chord with the general public and going through no fewer than three print-runs between 1929 and 1935.[63] Sub-titled, *Examples from the Plant World Photographed Direct from Nature*, the book (and its re-editions) featured a suite of large (320 × 250 mm),

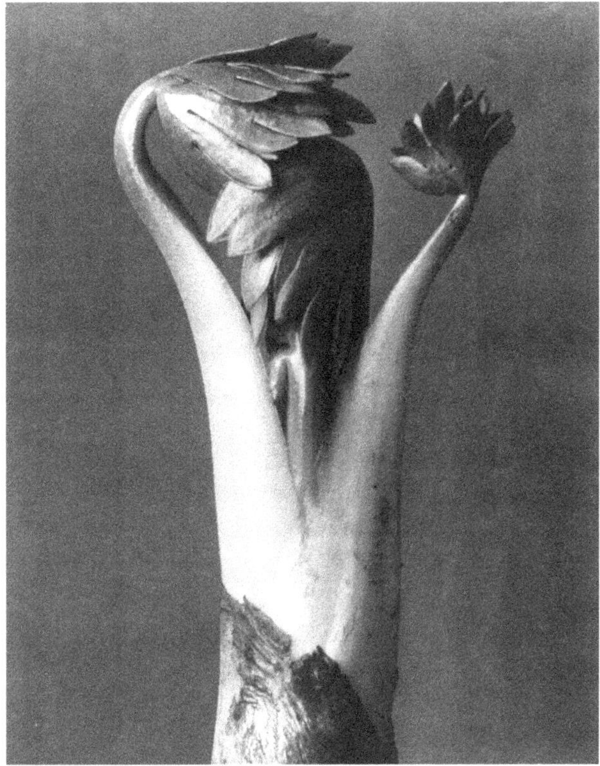

Figure 5.6

Karl Blossfeldt, '*Aconitum* (Monkshood)', 1900–26

full-page, black-and-white macrophotographic plates of 'plant forms' (typically, flower heads, stems, leaves, burrs and tendrils), most magnified between three and fifteen times (Figure 5.6).[64] The response to Blossfeldt's photo-book was unprecedented: an editorial in *The Times*, three pages of coverage in the *Illustrated London News* and substantial reviews in major publications such as the *TLS*, *The Listener* and the *Architectural Review* (not to mention smaller newspapers such as the *Manchester Guardian*). The journal *Architectural Review* described it – in nature-centric terms – as an 'enchanting collection of plates [that] gives us a rich compendium of "Historic Ornament" from a new angle',[65] while *The Listener* called it 'a book to pour over', the photographs inspiring 'not only admiration from the nature-lovers, but also study from many an artist'.[66] The 1932 co-edition elicited comparably enthusiastic praise from pundits: *Close-Up* promoted it to Modernist audiences as 'a superb affair. Better than the first, for the photos are more spread in their jewelry'.[67] As Guitemie Maldonado has remarked, the popularity of Blossfeldt's photographs represented part of a broader Modernist fascination with botanical photography – promoted initially under the auspices of the New Objectivity movement in Germany – which led to magnified vegetable forms being published and exhibited widely across Europe.[68]

Critics habitually interpreted Blossfeldt's photographs as invaluable blueprints for design, *The Listener* entreating that '*Art Forms in Nature* should find its way into the reference library of every art-school worthy of the name'.[69] In this respect, the reception of Blossfeldt's book recapitulated a long-standing aesthetic discourse, stretching back to Art Nouveau and the Arts and Crafts movement, that encouraged artists to study nature as a means of by-passing classical and Renaissance precedent and academic tradition.[70] Modernist commentators, in particular, stressed the closeness of Blossfeldt's plant forms to embroidery and ironwork, and how they positively contributed to design theory.

The pedagogic utility of *Art Forms in Nature* stemmed from the character of Blossfeldt's practice itself, the original aim of which was didactic, in providing photographs as blueprints for his students' plant-modelling workshops.[71] In this respect, it had an important antecedent in Haeckel's decorative lithographic plates of micro-organisms and plant life, which, a quarter of a century earlier, had acted as a design sourcebook for the Art Nouveau movement[72] – a point not lost on Blossfeldt's London publisher, who used a direct translation of Haeckel's title for the English edition of *Urformen der Kunst*.

Art Forms in Nature formed part of a tranche of books produced mainly by Bauhaus faculty members that were published in Britain throughout the 1930s. Many of these texts (such as László Moholy-Nagy's *The New Vision: Fundamentals of Bauhaus Design, Printing, Sculpture, and Architecture*) were intended as educational aids and often included photomicrographs among their illustrations. Generally, these publications dealt with aspects of the contemporary form–function debate by laying emphasis upon the acquisition of a knowledge of materials in order to implement more successful design techniques. As a result, characteristically scientistic values often found expression in Bauhaus texts that sought to nurture an empirical attitude so as to develop an all-round aesthetic sensibility in students. Albers' plea that 'through the study of the problems of materials, we acquire exact observation and new vision' was intended to focus student concerns onto the structural characteristics of physical objects.[73] Photomicrography was promoted as an indispensable tool in this respect as it enabled students to witness the smallest units of creation. Moholy-Nagy spoke of the 'microscope and microphotography disclos[ing] a new world. They reveal, in this age of haste and superficiality, the marvel of the smallest units of construction'. For Moholy-Nagy, photomicrography afforded students an accessible and effective opening into 'a new education in materials'.[74]

The acuity of vision proffered by *Art Forms in Nature* appeared to many to be symptomatic of the range of sight opened up by the new visualising technologies of the interwar years that had been announced to considerable fanfare by the press. Reviewers emphasised how the photographs were dependent upon new technology, and in being so, 'produce[d] an amazing sensation of looking into a new world, curiously familiar, yet remote from human handiwork'.[75] Likewise, the artist-critic Paul Nash claimed that *Art Forms in Nature* was 'an intensely interesting example of the peculiar power of the camera to discover formal beauty which ordinarily is hidden from the human eye'.[76] For Nash, the very significance of the photographs lay in the technology itself, without which humankind would remain forever ignorant of the underlying intricacy of the plant world:

> For it is the camera eye directed by acute human perception which is responsible for these remarkable observations. Actually these important forms do not exist for our vision except by virtue of a mechanical scientific process.[77]

In a pantheistic eulogy that re-rehearsed some of the metaphysical questions raised by popular accounts of the new physics, *The Times* concluded, in light of *Art Forms in Nature*, that:

> The bright star may have been proved not to be steadfast after all; the flower in the crannied wall may be on the point of yielding up its last secret; and, after neighbouring the invisible so close for centuries, man may see the spirits leap from trees and flowers only to find them very different from what he expected.[78]

Readings of this sort were encouraged by Karl Nierendorf's preface to the photographs, which spoke of a growing awareness of nature brought about through sport, the new architecture and 'film, [which] thanks to the time-lapse cine-camera, [allows humankind to] watch the swelling and shrinking, the breathing and the growth of plants. The microscope reveals whole systems of life in drops of water, and the instruments of the observatory open up the infinity of the universe'.[79] The pseudo-scientific qualities of Blossfeldt's images were heightened by the book's subtitle, which stressed an empirical agenda and by the fact that through carefully selecting and enlarging various morphological details so as to create a typology of plant forms, Blossfeldt's photographic practice outwardly linked to Enlightenment classificatory systems. There is a strong visual resemblance, for example, between Blossfeldt's work and the visual glossary presented in Carolus von Linnaeus' *Philosophia botanica* of 1751, in which the eleven plates collectively produce a typology of all the structural variations of each part of the plant.[80]

Between 1929 and 1935 a spate of 'art books' on the subject of photo-micrography were issued as publishers sought to capitalise upon popular interest in the physical world seen in close-up. Two notable publications, in this respect, were *World Beneath the Microscope* (1935) and *Snow Crystals* (1931). Both publications adopted a similar format to Blossfeldt's book – albumen or silver gelatin prints of magnified natural forms on high-quality, glossy and largish (250 × 180 mm) paper – even if the subject matter ultimately varied (focusing on micro-organisms and enlarged snowflakes, respectively) (Figure 5.7).[81] Ranging from the averagely priced (starting at 5*s* for *World Beneath the Microscope*) to the relatively expensive (£2 2*s* for *Art Forms in Nature*), the publication of these photo-albums suggested

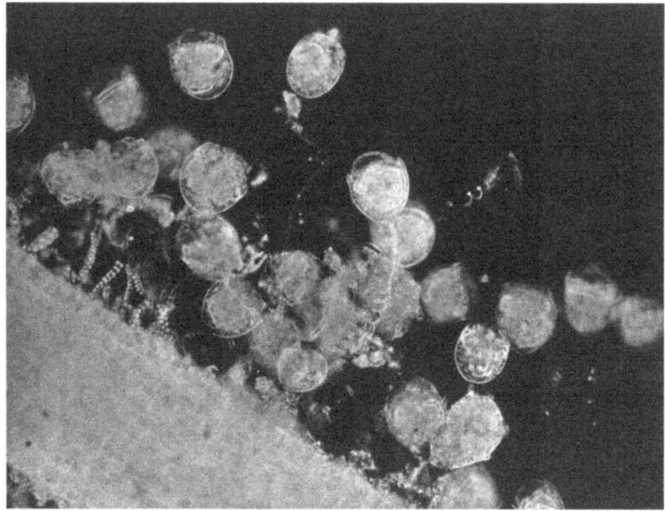

Figure 5.7

W. Watson-Baker, 'Photomicrograph of *Vorticella*', Plate 29 in *World Beneath the Microscope* (London: The Studio, 1935)

the willingness of publishers to gamble on photomicrography's growing appeal to the public. To be sure, the publisher Zwemmer put a great deal of time and effort into publishing and promoting the English co-edition of *Art Forms in Nature* by liaising with the book's German publisher as well as cultivating contacts at newspapers in the run-up to its release.[82]

Like *Art Forms in Nature*, *World Beneath the Microscope* and *Snow Crystals* were curious hybrids – superficially 'art' books targeted at amateurs and yet simultaneously books of 'science' that aimed at wowing readers with new microscopical discoveries. *World Beneath the Microscope*, for instance, included an introduction by W. Gaunt that stressed the usefulness of photomicrography to art alongside a highly technical essay by W. Watson-Baker that detailed the practical aspects of photomicrography. In his introduction, Gaunt accentuated the bioromantic possibilities of photomicrographic imagery, entreating the artist to approach Watson-Baker's photo-album as a sourcebook for contemporary design:

> The artist feels what the scientist calculates. As serious writers have abandoned the superficial devices of the picaresque novel and begun to explore the mind [...], the visual artist is urged to exchange the delineation of outward appearance for a nearer examination of form. The symmetrical exemplars of textbooks on design prove to be of the utmost banality when compared with the enormous repertoire of design in nature, a dimensional book which discloses repertoire within repertoire, cosmos within cosmos.[83]

Throughout *World Beneath the Microscope*, explanatory captions hinted at aesthetic parallels that could be drawn between the products of nature (as visualised

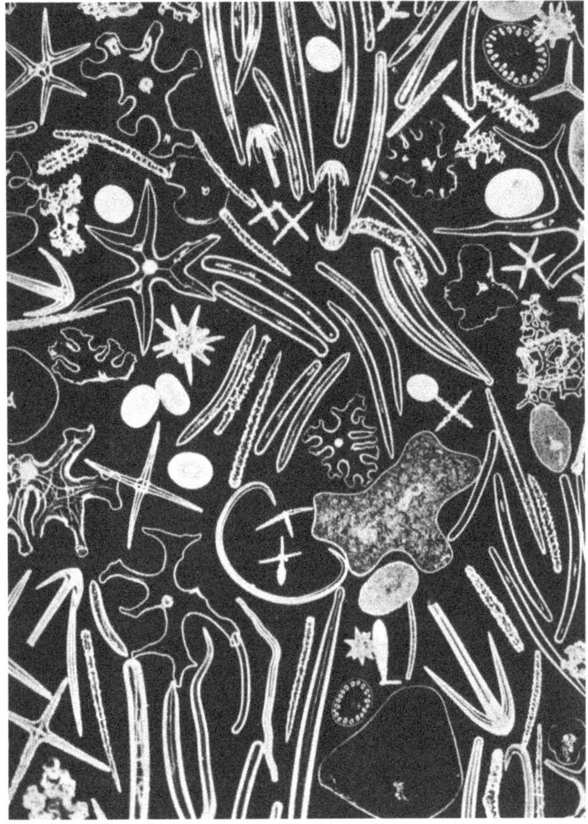

Figure 5.8

W. Watson-Baker, 'Photomicrograph of "Sponge Spicules"', Plate 7 in *World Beneath the Microscope* (London: The Studio, 1935)

by science) and the works of the artist. An image of some 'Sponge Spicules' (Figure 5.8), for example, included the exhortation that '[the] lavish invention of nature is manifest in a granule of sponge. Has the abstract painter of today achieved anything more interesting than this evolutionary design?'[84] This epistemological ambiguity led to some criticism of the 'pompous' analogies made by publishers between close-up photographs of nature and modern art,[85] yet, at the same time, reviewers frequently voiced frustration at the publications not appearing 'scientific' enough – the *TLS* bemoaning that 'though the wish to keep the plates free from typography is understandable, it would have been a help to appreciation if the subject and degree of magnification had been printed below them'.[86] Overall, however, reviewers saw no inherent contradiction in extolling the aesthetic virtues of these 'art-house' publications while simultaneously marvelling at their reputed scientific insights.

The power of the camera's eye: Blossfeldt, Nash and Wilenski

The critical reception that greeted the British publication of Karl Blossfeldt's macrophotographs served to galvanise Modernist interest in the hidden architecture of nature as it was revealed by scientific technology and, in so doing, greatly assisted in the popularisation of bioromantic concepts within the British avant-garde community. Although the 1929 edition was well received, it was only after Zwemmer published, in 1932, Blossfeldt's *Wundergarten der Natur* as the second volume of *Art Forms in Nature* and Robert Wellington curated an exhibition at the Zwemmer Gallery of Blossfeldt's photographs alongside reproductions of plant forms in the applied arts that Modernists – most especially Paul Nash and the critic Reginald Wilenski – began to theorise on the aesthetic rapport between modern art and natural form.[87] While exhibitions such as the 'Neue Wage der Photographie' organised in Jena in 1928 and the Deutsche Werkbund's 1929 'Film und Photo' had already introduced the new scientific imagery to European audiences,[88] the exhibition of Blossfeldt's plates in London was the first show of its kind in Britain and acted as a revelation in exposing, by technological means, the concealed aesthetic properties of natural form. In a 1932 critique of Blossfeldt's macrophotographs – which accompanied their British publication and exhibition – Nash forwarded a bioromantic hypothesis of the new scientific imagery, perceiving the basis for a meaningful encounter between technology and modern art:

> Are we to suppose that the old artists derived inspiration from minute examination of natural phenomena? Obviously, in certain cases, natural forms have supplied a *motif*, but in many it would have been impossible to detect the significance of natural design without the aid of a mechanical process. This is where the camera's 'eye' proves its incalculable power, but not as an archaeological, botanical or merely curious discoverer of 'interesting' comparisons between art and nature; its importance lies, surely, in the wealth of matter it places at the disposal of the modern sculptor or painter, which may prove stimulating to [...] a sound expansion in the realm of art.[89]

Nash's reaction to *Art Forms in Nature* was informed by the publication, just a few months earlier, of Wilenski's associationist treatise on Modernist sculpture, *The Meaning of Modern Sculpture*.[90] The central, idealist premise of this text was that the ordered appearance of the plant forms in Blossfeldt's macrophotographs had discredited the romantic paradigm of a 'wild', undisciplined nature by showing it to be inherently geometrical:

> It is impossible to sustain the Romantic notion of a wild, free, ragged, 'nature' when we examine the enlarged photographs of plant forms in Professor Blossfeldt's [...] *Art Forms in Nature* [...]. These photographs transform the apparently ragged constituents of a tangled hedgerow into a series of structures informed with a most definite shape, with most evident order and with most evident logic. The study of these photographs makes it quite clear that the artist who reacts to the superficial tangled

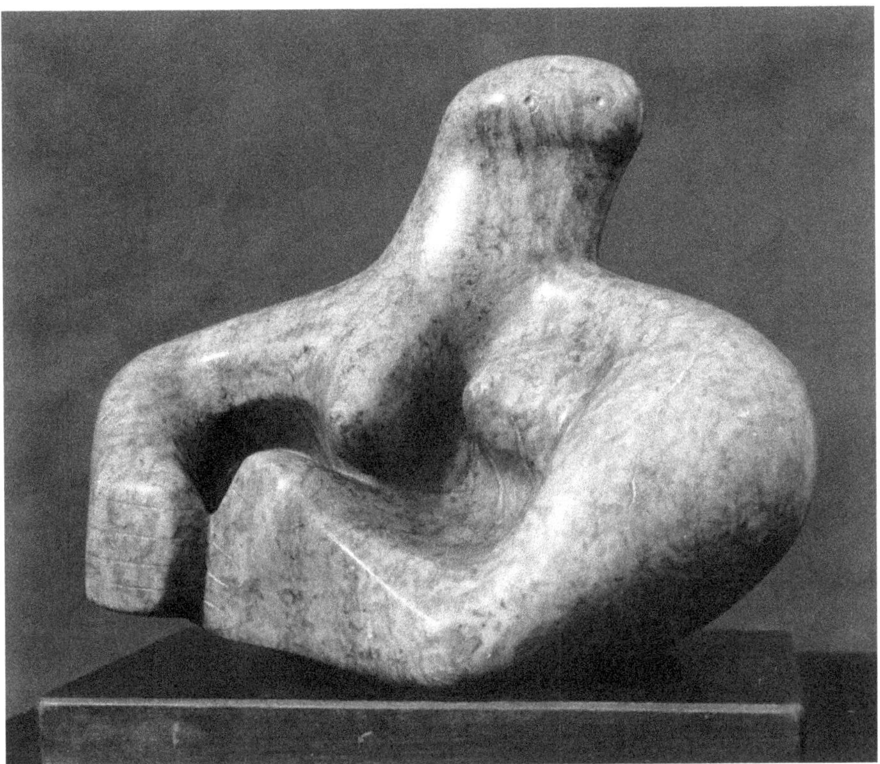

Figure 5.9

Henry Moore, *Composition*, 1931. Henry Moore Foundation

appearance of nature [...] is producing the kind of art which Socrates described as 'only conjecture [...]'; and that the artist who gives us truth to form is the artist who symbolizes the forms contained in the hedgerow by geometric forms which Socrates *for that reason* described as 'eternally and absolutely beautiful'.[91]

Applying a psychobiological approach to his conception of artistic form, Wilenski concluded, in light of Blossfeldt's photographs, that:

[Even] the most drastically nonrepresentational work of art, if produced by an artist who consciously or unconsciously apprehends the formal character of natural structures, may be 'true to nature' in this larger sense of the symbolization of natural form.[92]

Wilenski's argument was based upon the similarity of form between Blossfeldt's enlarged photographs of blooms and seed heads and the stylised regularity of Modernist sculpture, most especially the biomorphic work of Moore (Figure 5.9) and Bedford (Figure 5.10).[93] This method of analogising between macrophotography

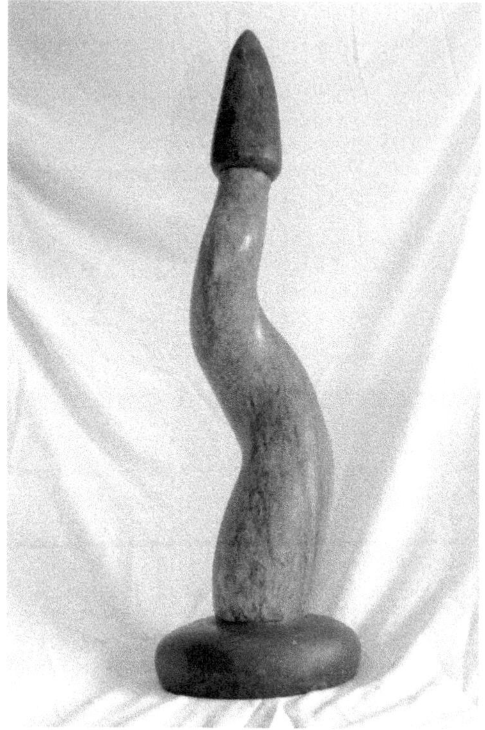

Figure 5.10

Richard Bedford, *Black Tulip*, c.1936. Private collection

and modern art was merely one offshoot of a larger, European phenomenon – originating in the writings of biocentric artists like Wassily Kandinsky and László Moholy-Nagy – that Oliver Botar has described as 'the scientific image analogy'. A scientistic conveyance of Kállai's notion of bioromanticism, this analogical approach to the relationship between the new scientific imagery and artistic Modernism emphasised the value of technology to art by comparing microscopic, macroscopic, X-ray, telescopic and aerial images to abstract art.[94] Kandinsky pioneered this analogical conception of scientific and artistic imagery during the 1920s and early 1930s by collecting images of 'animals, airplanes [...], objects shown under high magnifications and subjects that could generally be characterized as "technology" and "nature"' in scrapbooks and comparing artistic and scientific imagery in his Bauhaus lecture series of 1931.[95]

Wilenski's strategy of comparing Blossfeldt's macrophotographs to Modernist sculpture had a parallel in earlier critiques of Blossfeldt's work, which first emerged in continental Europe following the German publication, in 1928, of *Urformen der Kunst*. In a 1928 review, for example, the critic Walter Benjamin not only analogised between Blossfeldt's photographic enlargements and architectural design but also explicitly linked biomorphic Modernism to the revelation of microscopical imagery:

> Even the most impassive observer would be thrilled to see that the enlargement of parts of plants visible to the eye could be as extraordinary as plant cells glimpsed through a microscope. When we remember that Klee and, even more, Kandinsky worked for so long on the elaboration of forms which only the intervention of the microscope could – brusquely and violently – reveal to us, we notice that these enlargements of plants also contain original stylistic forms. In the crosier depicted by the fern, in the larkspur and in the blooms of saxifrage, we see forms reminiscent of the tracery in the rose windows of cathedrals.[96]

The willingness of Modernist critics to interpret the plant forms revealed by Blossfeldt as emblematic of the prototypical forms from which all objects derived had a foundation in Blossfeldt's practice. Echoing Goethe's morphological interest in finding the archetypal plant, Blossfeldt's interest lay in discovering an 'archetypal art' which corresponded to the nineteenth-century architectural theory of 'basic form' propounded by Gottfried Semper, a holistic concept which explained the states wherein the form and function of an object were one.[97] Blossfeldt saw this unity in plants, noting – in an unpublished essay of 1929 – that plants 'build and shape themselves functionally, converting everything into artistic form'.[98] In this fashion, the reception of his photographic enlargements overlapped with Modernist interest in Raoul Francé's theory of biotechnics, in which the archetypes of every human technology were believed to be present in nature and which suggested that designers and artists should thus look to the organic world for creative inspiration.[99] As Moholy-Nagy explained in *The New Vision*: 'How naïve is the fear of the "cold intellectualism" of form merely according to purpose, if we but learn to observe the meaningful forms according to purpose in nature [...]. An understanding becomes easier as soon as we attempt to grasp the essence of an organic form, which is the crystallization of its function'.[100]

Despite sharing with Wilenski a propensity to analogise between macro-photography and art, Nash disregarded Wilenski's theory that Blossfeldt's photographs had opened a window onto nature's design sourcebook and divulged a logical, fundamental order that contemporary art instinctively obeyed. While Wilenski's idealist, associationist interpretation of Blossfeldt was 'sympathetic' to Nash's interests, Nash chose instead to forward a highly anti-rational reading of Blossfeldt's photographs that more closely followed that of a 'dissident' Surrealist like the philosopher Georges Bataille.[101] Far from providing 'indisputable evidence of definite, sculptural order most un-wild', Nash saw 'the manifestations of modern photography not only [supporting] the statements of many so-called "perverse" sculptors and painters, but [running] parallel to and, to a great degree [influencing] the course of modern art'.[102] Here 'perverse' functioned as a kind of coded reference for Surrealism, which contemporary critics often spoke of as being in some way psychically distorted or imbalanced.[103] Describing the 'irrational lop-sidedness' of Surrealism, for example, the abstract painter Eileen

Holding claimed that it was 'as destructive of aesthetic value as the equally unbalanced intellectuality that it sets out to neutralize'.[104] In this manner, Nash's technophilic, bioromantic analysis anticipated J. and M. Thwaites' infamous 1936 assertion that Modernists such as Arp and Moore had 'arrived at Surrealism by the back door' precisely through developing a form of sculptural practice that eschewed psychic automatism in favour of a vital form of plastic and material expression.[105] Certainly, by making this argument, Nash may well have been thinking of Bataille's celebrated 1929 *Documents* essay on Blossfeldt's macrophotographs – 'The Language of Flowers' – in which the philosopher posited that, far from being graceful paradigms of virtue, flowers – when viewed under a magnifying lens – were actually quite repugnant. 'The most beautiful flowers', Bataille wrote, 'are spoilt at the centre by the hairy stain of sexual organs. It is thus that the interior of a rose corresponds in no part to the beauty of its exterior, as if one pulls away the petals to reveal the corolla, there remains nothing more than a tuft with a sordid appearance'.[106]

Similarly, in an essay on 'The Big Toe' (1929), Bataille sought to undermine systems of aesthetic value precisely by highlighting the perverse dichotomy of 'high' and 'low', in which everything that is regarded as lofty and elevated is idealised, whereas all that is 'base' is maligned and subject to scorn. Explaining how an inconsequential and ignored part of the anatomy could become the focus in humanity of a 'rage of seeing oneself as a back and forth movement from refuse to the ideal, and from the ideal to refuse', Bataille spotlighted how a small organ could unleash a torrent of disgust and self-loathing simply due to its 'base' position on the lower extremity of the body – a discomfiting reminder, as it were, of humankind's lowly origins in the soil.[107] Jacques-André Boiffard's accompanying macrophotographs rammed this philosophical point home (Figure 5.11). Featuring a big toe that had been visually isolated from the body and significantly enlarged, these photographs emphasised the potentially disturbing nature of magnification by picturing a toe that appeared malformed, with a cracked and grossly discoloured toenail – precisely the somewhat nauseating qualities that Bataille understood to cause rage to be 'directed against an organ as *base* as the foot'.[108] The harshly cropped and brightly lit character of the images equally acted as a parody of the New Objectivity school of photography – for while the New Objectivity school valued such techniques for their analytical clarity and 'anti-emotional nature'[109] – Boiffard's use of the close-up deliberately aimed to provoke the viewer into an emotive response due to the disturbing effect of the very small (and hitherto unnoticed) rendered disproportionally large.

Exercising an incomparable interest in the disquieting potential of the photographic close-up, *Documents* pursued an aggressive policy of de-familiarisation in which its readers were introduced to an iconography of extreme magnification as a means of disrupting the aesthetic codes conventionally adopted in the viewing of pictures.[110] Yet this de-sublimatory strategy was also extended to contemporary sculpture, which was routinely discussed in highly anti-formal terms as indicative

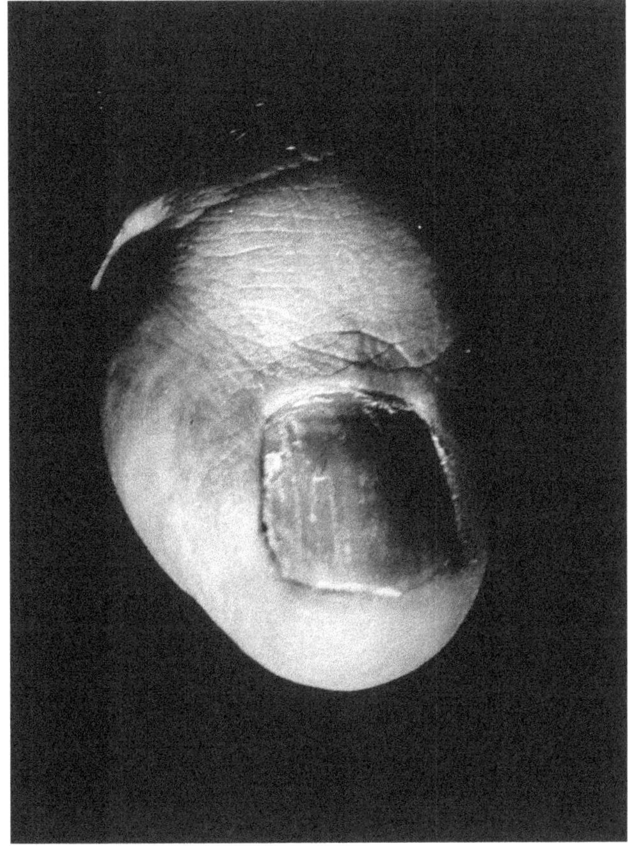

Figure 5.11

Jacques-André Boiffard, *Gros Orteil: Suject Masculin 30 ans*, 1929.
Musée national d'Art moderne – Centre Georges Pompidou, Paris

of a wantonly deformative and capricious aesthetic.[111] Critiquing a 1929 exhibition of Hans Arp's string and wood reliefs, for example, Michel Leiris spotted a tendency towards morphological mutability and classificatory incoherence which led to forms so alike one another in size and shape that any sense of taxonomic hierarchy was overturned.[112]

Nash's spotlighting of how the 'perversity' of modern sculpture explicitly rejected rational order hence suggests, at the very least, a cognisance of *Documents'* philosophical tenets. Indeed, elsewhere he appeared precisely drawn to those disquieting, de-sublimatory aspects of scale that appealed to the Surrealists. In the photo-essay *Monster Field* (1939), for instance, he demonstrated an inquisitiveness towards nature seen in extreme close-up, evoking, in a pungent description of decaying arboreal forms, the distasteful side of nature that Bataille had witnessed in Blossfeldt's macrophotographs of flowers. 'At a distance', Nash wrote, '[the trees] looked literally dead-white, but, at close range, their surfaces disclosed many inequalities of tone and subtle variety of ashen tints. Also in many places the bark still clung, a rich, dark, plum-coloured brown. Here and there the smooth bole, gouged by the inveterate beetle, let out a trickle of yellow dust which mingled with the red earth of the field'.[113]

The opposition thus staged between Wilenski's idealist critique and Nash's de-sublimatory, Bataillean analysis illustrates the philosophical breadth of biologistic interpretations of Modernist sculpture in the 1930s, whereby the new scientific imagery was employed by differing ideological factions to co-opt artists into divergent aesthetic groupings.[114]

Biology in the blood: photomicrography and Modernist sculpture

The syncretic attempt to find a 'biomorphic' or 'organic' middle ground between the opposing schools of Constructivist abstraction and Surrealism that Modernist critics – most especially Grigson and Read – undertook during the mid-1930s expanded the bioromantic discourse that had emerged in response to Blossfeldt's macrophotographs by yoking the philosophical tenets of neo-vitalism to the aesthetic values of a specific kind of micrographic imagery.[115] Indeed, the new age of mechanical objectivity – spearheaded by European scientists such as the histologist Erwin Christeller – engendered microscopical apparatuses (such as the electron microscope) which permitted the enlargement of organic structures by the factor of several hundred.[116] The apprehension of microscopical form, magnified to a startling degree of clarity, allowed artistic Modernists to hypothesise a relationship between biomorphic Modernism and microscopical imagery that gave credence to the neo-romantic, biophilic notion that biomorphism, as a style, openly drew inspiration from the natural world. Indeed, it was the impression that biomorphist imagery derived, in part, from the encyclopaedias of the New Biology that gave the style label its specifically modern undertones and thus served to connect it, iconographically, to the paradigmatic science of the period.[117]

The evocative, abstract language of pleats, curlicues and curves that characterised biomorphic Modernism appeared – to critics and artists – to mirror the shapes of micro-organisms, cells and membranous structures and therefore symbolised a return, in some way, to embryonic origins and primordial nature.[118] While explaining how Moore's sculptures had a precedent in Constantin Brancusi's practice, Grigson noted that the latter's irregular, ovoid forms appeared to have a subtle kinship with single-celled organisms: 'It is Brancusi whose polished unicellular forms have been the basis for such different figures, more complex, more "impure" as those of Mr Henry Moore'.[119] Similarly, in a catalogue of 1936, the director of the Museum of Modern Art in New York, Alfred Barr, described Arp's sculptures in biologistic terms as 'a kind of sculptural protoplasm'.[120] For Barr, the *topos* of biomorphic Modernism was therefore the amoebic figure, a shape whose rippling, asymmetrical form most closely approximated the 'near-abstractions' that he classified as biomorphism's defining biophilic logo.[121]

The popularisation of the New Biology through such conduits as the publishing houses of Oxford University Press and Cassel & Co., photography journals and the cinema led to the dissemination of photomicrographic imagery on a previously unimaginable scale and – by heightening popular interest in the new scientific imagery – added vigour to comparisons between Modernist sculpture and microscopical form.[122] Analysing the 'organic' qualities of contemporary sculpture in an *Apollo* article of 1930, for example, the Scottish documentary filmmaker John Grierson considered the wider impact of microcinematography on the public psyche: 'I am apt to think that the cinema has done something to open our eyes [to organic life] in this respect, with its power of revealing the constructions of plant life, animal life, and all life together in motion. It would still be more accurate to say that biology is getting into our blood'.[123]

Grigson was one of the first British critics to explicitly analogise between artistic Modernism and the new scientific imagery,[124] and, like Grierson, pronounced that awareness of 'life' was changing due to the popularisation of biological imagery. The crude 'knowledge of life' which had characterised earlier periods in history had been enhanced by the New Biology, making it now a commonplace for a 'pictorial knowledge of life' to be acquired 'in the plates and diagrams of a biological textbook'.[125] In Grigson's interpretation, Moore's role as a 'biomorphist producing viable work' emphatically represented the degree to which biology had newly revealed the hidden forms of nature:[126]

> Rounded shapes by Moore may be related to a breast, or a pear, or a bone, or a hill, or a pebble shaped among other pebbles on a shingle bar. But they might also relate to the curves of a human embryo, to an ovary, a sac, or to a single-celled primitive organism. Revealed by anatomy or seen with a microscope, such things are included now in our visual knowledge. Art, or the forms of art, change with such knowledge.[127]

In his writings on art, Moore accentuated the modernity of his own work by presenting Modernist sculpture as a purposive response to the new scientific imagery. In a piece of writing for the *Architectural Association Journal* of 1930, he noted how the modern sculptor's work 'may be comparatively representational or may be as Music or Architecture are, nonrepresentational, but mechanical copying of objects and surrounding life will leave him dissatisfied – the Camera and the Cameo-graph have nullified this as his aim'.[128] By 1934, membership of the neo-romantic, avant-garde group Unit 1 caused Moore to justify Modernist practice in starker, technological terms by singling out the influence of specific, lens-based technologies on the contemporary sculptor: 'There is in Nature a limitless variety of shapes and rhythms (and the telescope and microscope have enlarged the field) from which the sculptor can enlarge his form-knowledge experience'.[129]

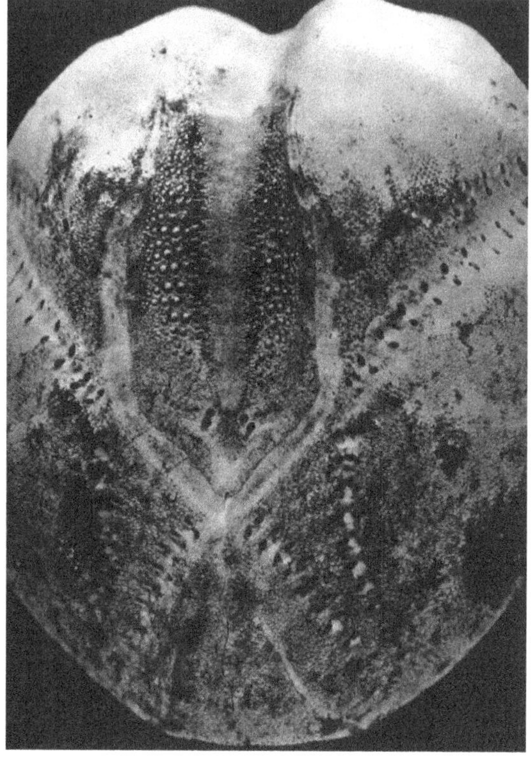

Figure 5.12

W. Watson-Baker, 'Photomicrograph of a Sea Urchin Shell', Plate 3 in *World Beneath the Microscope* (London: The Studio, 1935)

Moore's nature-centric, technophilic comment referred to a long-standing tradition in the visual arts that stretched back to the nature philosophies of the Enlightenment and to works such as Edward Young's *Conjectures on Original Composition* of 1759 – in which nature was seen as the model and justification for artistic originality.[130] Nevertheless, the opinion that the new scientific imagery had expanded the horizons of human perception and – as Haeckel's lithographs had at the turn of the century – engendered a dialogue between art and science encouraged commentators to openly analogise between the imagery of the New Biology and artistic Modernism.[131] A caption in *World Beneath the Microscope*, for example, that accompanied a photomicrograph of a sea-urchin shell, compared its swollen, bulbous form to a Modernist sculpture (Figure 5.12): 'The modern sculptor must envy the massiveness of form, the grandeur of contour, of this small shell, whose dovetailing makes a strange and interesting pattern'.[132] For the abstract painter John Piper, such imagery opened up a design sourcebook in which the motifs of modern artists were laid bare: 'It is amusing in fact to turn the pages [of *World*

Beneath the Microscope] and notice the artists suggested by the photographs: Klee (anchors and plates of Synapta), [Max] Ernst (a great many times), [Joan] Miró (sponge spicules), Giacometti (chemical crystals), and so on'.[133]

The sense that microscopy had collapsed the perceptual gap between artistic composition and biological imagery also led artists to employ microscopical metaphors as a means of emphasising the cognitive acuity of Modernist practice.[134] For instance, in his *Circle* essay – 'The Constructive Idea in Art' – Naum Gabo likened Cubism's fragmented aesthetic to a method of microscopical analysis which shunned the vagaries of external appearances in order to investigate – in biologistic terms – the 'inner mechanism' of objects:

> Cubism was a revolution [...]. The borderline which separated the external world from the artist and distinguished it in forms of objects disappeared; the objects themselves disintegrated into their component parts and a picture ceased to be an image of the visible forms of an object as a unit, a world in itself, but appeared as a mere pictural [*sic*] analysis of the inner mechanism of cells. The medium between the inner world of the artist and the external world has lost its extension and between the inner world of the perceptions of the artist and the outer world of existing things there was no longer any substantial medium left which could be measured either by distance or by mind.[135]

Such metaphors of 'penetrative' insight were opportune re-articulations of philosophical concepts first coined in the medical and scientific literature of the Enlightenment. The idea of unravelling the secrets of nature related to eighteenth-century medical discourse, which empowered a 'masculine' science to disrobe and scrutinise a 'female' nature – a concept illustrated by the use of wax anatomical models of women's bodies throughout this period as objects of medical enquiry.[136] The notion that a form of empirical knowledge was essential to the comprehension of nature subsequently legitimised the stereotyping of women as living objects 'brought under the penetrating enquiry of male reason'.[137] The legacy of this misogynistic bias in microscopical discourse is evident in the gendered description of nature given by J. Milton Offord in his 1933 presidential address to the Quekett Microscopical Club: 'Nature is an inexhaustible mine. There is far more to be discovered than has been hewn out of her secret recesses'.[138] By describing Cubist practice as a penetrative form of analysis, Gabo therefore aligned Constructivist theory with a well established rhetorical trope in scientific discourse, one which associated perspicacity of vision with misogynistic empowerment: '[T]he Cubist transfers the entire inner world of his perceptions with all its component parts [...] into the interior of the object penetrating through its whole structure, stretching its substance to such an extent that the outside integument explodes and the object itself appears destroyed and unrecognizable'.[139]

The intricate nature of Gabo's constructions – which comprised complex intertwinements of nylon and string filaments in such a way as to abstractly emulate natural structures – evinces the degree to which his art was influenced by micrographic imagery, most especially the illustrations in Thompson's *On Growth*

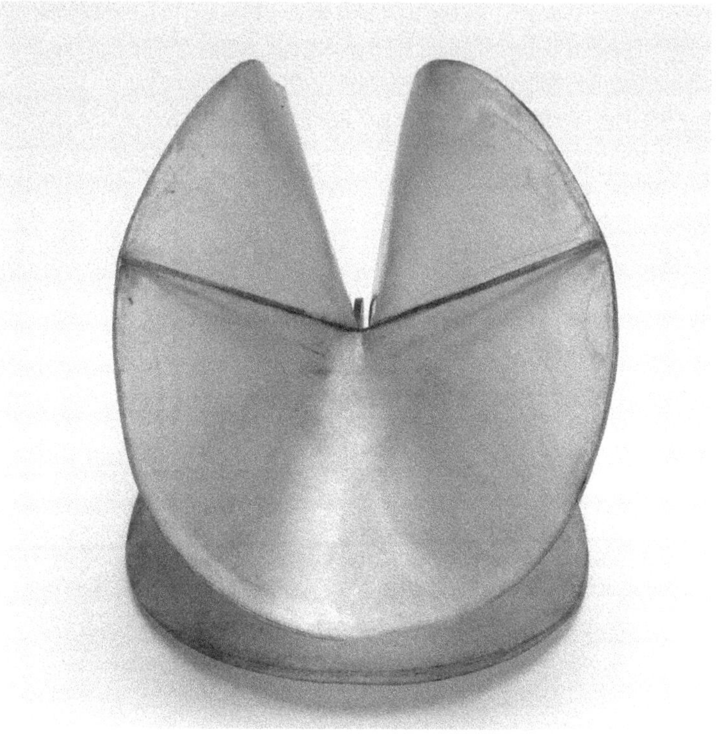

Figure 5.13

Naum Gabo, Model for *Spheric Theme*, c.1937. Tate, London

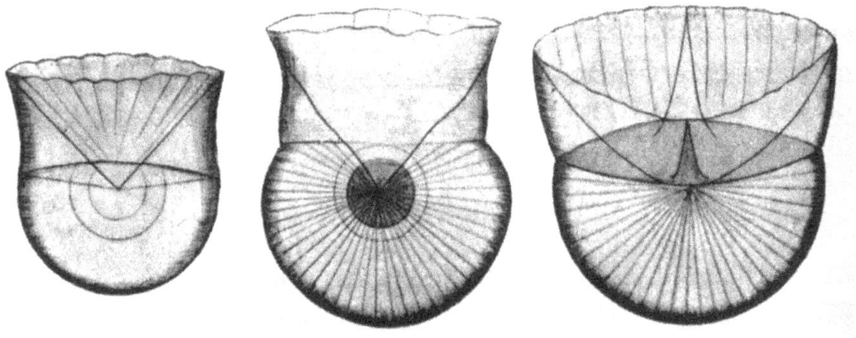

Figure 5.14

D'Arcy Wentworth Thompson, 'Illustration of Conostats', Fig. 302 in *On Growth and Form* (Cambridge: Cambridge University Press, 1942)

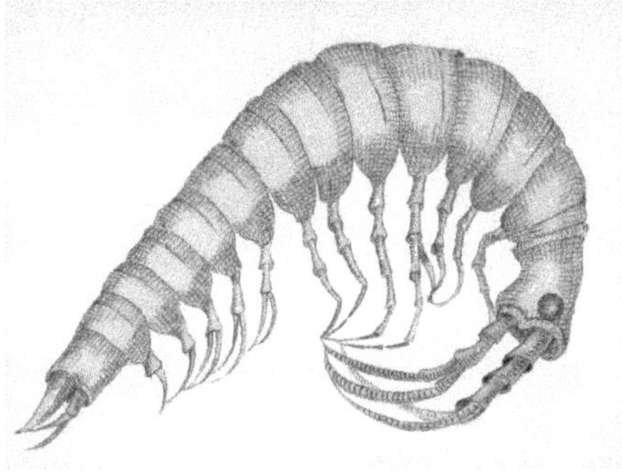

Figure 5.15

Richard Bedford, *Strange Creature 1 – Shrimp*, c.1938. Victor Batte-Lay Foundation Collection

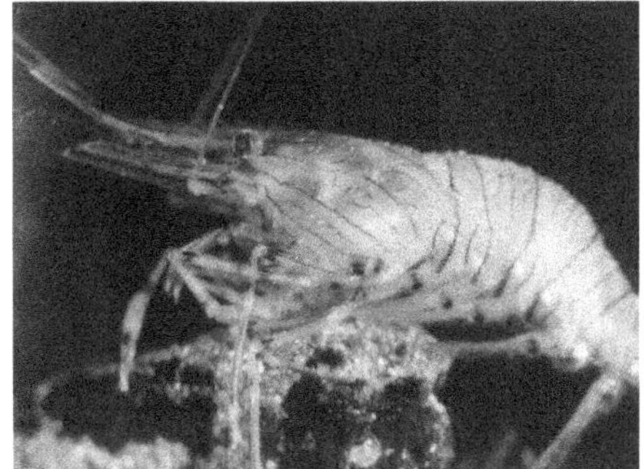

Figure 5.16

Jean Painlevé, *Crevette*, 1929. Les documents cinématographiques, Paris

and Form. The representation of volume in *Spheric Theme* (c.1937) (Figure 5.13), for example, echoes the veined, membranous structures Thompson used to illustrate the phenomena of spheroidal calcospheres and conostats (Figure 5.14), whereas *Linear Construction No. 1* (1942–49) (Figure 4.8, p. 147) closely resembles Thompson's images of the ventricular formation of radiolarian skeletons.[140]

Whereas the geometric character of Gabo's constructions straight-forwardly aligned his practice with biocentric Constructivism, the impact of photomicrography on Modernist practice was also apparent in the biomorphic, quasi-abstract work of Richard Bedford. Bedford produced a suite of watercolour drawings during the 1930s – entitled 'Portfolio of Shapes and Plants' – illustrating small creatures (such as insects, marine organisms and reptiles) rendered strange through a process of graphic enlargement. *Strange Creature 1 – Shrimp* (c.1938) (Figure 5.15), for instance, depicts the grotesque appearance of a shrimp or water-louse when perceived in extreme close-up – a representational strategy that explicitly evokes the microcinematographic footage of undersea life popularised by contemporary filmmakers such as Jean Painlevé and Percy Smith (Figure 5.16). Painlevé's work was much celebrated by the avant-garde during the 1930s, with Man Ray, in particular, using Painlevé's footage of starfish in his film *L'Etoile de mer* and Georges Bataille publishing Painlevé's stills of crustaceans in his journal *Documents*.[141] In this way, the knotted, snakelike form that appears in Bedford's *Creature (Snakelike)* (c.1938) (Figure 5.17) and the supple, tendrilous shapes of his biomorphic sculptures such as *Horned Lizard* (1936) (Plate 9) mirror the outlines of infusoria as they were shown in films such as *World in a Wineglass* (1931) (Figure 5.18). It was precisely this impression of a 'quickened consciousness of organic life', displayed by Modernist sculptors like Bedford, that led Grierson to speculate – in biologistic, neo-romantic terms – that, through the cinema, 'biology is getting into our blood'.[142] Similarly, while considering Moore's sculptures such

197

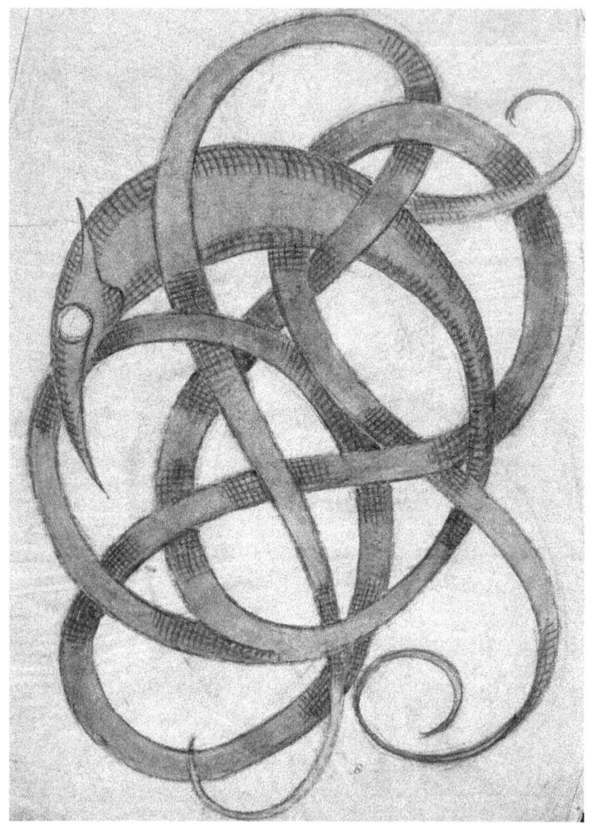

Figure 5.17

Richard Bedford, *Creature (Snakelike)*, c.1938.
Victor Batte-Lay Foundation Collection

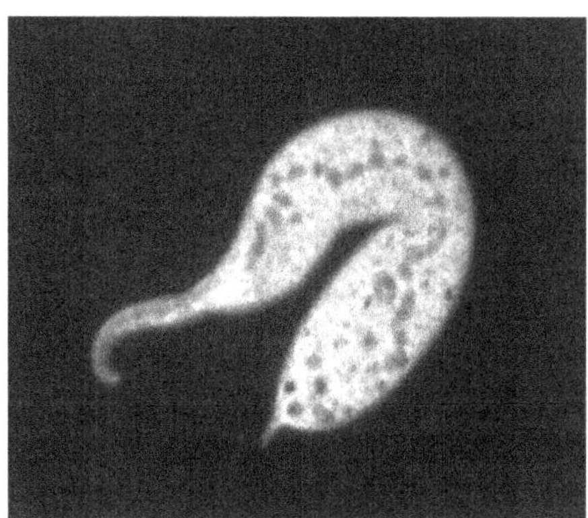

Figure 5.18

Film still from 'Secrets of Nature', *The World in a Wineglass*, 1931

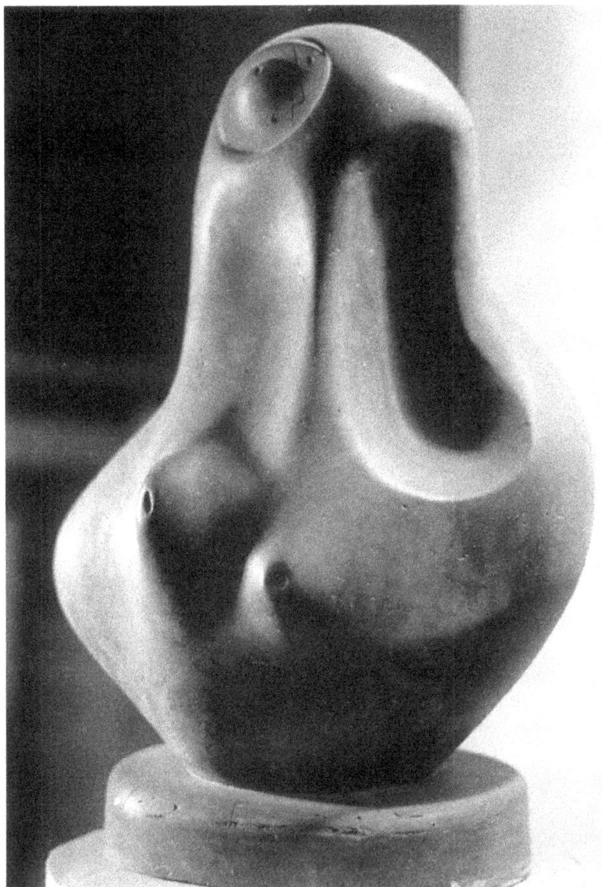

Figure 5.19

Henry Moore, *Composition*, 1933. British Council

as *Two Forms* (1934) (Plate 10) and *Composition* (1933) (Figure 5.19), Grigson was compelled to admit a close parallel between Moore's brand of biomorphic Modernism and the turgescent structures of micro-organisms: 'When I look at [Moore's] carvings I sometimes have to reflect that so much of our visual experience of the anatomical details and microscopical forms of life comes to us, not direct, but through the biologist'.[143] It was this apparent proximity to the imagery of the New Biology that gave biomorphic Modernism a veneer of contemporaneity and which distanced it from more traditional, neo-romantic conceptions of art's relationship to nature.[144] Thus, the new age of mechanical objectivity – ushered in by the new visualising technologies – allowed the neo-vitalist, biologistic sensibilities which informed British Modernist sculpture to find – in terms framed by Kállai's bioromanticism and Wilenski's biocentric associationism – a visual equivalent in the micrographic imagery of the New Biology.

Notes

1 A. Ozenfant, *The Foundations of Modern Art* (1931) (New York: Dover, 1952), p. 289.
2 A. Breton, *Surrealism and Painting* (1928) (Boston: MFA Publications, 2002), p. 1.
3 W. Gaunt, 'Introduction', in W. Watson-Baker, *World Beneath the Microscope* (London: The Studio, 1935), p. 9.
4 Hooke's text was originally intended as a popularly accessible book: see J. Maienschein, 'Cell Theory and Development', in R.C. Olby, G.N. Cantor, J.R.R. Christie & M.J.S. Hodge (eds), *Companion to the History of Science* (London: Routledge, 1990), p. 357. See also R. Hooke, *Micrographia or Some Physiological Descriptions of Minute Bodies Made by Magnifying Glasses with Observations and Inquiries Thereupon* (1665) (New York: Dover, 1961).
5 Pickstone, *Ways of Knowing*, pp. 145–6.
6 Daston & Galison, *Objectivity*, p. 120.
7 *Ibid.*, pp. 120–1.
8 P. Broks, *Media Science Before the Great War* (London: Macmillan, 1996), pp. 23–30.
9 On this subject see I. Eibl-Eibesfeldt, 'Ernst Haeckel: The Artist in the Scientist', in *Art Forms in Nature: The Prints of Ernst Haeckel* (Munich: Prestel, 1998), pp. 19–29.
10 W. Breitenbach & E. Haeckel, 'Die Natur as Künstlerin', in F. Goerke (ed.), *Die Natur als Künstlerin* (Berlin, 1913), quoted in I. Eibl-Eibesfeldt, 'Ernst Haeckel', p. 24.
11 Thomas, 'The Search for Pattern', p. 101.
12 Daston & Galison, *Objectivity*, p. 247.
13 *Ibid.*, p. 247.
14 *Ibid.*, p. 247; Eibl-Eibesfeldt, 'Ernst Haeckel', p. 19.
15 Gamwell, *Exploring the Invisible*, pp. 77–87.
16 Botar, 'Defining Biocentrism', p. 16.
17 Daston & Galison, *Objectivity*, pp. 138–190; Gamwell, *Exploring the Invisible*, p. 210.
18 Mundy, *Biomorphism*, pp. 113–14.
19 B. Berg, 'Contradictory Forces: Jean Painlevé, 1902–1989', in A.M. Bellows & M. McDougall (eds), *Science Is Fiction: The Films of Jean Painlevé* (Cambridge: MIT Press, 2000), pp. 17–18.
20 E. Faure, *De la cinéplastique* (Paris: Editions Séguier, 1995), quoted in Berg, 'Contradictory Forces', p. 19.
21 See Ades, 'Little Things', pp. 9–60.
22 Mundy, *Biomorphism*, pp. 111–12.
23 Grigson, *Henry Moore*, p. 9.
24 See Botar, *Prolegomena to the Study of Biomorphic Modernism*, pp. 569–70. Although Kállai was a relatively unknown figure in British artistic circles in the early 1930s, Read's knowledge of German art theory and professional interest in the Bauhaus, Kállai's friendship with the Constructivist Naum Gabo and the unparalleled success of Blossfeldt's photo-books in Britain during the late 1920s and early 1930s all contributed to a nature-centric, scientific climate in which bioromantic ideas could be understood to flourish.
25 Botar, 'Ernö Kállai and the Hidden Face of Nature', p. 77.
26 See also Mundy, 'Form and Creation', pp. 16–23.
27 See Anthony Blunt's review of the second edition of Blossfeldt's *Art Forms in Nature*: A. Blunt, 'Nature and Design', in *The Listener*, Vol. XI, No. 274, 11 April 1934, p. 612.
28 Anon., 'Art and Nature: A Review of "Art Forms in Nature"', in *Times Literary Supplement*, No. 1148, 31 October 1929, p. 871.
29 Aspects of my discussion of photomicrography and microcinematography have previously been published as: E. Juler, 'The Key to a Hidden World: Photomicrography and Close-Up Nature Photography in Interwar Britain', in *History of Photography*, Vol. XXXVI, No.1, February 2012, pp. 87–98 (www.tandfonline.com/doi/full/10.1080/03087298.2012.633442). I am grateful to the Taylor & Francis Group for granting me permission to reproduce elements of my *History of Photography* article in this book.
30 See Whitworth, 'The Clothbound Universe', pp. 53–4.
31 L.J. Schaaf, 'Invention and Discovery: First Images', in A. Thomas (ed.), *Beauty of Another Order: Photography in Science* (New Haven: Yale University Press, 1997), p. 26.

32 Thomas, 'The Search for Pattern', pp. 76–7.

33 Anon., 'The Progress of Physical Science: Films of Living Cells, Study of Growth', in *The Times*, 6 July 1931, p. 19.

34 Anon., 'Key to a Hidden World: New Light on the Infinitesimal, the Super-Microscope', in *The Times*, 26 November 1938, p. 9.

35 *Ibid.*, p. 9.

36 See *The Times*, 26 November 1938, p. 16. 'The Nature of Things' by William Bragg in *The Listener* included photomicrographs of snow crystals alongside schematic models of atomic structures, and photomicrographs were often used as eye-catching illustrations in his articles for *Nature*, for example in W. Bragg, 'Liquid Crystals', in *Nature*, No. 3360, 24 March 1934, pp. 445–56.

37 See Anon., 'Pond Life on the Screen', in *The Times*, 18 April 1939, p. 10.

38 G.H. Saxon Mills, 'Modern Photography: Its Development, Scope and Possibilities', in *Modern Photography* (London: The Studio, 1931), p. 4.

39 'Editor's Introduction', in *Modern Photography* (London: The Studio, 1931), p. 1.

40 *Photography Yearbook* (London: Fountain Publications, 1935), pp. 266–9.

41 C. Leeston Smith, 'Present Day Technical Apparatus and Its Applications', in *Modern Photography* (London: The Studio, 1931), pp. 111–18.

42 T. Weiss, 'The International Exhibition of Photography in Brussels', in *Close-Up*, Vol. IX, No. 3, September 1932, pp. 188–9.

43 See: Anon., 'The 'Positive Electron', in *The Times*, 11 March 1933, p. 12; Anon., 'Photographs of Atom Tracks', in *The Times*, 20 November 1937, p. 17.

44 R. Low, *Documentary and Educational Films of the 1930s* (London: George Allen & Unwin, 1979), p. 2.

45 *Ibid.*, p. 4.

46 See R. Low, *The History of British Film, 1918–1929* (London: George Allen & Unwin, 1971), p. 286.

47 P. Smith & M. Field, *Secrets of Nature* (London: Faber, 1934).

48 See T. Boon, *Films of Fact: A History of Science in Documentary Films and Television* (London: Wallflower Press, 2008), pp. 85–7. Gaumont–British Instructional (G-BI) was the successor company to British Instructional Films (BIF).

49 Indeed, *The Film in National Life*, the report of the Commission on Educational and Cultural Films, noted that: 'Accelerated photography (very slow mechanical exposure) has given a new significance to the gradual processes of plant growth and decay. Microcinematography joins the perception of the microscope to that of the film camera'. Commission on Educational and Cultural Films, *The Film in National Life* (London: George Allen & Unwin, 1932), p. 122. See also Low, *The History of British Film*, pp. 286–7.

50 Low, *Documentary and Educational Films of the 1930s*, p. 3.

51 See Werskey, *The Visible College*.

52 See *The Cinema in Education* (1935) for a contemporary discussion of the pedagogic utility of microcinematographic films as classroom aids: C.D. Otterly, *The Cinema in Education: A Handbook for Teachers* (London: George Routledge & Sons, 1935), esp. pp. 5, 19, 23.

53 Anon., 'The Film World and Travel Pictures', in *The Times*, 18 December 1929, p. 12. Similar praise for BIF's science films was given in a review of 1930: Anon, 'The Film World and Travel Pictures', in *The Times*, 22 October 1930, p. 14.

54 The *Monthly Film Bulletin*'s Science Committee rebuked the 'popular' and 'facetious' nature of the commentary for *Magic Myxies*, as well as some of its scientific conclusions. Anon., 'World in a Wineglass', *Monthly Film Bulletin*, Vol. VI, No. 69, 30 September 1939, p. 196.

55 See Low, *The History of British Film*, p. 130.

56 Anon., 'The Progress of Science: Films of Living Cells, Study of Growth', in *The Times*, 6 July 1931, p. 19.

57 For an overview of *Close-Up*'s influence and cultural strategy, see A. Friedberg, 'Reading *Close-Up*, 1927–1933', in J. Donald, A. Friedberg & L. Marcus, *Close-Up, 1927–1933: Cinema and Modernism* (London: Cassell, 1998), pp. 1–26.

58 After 1935 Smith made natural history films for G-BI under the new series title of 'Secrets of Life'. It is also worth noting here the possible influence of the American pioneer of time-lapse photography, John Ott, who worked extensively on photomicroscopy in the interwar period and

lectured widely, including in Britain. See also O. Blakeston, 'Personally About Percy Smith', in *Close-Up*, Vol. VIII, No. 2, June 1931, pp. 143–6.

59 Friedberg claims that *Close-Up*'s writers were 'determined to transform the cultural topography of cinema and its future'. See Friedberg, 'Reading *Close-Up*, 1927–1933', p. 3.

60 T. Weiss, 'Film Review: *Das Keimende Leben*', in *Close-Up*, Vol. IX, No. 3, September 1932, p. 208.

61 For an overview of this topic see Overy, *The Morbid Age*, pp. 93–135.

62 *Monthly Film Bulletin*, No. 6, 1939, p. 28, cited in Boon, *Films of Fact*, p. 87.

63 See Halliday, *More Than a Bookshop*, pp. 75–6. The first English co-edition was printed in 1929 with an introduction by the art-dealer Karl Nierendorf and featuring 120 full-page, black-and-white plates; a 'second series' co-edition was published in 1932 with a one-page introduction by Blossfeldt and 120 new images; and in 1935 a second edition was published with an abridged selection of 96 plates.

64 K. Blossfeldt, *Art Forms in Nature: Examples from the Plant World Photographed Direct from Nature*, with an introduction by Karl Nierendorf (London: Zwemmer, 1929).

65 F. Etchells, 'Nature and Art', in the *Architectural Review*, December 1929, p. 299.

66 Anon., 'What Nature Knows of Art', in *The Listener*, Vol. II, No. 36, 18 September 1929, p. 370.

67 Anon., 'Book Reviews – New Books', in *Close-Up*, Vol. IX, No. 3, September 1932, p. 208.

68 Maldonado, *Le cercle et l'amibe*, pp. 214–15.

69 Anon., 'What Nature Knows of Art', p. 370.

70 M. Lavallée, 'Art Nouveau', *Grove Art* online, Oxford Art online, Oxford University Press, www.oxfordartonline.com/subscriber/article/grove/art/T004438 (accessed 4 March 2013).

71 U.M. Stump, 'Karl Blossfeldt's Working Collages – A Photographic Sketchbook', in *Karl Blossfeldt: Working Collages* (Cambridge: MIT, 2001), p. 7.

72 See Mundy, 'Form and Creation', p. 17.

73 Albers, 'Concerning Fundamental Design', p. 120.

74 L. Moholy-Nagy, *The New Vision: Fundamentals of Bauhaus Design, Painting, Sculpture, and Architecture* (London: Faber, 1934) (unpaginated introduction). The first English-language edition of this text was published by Brewer, Warren & Putnam of New York in 1930.

75 Anon., 'What Nature Knows of Art', p. 370.

76 P. Nash, 'Photography and Modern Art', in *The Listener*, Vol. VIII, No. 185, 27 July 1932, p. 130.

77 *Ibid.*, p. 130.

78 Anon., 'Art Forms in Nature', in *The Times*, 5 October 1929, p. 13.

79 K. Nierendorf, 'Preface to Karl Blossfeldt, *Urformen Der Kunst*' (1928), in D. Mellor (ed.), *Germany: The New Photography 1927–33* (London: Arts Council, 1978), p. 18.

80 The plates from Linnaeus' book were copied or adapted in almost every botanical textbook of the Enlightenment era, and provided a visual taxonomy shared by botanists throughout Europe. As such, the visual correspondences between Blossfeldt's photographs and Linnaeus' plates suggest something more than pure coincidence and point towards a deeply ingrained method of classifying the plant world. See D. Bleichmar, 'Training the Naturalist's Eye in the Eighteenth Century: Perfect Global Visions and Local Blind Spots', in A. Graciano (ed.), *Visualising the Unseen, Imagining the Unknown, Perfecting the Natural: Art and Science in the 18th and 19th Centuries* (Newcastle: Cambridge Scholars, 2008), p. 6.

81 See Watson-Baker, *World Beneath the Microscope*; W.A. Bentley & W.J. Humphreys, *Snow Crystals* (New York: McGraw-Hill Book Company, 1931).

82 Halliday, *More Than a Bookshop*, pp. 75–6.

83 Gaunt, 'Introduction', p. 9.

84 Watson-Baker, *World Beneath the Microscope*: text for Plate 7 (the image is reproduced here as Figure 5.8, p. 185).

85 For instance J. Piper, 'Review of *World Beneath the Microscope*', in *Axis*, No. 4, 1935, p. 28.

86 Anon., 'Art and Nature: A Review of "Art Forms in Nature"', p. 871.

87 Paul Nash provides details of this exhibition in his 1932 *Listener* review: Nash, 'Photography and Modern Art', p. 130. See also Halliday, *More Than a Bookshop*, p. 76; Botar, *Prolegomena to the Study of Biomorphic Modernism*, p. 48.

88 Mundy, *Biomorphism*, pp. 111–13.

89 Nash, 'Photography and Modern Art', p. 130.

90 Nash specifically mentions Wilenski's 'admirable treatise on modern sculpture' in his analysis of Blossfeldt's photographs. See Nash, 'Photography and Modern Art', p. 130.

91 Wilenski, *The Meaning of Modern Sculpture*, p. 158.

92 *Ibid.*, pp. 158–9.

93 Wilenski's particular use of photographic reproduction assisted such iconographic metonymy. Every reproduction in *The Meaning of Modern Sculpture* adopted the same format, with the photographs of Modernist sculpture, especially, appearing like mocked-up Blossfeldt enlargements: closely cropped and shot against a neutrally grey background they mimicked the analytical pretensions of the New Objectivity school of photography.

94 See Botar, *Prolegomena to the Study of Biomorphic Modernism*, pp. 65–6, 337–8.

95 See Barnett, 'Kandinsky and Science', p. 219.

96 W. Benjamin, 'New Things About Plants' (1928), in D. Mellor (ed.), *Germany: The New Photography, 1927–33* (London: Arts Council, 1978), p. 21.

97 See U.M. Stump, 'Karl Blossfeldt's Working Collages – A Photographic Sketchbook', in A. Wilde & J. Wilde (eds), *Karl Blossfeldt: Working Collages* (Cambridge: MIT Press, 2001), pp. 14–15.

98 Karl Blossfeldt in an unpublished essay of 1929 quoted in Stump, 'Karl Blossfeldt's Working Collages', p. 15.

99 Botar, 'The Origins of László Moholy-Nagy's Biocentric Constructivism', p. 315.

100 Moholy-Nagy, *The New Vision*, p. 62.

101 Nash, 'Photography and Modern Art', p. 130.

102 *Ibid.*, p. 130.

103 Nash was, in this respect, doubtless aware of the degree to which Moore's work had been appropriated by *Documents* (and other Surrealist journals) as symbolic of a psychologically disturbing expression of anti-classicism. See J. Kelly, 'The Unfamiliar Figure: Henry Moore in French Periodicals of the 1930s', in J. Beckett & F. Russell (eds), *Henry Moore: Critical Essays* (Aldershot: Ashgate, 2003), p. 43.

104 E. Holding, 'London Shows', in *Axis*, No. 2, April 1935, p. 30.

105 J. Thwaites & M. Thwaites, 'Surrealism and Abstraction – The Search for Subjective Form', in *Axis*, No. 6, summer 1936, pp. 24–5.

106 '[Les] fleurs les plus belles sont déparées au centre par la tache velue des organs sexués. C'est ainsi que l'intérieur d'une rose ne répond nullement à sa beauté extérieure, que si l'on arrache jusqu'au dernier les petals de la corolla, in ne reste plus qu'une touffe d'aspect sordide.' G. Bataille, 'Le langage des fleurs', in *Documents*, No. 3, June 1929, pp. 162–3.

107 'La vie humaine comporte en fait la rage de voir qu'il s'agit d'un movement de va-et-vient de l'ordre à l'idéal et de l'idéal à l'ordure [...]'. G. Bataille, 'Le gros orteil', in *Documents*, No. 6, November 1929, p. 297.

108 '[...] passer sur un organe aussi *bas* qu'un pied'. *Ibid.*, p. 297, original emphasis.

109 Lajos Kassak quoted in Metken (ed.), *Realismus zwischen Revolution und Reaktion 1919–1939* (Munich, 1981), p. 154, reprinted in S. Michalski, *New Objectivity: Neue Sachlichkeit – Painting in Germany in the 1920s* (Cologne: Taschen, 2003), p. 192.

110 See S. Baker, 'Doctrine (The Appearance of Things)', in D. Ades & S. Baker (eds), *Undercover Surrealism: Georges Bataille and Documents* (London: Hayward Gallery, 2006), pp. 34–40.

111 This de-sublimatory strategy suggests some sort of engagement with post-Enlightenment philosophies of perception. There was, after all, a strand in post-Enlightenment philosophy that foreshadowed romantic aesthetics by noting how the disproportionate enlargement of an object grossly distorted conventional standards of beauty, ushering in a sensation of sublimity because of the fearfulness induced by the magnitude of its deformation. Thus writing on the cusp of the transition between the aesthetic values of the Enlightenment and those of the romantic era, the philosopher and statesman Edmund Burke wrote in *A Philosophical Enquiry into the Origin of Our Ideas of the Sublime and the Beautiful* of 1757 that natural objects 'which greatly exceed are by that excess, provided the species be not very small, rather great and terrible than beautiful'. Edmund Burke, *A Philosophical Enquiry into the Origin of our Ideas of the Sublime and Beautiful* (Oxford: Oxford University Press, 1998), p. 142. This notion of the sublime had important implications for microscopical imagery, as the philosopher George (Bishop) Berkeley explained in his work during the period 1709–10 that, by obtaining hitherto unseen views of objects, the microscope severed the normal phenomenological ties between things viewed and

the body of the viewer. This sense of perceptual dislocation was seen to carry connotations of unfixed dimensionality, a condition that Burke would later link to sublimity, especially in the case of tiny objects as: '[The] great extreme of dimension is sublime, so the last extreme of littleness is in some measure sublime likewise; when […] we pursue animal life into these excessively small, and yet organized beings, that escape the nicest inquisition of the sense, we become amazed and confounded at the wonders of minuteness; nor can we distinguish in its effect this extreme of littleness from the vast itself'. G. Berkeley, *Philosophical Works – Including the Works on Vision* (London: Everyman's Library, 1983), pp. 33–4.

112 'Il fait se gondoler les formes et, systématiquement, rendant tout presque semblable à tout, boulverse les classifications illusoires et l'échelle meme les choses créés'. Leiris, 'Hans Arp Exhibition', p. 340.

113 P. Nash, *Monster Field* (1939) (Oxford: Counterpoint Publications, 1946), pp. 244–7, reprinted in A. Causey (ed.), *Paul Nash: Writings on Art* (Oxford: Oxford University Press, 2000), p. 150.

114 Although, by interwar British standards, Nash's reading of Blossfeldt's 'perverse' influence on modern sculpture was unorthodox, it nonetheless illuminated a split in artistic responses to scientific photography – one that divided opinion between the visual impact of this medium being felt in a structured, rational way or in irregular, neo-romantic terms. We can here note that this division was partly generational. The emphasis that Wilenski laid throughout his *The Meaning of Modern Sculpture* upon the bankruptcy of the European classical tradition and the vitality of primitive sculpture was heavily indebted to the formalist dogma of Roger Fry and the anti-humanist aesthetics of T.E. Hulme. While maintaining such a standpoint had been a question of avant-garde loyalty throughout the 1920s, by 1932 it appeared old-fashioned to Modernists who had reached artistic maturity long after the prewar heyday of Vorticism. That Wilenski was primarily concerned with British sculpture and had little time for Surrealism or ethnography meant that *The Meaning of Modern Sculpture* was also somewhat shallow in its contemporary frame of reference – a point that was evidently picked up on by Nash, whose censure of Wilenski's thesis operated on the basis that it displayed scant knowledge of the 'so-called "perverse" sculptors' of the European avant-garde. See Causey, 'R.H. Wilenski and *The Meaning of Modern Sculpture*', pp. 267–72.

115 Botar, *Prolegomena to the Study of Biomorphic Modernism*, p. 50.

116 On Christeller's role in the development of objective photomicrographic technology, see Daston & Galison, *Objectivity*, pp. 171–3.

117 Mundy, *Biomorphism*, pp. 111–12.

118 Botar & Wünsche, 'Introduction', p. 3.

119 Grigson, 'Painting and Sculpture Today', pp. 89–90.

120 Like Grigson, Barr saw this protoplasmic language of sculptural form as hugely influential in the development of British Modernist sculpture: 'The Arp "shape", a soft, irregular, curving silhouette, halfway between a circle and the object represented, appears again and again in the work of […] Moore'. Barr, *Cubism and Abstract Art*, p. 186.

121 *Ibid.*, pp. 12, 19.

122 See also Mundy, *Biomorphism*, pp. 111–12.

123 Grierson, 'The New Generation in Sculpture', p. 350. Grierson's 1930 article is possibly the first example of a British writer openly analogising between Modernist sculpture and the new scientific imagery. As far as I am aware, it anticipates Grigson's writings on the subject by at least four years.

124 See Botar, *Prolegomena to the Study of Biomorphic Modernism*, p. 50. Botar claims that as 'early as 1934 Nash and Grigson were calling for the production of a nature-centred biomorphic Modernist art, i.e. Bioromanticism. To my knowledge, they were the only art professionals at the time to actively promote such art'. It would seem, however, that – while Grigson was undoubtedly a pioneer in this regard – his bioromantic aesthetic was developed only several years after Grierson published his *Apollo* article.

125 Grigson, *Henry Moore*, p. 8.

126 Grigson, 'Comment on England', p. 10. Botar describes Grigson's *Henry Moore* as 'the earliest critical text I know of in English to discuss biomorphic Modernist art systematically in conjunction with contemporary biology, Vitalism *and* the scientific image analogy'. Botar, *Prolegomena to the Study of Biomorphic Modernism*, p. 50.

127 Grigson, *Henry Moore*, p. 8.
128 Moore, 'A View of Sculpture', p. 408. The word 'cameograph' does not appear in the *Oxford English Dictionary* and it has been suggested by Phillip James that Moore meant the pantograph. P. James (ed.), *Henry Moore on Sculpture* (London: MacDonald, 1966), p. 57. A pantograph is an instrument that enables the mechanical copying of a plan, diagram, pattern, etc., especially on a different scale, typically with two drawing points connected by an adjusted parallelogram of jointed rods. See *Oxford English Dictionary* online, 'pantograph, *n.*', www.oed.com/view/Entry/137027 (accessed 28 January 2011).
129 Moore, Untitled statement, p. 30.
130 See van Eck, *Organicism in Nineteenth-Century Architecture*, p. 181.
131 See Mundy, *Biomorphism*, pp. 114–15.
132 Watson-Baker, *World Beneath the Microscope*, caption for Plate 3.
133 Piper, 'Review of *World Beneath the Microscope*', p. 28.
134 Scholars have claimed that while the discourse of pure science has typically been dealt with as a 'textual object', technology, by contrast, has been characteristically viewed as irreducible, owing to its manifestly non-textual nature. For although the texts surrounding new technologies have often been fruitfully analysed, technology – like art – has brought up problems for the history of ideas, not least because its very *physicality* signifies its ontological divergence from linguistic discourses. Of course, in terms of metaphor analysis, the study of textual paradigms can be edifying when attempting to reconstruct the speech patterns through which new technologies were understood. See Whitworth, *Einstein's Wake*, p. 7.
135 Gabo, 'The Constructive Idea in Art', pp. 3–4.
136 L. Jordanova, 'Natural Facts: A Historical Perspective on Science and Sexuality', in C. MacCormack & M. Strathern (eds), *Nature, Culture and Gender* (Cambridge: Cambridge University Press, 1980), pp. 54–8.
137 *Ibid.*, pp. 45, 58.
138 J. Milton Offord, 'President's Address' (14 February 1933), in *Journal of the Quekett Microscopical Club*, Third Series, Vol. I, 1934–37, p. 5.
139 Gabo, 'The Constructive Idea in Art', p. 4.
140 See Hammer & Lodder, *Constructing Modernity*, p. 387.
141 Berg, 'Contradictory Forces', p. 19.
142 Grierson, 'The New Generation in Sculpture', p. 350.
143 Grigson, *Henry Moore*, p. 9.
144 Mundy, *Biomorphism*, pp. 111–12.

6. Epilogue

In many ways, the 1930s was the decade in which modernity registered most powerfully in twentieth-century British culture, not just in terms of the inception of new technologies, the naissance of artistic Modernism and the growing professionalisation of society, but also in the sense of the emergence of a genuinely mass society that was governed by capitalism 'red in tooth and claw'.[1] Biology was perceived to be an integral aspect of this new society, as – to paraphrase Julian Huxley – the application of biological knowledge to practical affairs was held to have noteworthy sociological repercussions, extending from contemporary medical advances to the possibility of altering the future course of human evolution itself.[2] As a potent symbol of modernity, biology thus provisioned Modernist aestheticians with a range of rhetorical devices through which the novel conceptions of space, time and form engendered by modernity could be successfully navigated and consolidated. In this sense, the exchange of ideas that took place between Modernist sculpture and the New Biology signalled more broadly the effects of modernity on British society, in which subjects as diverse as scientism, biologism, technology, philosophy and religion were put under review as symbolic of the changes wrought upon society by the unbridled forces of a Promethean Modernism.[3]

We have observed how the boom in popular science that took place in the years between the wars made easy the transition of ideas between art and science, providing biology with a gloss of accessibility by means of its appeal to a literary readership, allowing it to become, in the words of Herbert Read, 'layman's property and be given general currency'.[4] Nonetheless, the concepts that Modernist sculptors and their critics picked up on were characteristically those that possessed the greatest popular resonance: organicism, for example, heralded not only a new understanding of *space* but also related more broadly to an anti-mechanistic philosophical drive which critiqued mainstream biology, economics, society and politics as overly reductive and deterministic.

If C.P. Snow's 'Two Cultures' paradigm conventionally sites art and science at the opposite ends of the cultural spectrum, then one aim of this study has been to demonstrate the untenability of that position. After all, the formation of a *biologistic* Modernist sculptural discourse literally bespeaks of an ongoing dialogue between biology and modern sculpture in which nominal disciplinary boundaries continually undergo revision and renegotiation. Once started, such a conversation – especially one that pertains so closely to the philosophical and material conditions of modernity itself – may not be easily ended, even when epochal events, such as the Second World War, threaten to destroy the socio-economic infrastructure upon which all art, to some extent, depends. Unlike the Great War, which witnessed the obliteration of a prewar avant-garde swept up by the impassioned bellicosity of wartime propaganda, the Second World War experienced no wholesale slaughter of British artistic talent, meaning that the postwar avant-garde did not suffer from the same lack of continuity that had hampered the ambitions of the surviving Vorticists.[5] As the great seer of prewar artistic radicalism, Wyndham Lewis, commented in 1915: '[The] War has stopped Art dead'.[6] Decamped to locations such as St Ives and Hertfordshire, however, the majority of the leading sculptors of the 1930s (Henry Moore, Naum Gabo and Barbara Hepworth among them) managed to avoid, by age, gender or political persuasion, the same fate as the previous generation and were able to continue developing their practice in relative quietude.[7] In this sense, many of the biological themes that occupied the sculptural avant-garde during the 1930s continued to enjoy cultural currency in the years following the Second World War, with Moore maintaining his interest in morphology and Read elaborating upon his 'organic aesthetic' until well into the 1960s.[8] Of no less importance was the efflorescence of certain keynote ideas on art and science that had been formulated in the 1930s but which found true expression in only the postwar period: Snow's 'Two Cultures' paradigm, for example, while conceived in the 1930s, was first delivered as a Rede Lecture in 1959. And while the ubiquitous postwar presence of Moore's brand of sculptural Modernism on the international art scene created something akin to a generational backlash against his reputation as a founding father of modern sculpture, many of the aesthetic repercussions stemming from the interchange evinced between the New Biology and Modernist sculpture during the 1930s continued to be felt until well into the 1950s.[9]

That biology should continue to exert influence in postwar artistic circles was partly due to the extraordinary cultural effort that went into encouraging greater cooperation between the arts and sciences in the years following the Second World War. Undoubtedly, the most effective expression of this pan-disciplinary tendency was the Festival of Britain, which opened on London's South Bank in May 1951. The political background to the Festival was the postwar Labour government and the creation of the welfare state, and the idea of the Festival (emerging as a proposal in 1947–48) had been first mooted as a commemoration of the centenary of the 1851 Great Exhibition and as part of an attempt to relieve the country during a period of unmatched hardship.[10] First and foremost, the Festival performed a double function by, on the one hand, serving to advertise national science and technology, and, on the other, promoting British goods and 'ideas about the need for good design, particularly in those goods destined for export'.[11] This enterprise had extra resonance during the Cold War, as the Festival planners argued that Britain was uniquely placed between the USA and USSR to create 'a proper coordination of the fine arts, machine-work, and "handwork", because this sort of "balance" [is] essential if Western Civilization [is] to survive'.[12] Part of this initiative involved reconciling the sciences and the arts – a project that was undoubtedly influenced by the atmosphere of postwar reconciliation pervading cultural politics. Jacob Bronowski, a polymathic organiser of the Festival, commented, for example, that any disjuncture between the two disciplines was symptomatic of a wider dialogic breakdown in society:

> Science and the arts today are not as discordant as many people think. The difficulties which we all have as intelligent amateurs in following modern literature and music and painting are not unimportant. They are one sign of the lack of a broad and general language in our culture. The difficulties which we have in understanding the basic ideas of modern science are signs of the same lack.[13]

Cooperation between science and art was deemed advantageous to publicising the modernity of the Festival, and science was consequently recruited as the footing upon which a new Modernist aesthetic could be established. Just some of the ways in which this design philosophy found expression were the South Bank Regatta restaurant (which was decorated with furnishings employing molecular and crystalline patterns), the aluminium-coated Dome of Discovery, and the aerodynamic Skylon tower, all of which possessed compelling futuristic overtones.[14] Although the use of abstract symbols derived from organic shapes became popular in commercial design throughout the 1950s,[15] the Festival of Britain contributed to this development in particular ways, specifically through the activities of the Festival Pattern Group. As Mark Hartland Thomas noted at the time, the chief purpose behind the creation of the Group had been to make public the aesthetic nature of scientific activity, the 'beauty of the patterns in the basic structure of nature [as] revealed [by science]'.[16] Like the Constructivists of the interwar period, Thomas judged that crystal patterns succeeded as contemporary design, firstly because 'they were essentially modern because the technique that constructed them was quite recent', and secondly because 'like all successful decoration of the past they derived from nature – although it was nature at a sub-microscopic scale not previously revealed'.[17] Through application to various textiles and furnishings, the crystal motifs appeared on an assortment of products (lace veils, ties, wallpaper, crockery, etc.) and were forcefully promoted at the Festival's Regatta restaurant with the hope of stimulating international interest in British products (Figure 6.1).

Another upshot of the Festival was the involvement of the Independent Group in the exhibition 'Growth and Form', an avant-garde happening that aimed to continue the dialogue between art and science already established by the principal Festival exhibits (Figure 6.2).[18] The exhibition operated as a response to the Festival science display, which attempted – through a variety of enlarged models and photographs – to 'look inside nature [and] show the processes, living and dead, by which nature works', in order to augment public knowledge of

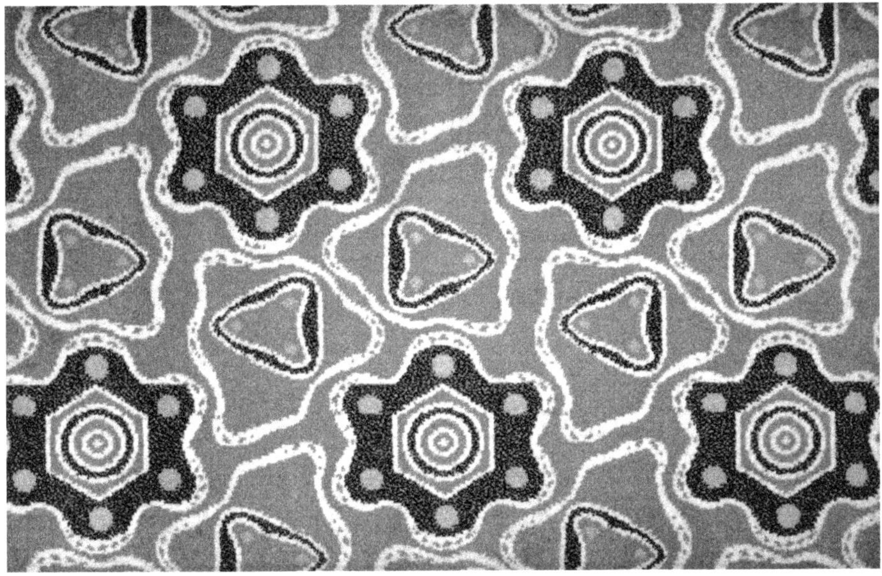

Figure 6.1

Carpet pattern derived from the crystal structure of insulin. Designed by G. Brown for James Templeton & Co. Ltd as part of the Festival Pattern Group, 1950–51. Design Council Slide Collection at Manchester Metropolitan University

natural science.[19] Like the popularising undertakings of the 1930s, the Festival had a powerful pedagogic role and, through an interest in marketing strategies, the Independent Group similarly attempted to educate public conceptions of 'science' through exhibitory display – an objective enhanced by the closeness of the exhibition to the Festival's official expository agenda, one of its aims being to bridge the 'Two Cultures' gap.[20]

The inspiration for the exhibit was a book well known to interwar aestheticians: D'Arcy Wentworth Thompson's *On Growth and Form*, which the exhibition organisers attempted to visually reconstruct so as to inform the public of the differences between artistic and natural creation.[21] Even though the disquieting way in which 'Growth and Form' brought together magnified photographs of skeletal hands, sea-urchin larvae and cells (within an exhibitory environment that incorporated free-standing frames in conjunction with floor and ceiling displays) shared much with prewar Surrealist projects,[22] the exhibition also borrowed from the techniques used by Festival organisers (such as the use of enlarged models to suggest dramatic shifts in scale), as well as using images predominantly produced by scientists: all of which suggested broad complicity with the primary goals of the Festival.[23] While the curator, Richard Hamilton, was drawn to the nonconformist aspects of Thompson's morphological theory – which contradicted the

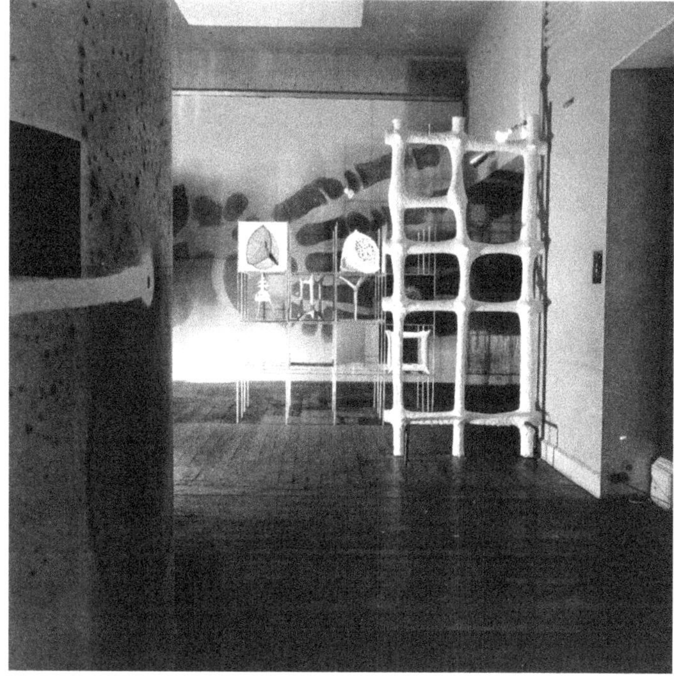

Figure 6.2

Nigel Henderson, Installation shot of Richard Hamilton's work (title unknown)
in the exhibition 'Growth and Form' at the Institute of Contemporary Art, 1951.
Tate, London

linear lines of evolutionary descent presented in the Festival display[24] – overall,
the stress in 'Growth and Form' on 'instructive' spectatorship corresponded with
the pedagogic aims of the Festival, not least in the attempt to induce 'an appreciation of the forms in nature beyond [...] the immediate visual environment
[of the viewer]', even if this worked unconventionally through a process of visual
de-familiarisation.[25]

That postwar exhibitory activity suggested continuity with interwar pedagogic projects was substantiated by the continuing dialogue effected between art
and science at this time. Just as interwar expositions of popular science typically
highlighted relationships between philosophy and scientism, so, too, much postwar
discoursing reiterated perceived interdisciplinary ties and, as such, 'Growth and
Form' allowed for further speculation as to the linkages between art and science.
The symposium 'Aspects of Form' which followed on from the Independent
Group show drew together various speakers from the arts and sciences with the
purpose of demonstrating how 'form' was fundamental to science, and how art
and science corresponded. For instance, in the preface to the ensuing publication,

Herbert Read noted: 'The increasing significance given to *form* or *pattern* in various branches of science suggest[s] the possibility of a certain parallelism [...] in the structures of natural phenomena and of authentic works of art'. As a result: 'Aesthetics [could be] no longer an isolated science of beauty [and] science [could] no longer neglect aesthetic factors'.[26] Throughout the symposium, emphasis was thus laid upon realising the 'unity of spatial form in the complex processes of physics, biology, psychology and art',[27] implying ideological continuity with the pan-epistemological enterprises of the 1930s.

Alongside the exhibitory agenda of 'Growth and Form', biology provided artists with a philosophical grounding upon which to base postwar experimental art-making. Although the influences of Tachisme and action painting undoubtedly stimulated curiosity in gestural types of artistic production,[28] the bias towards *process*-based art forms in the postwar period suggests continuity with interwar aesthetic projects, not least because of the particular way in which biology found expression as a metaphor for non-representational art manufacture. While speaking of his practice, for instance, the painter Adrian Heath noted that the issue for the contemporary artist was 'to exercise some logical or objective control on [the] gradual evolution [of the image], for to abandon oneself in a purely subjective way to a sensuous medium makes further progress difficult'.[29] Indeed, in contrast to the subjective excesses of Tachisme, the process-dominant work of the Constructionists emphasised objectivism over expressiveness and, as such, morphology provisioned a helpful methodology through which non-figurative images could be understood to intuitively 'develop' without abandoning positivist principles (Figure 6.3).

For Heath, geometry supplied some measure of control, allowing him to progressively flesh out the composition at regular intervals without any implications of subjectivity. Yet at the same time, '[i]t is the process', he noted, 'the method of development that is the life of the painting [and] [t]he forms themselves, their size and position are born from the process. In fact all forms should witness clearly to their evolution. The colour and shape of their past is responsible for their present which is richer for these transitions'.[30] In this sense, the non-figurative shapes depicted in each artwork symbolised the developmental prehistory of the image, meaning that Constructionist art operated as a cipher of process rather than as the representation of some predetermined idea.[31] As a consequence, the radicalism of postwar abstraction lay in the allegation that it was 'not the result of a process of abstraction in front of nature, but a method of construction emanating from within'.[32] Accordingly, morphology proved useful in expounding upon a methodology that prioritised changeability over finality while also appearing geometrically reductionist. Heath, for example, spoke of Constructionist artworks in terms suggestive of Thompson's description of an object as a 'diagram of forces', asking whether 'a final form [can] be achieved in a painting any more than it is in a natural growth such as a tree [...]? Both are shaped by the continuous play of opposing forces'.[33] At the same time, the elementary morphologies that

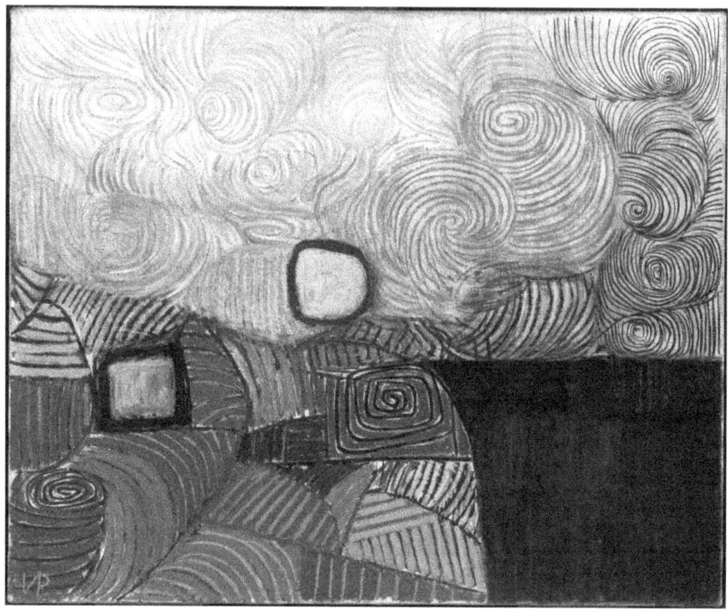

Figure 6.3

Victor Pasmore, *Spiral Motif in Green, Violet, Blue and Gold: The Coast of the Inland Sea*, 1950. Tate, London

so preoccupied the critics and sculptors of the 1930s found renewed expression in the simple geometric compositions of the Constructionist movement, which were similarly justified on the basis that certain formal elements (the spiral, circle, triangle, etc.) were inherent to the physical world and 'universally recognized [and had] become concrete elements in themselves'.[34] Producing both abstract tableaux and sculptural collages at this time, all of which contained loose constellations of geometric elements, the painter Victor Pasmore noted that the 'ancient maxim *art imitates nature*, must no longer be construed in the superficial sense [...] but in its deeper meaning – art imitates nature in its manner of operation'.[35]

The postwar resurgence of interest in morphology responded to a variety of concerns, although perhaps the most important of these was the growing pan-disciplinary temperament of cultural discourse. Interdisciplinary theory found expression within postwar aesthetics chiefly as a means through which creativity could be better comprehended, using science to expound upon the reasoning behind the emergence of new styles. The art historian J.P. Hodin confirmed as much by confessing that: 'The newer aesthetics is not exclusively concerned with problems of form, but devotes great attention to the creative processes themselves and the relationship between style and personality'.[36] This development had come about through prewar attempts to found 'a comparative science of styles [which had led to] the beginnings of a morphology of style; and the interrelations between aesthetics and other branches of learning were [thus] developed'.[37] While this interrelationship essentially succeeded owing to the rationality of modern art (which found its counterpart in scientific rationalism), by and large Hodin disapproved of this trend, claiming that the 'application of scientific methods to other realms of experience necessarily results in an impoverishment of intellectual activity. The analytical spirit and the creative functions of the mind are diametrically opposed'.[38] Yet – from the opposite end of the ideological spectrum – Herbert Read averred that postwar society actually required *greater* interdisciplinary fluidity, so as to 'take what exists – the detritus of a defunct civilization [and] build the foundations of a new civilization' by utilising science to elucidate the purpose of art – and, in so doing, implied that pan-disciplinary methodologies were indispensable to the rebuilding of postwar culture. 'From our historians', he wrote, 'we must expect a more exact analysis of the social conditions which have produced art in the past. From our psychologists we must expect analysis of the creative process in man'.[39] Consequently, in spite of explicit differences over the suitability of applying scientific methods to art theory, critics and art historians were broadly united in recognising the influence of science upon postwar aesthetics.

A resurgence of interest in the history of morphology occurred throughout the late 1940s and early 1950s – a phenomenon that sought to explain the primacy of formation in the physical world by means of exploring the historical origins of morphological thought. In this respect, the work of Goethe was seen as exemplary in demonstrating both the principles of morphology and the way in which the study of art could be treated scientifically. Read was particularly active in this field, writing an essay in 1950 in which he maintained that the founding of morphology had relied upon 'Goethe [realising] that form [...] was present, not only in the leaf and the cloud, but also in the painting and the poem'. Moreover, this discovery led to Goethe's realisation that 'there is no essential difference between any [...] manifestations of form – that the form discoverable in nature is the same as the form revealed by art. There is one creative process – formation and transformation, and at no point can we detect a caesura'.[40] Such opinions supplied the theoretical foundation upon which postwar Constructionism could be based, as Goethe was understood to have deprecated 'the slavish imitation of the external appearances of nature', arguing that, as an intuitive act, art had access to 'the realm of what we would nowadays call archetypal forms [which] is still a part of the natural world'.[41] As such, abstract pictorial representations of natural processes matched up to the anti-reductionist aspects of Goethean philosophy. To be sure, in an article of 1949, Lancelot Law Whyte contended that conventional science had failed 'to recognize that the formative aspect of process holds the clue to the unity of nature' due to its empirical prejudice. An intuitive philosophy of nature, on the other hand, enabled the recognition that 'all natural processes are variants of one universal formative process', and that impermanent (and, by definition, irreducible) phenomena such as 'growth, development, and the reproduction of characteristic forms, [were] richer than those expressing mere permanence', and therefore allowed for a more integrated and nuanced understanding of the natural world.[42]

Much of this interest in morphology was linked to postwar attempts at reconstruction and reconciliation, with prewar rationalism increasingly coming under fire for a misplaced utopianism that had failed to build a new world or to stop the growth of militant fascism.[43] In view of that, Goethean morphology was seen to provide a radically anti-determinist reading of nature, which, by attending to the irreducible qualities of growth, was praised by logicians as aiding in the revitalisation of postwar society. Hodin alleged that 'in many marked minds of the present day [...] [Goethe] has been the initial cause of a process of regeneration [...]. For the laws of natural growth are those of a slow organic process, and only what grows naturally will bear fruits – not what is organized and constructed, quickly and rationally spread by propaganda'.[44] Such speculations were readily transposed onto art theory, as critics sought to find ways in which to encourage cultural rejuvenation without resorting to prescriptive formulae. Read, for whom Goethe was an important point of reference, analogised between natural and artistic development, vouchsafing that, contrary to mechanistic thought, neither process could be limited to purely materialistic characteristics: 'There is, admittedly, no direct solution of cultural problems. Let me reaffirm once again the *radical* nature of cultural growths. Art is an organic phenomenon, a biological process. Like flowers and fruit [...], it is a product of the life-force itself. I am not trying to reduce art to materialistic factors'.[45] In this context, organicism reappeared as a compelling postwar alternative to mainstream rationalist philosophies, as it allowed for nature and art to be thought of as totalities that were not analysable via standard mechanistic theories, but instead were processes that occurred over time and, as such, demonstrated that deterministic systems of thought were counterintuitive. More to the point, as with its interwar antecedents, postwar organicism explicitly posited that organic systems were 'natural' as opposed to artificially mechanistic and so more conducive to cohering the disjointed (and hence denaturalised) structure of postwar society.

Organicism thus found application in the arts through reviews which stressed the contingent and dynamic character of postwar sculpture, and its relatedness – in both an iconographic and an ontological sense – to the forms of the natural world. A 1946 *Horizon* essay by Carola Giedion-Welcker argued that 'the most characteristic element in Arp's art' was 'what [could] perhaps be best described as structural growth': an impression that was upheld by the way in which his tableaux evinced dynamic interrelations between irregularly shaped components (forms which likewise suggested the underlying morphologies of the natural world).[46] The manner by which these compositions implied contingency (in a mode akin to that of multipart sculpture) and organic development (through bulging outlines and spatial extension) led the reviewer to conclude that Arp's works 'seem[ed] to belong to nature itself, their elementary plastic language [being] somehow permeated with the primary forces of growth, movement and change'.[47] In claiming that artworks of this kind possessed 'a mode of expression analogous to that of nature itself' rather than resorting to 'a mere copying of nature', these interpretative strategies acted as ideological precursors to Constructionism, which likewise attempted to physically embody rather than visually represent nature.

Collectively, these organismal approaches to morphology and art indicated a point of continuity between the biologistic undercurrents of Modernist sculptural discourse in the 1930s and the radically holistic and anti-deterministic character of postwar art theory. This was no doubt partly due to the cultural influence key interwar figures within the biocentric Modernist movement – such as Moore and Read – continued to wield within the postwar art world; yet it also underscored the extent to which biologistic philosophies and organicist concepts – first introduced by British aestheticians in the years between the wars – found renewed urgency in the ravaged cultural environment of late 1940s and early 1950s Britain. The neo-romantic, organicist impulse, which led to the establishment of a biologistic aesthetic in the 1930s, possessed legitimacy in a postwar cultural milieu where disaffection towards rationalism was widespread, and modernisation – allegedly predicated on an outdated mechanistic philosophy – was held responsible for the collective socio-political failings that had led to the traumas of the Second World War. Thus if, originally, the New Biology provided Modernist sculpture with a new, anti-reductive vocabulary through which to articulate the novel consciousness of space, time and form spawned by modernity, then by the late 1940s it offered a new generation of biophilic artists an authoritative model of radical contingency which powerfully countermanded the deterministic imperatives of a mechanistic culture still in thrall to reductionist mantra. These strands of connective tissue, based as they were upon the New Biology, Goethean morphology, bioromanticism and organicism, closely bound Modernist sculpture to its postwar counterpart and demonstrate the longevity that biologism, as an aesthetic philosophy, enjoyed within British sculptural discourse in the middle years of the twentieth century.

Notes

1 See Lawrence & Meyer, 'Regenerating England', pp. 1, 11.
2 See Huxley, 'Tissue Culture and Human Habits', p. 953.
3 See: Bowler, *Reconciling Science and Religion*, pp. 407–12; Lawrence & Meyer, 'Regenerating England', p. 2.
4 Read, 'The Universal Harmony', p. 187.
5 Henri Gaudier-Brzeska and T.E. Hulme, leading figures in the prewar avant-garde, both died in the Great War, leading to a postwar environment dominated by the reactionary Bloomsbury Group, whose members, by and large, had been pacifists and conscientious objectors and thus escaped much of the devastation experienced first-hand by the Vorticists. See Harrison, *English Art and Modernism*, p. 146.
6 Wyndham Lewis in a letter to Kate Lechmere, reprinted in W. Rose (ed.), *Letters of Wyndham Lewis* (London: New Directions, 1963), p. 69.
7 It is of course important to note that Henry Moore was a war artist during the Battle of Britain

(his 'Shelter Drawings' having become powerful representations of this experience) and that he stayed in London until bomb damage to his studio obliged him to move to rural Hertfordshire.

8 Many books exist which chronicle Moore's abiding artistic engagement with natural form. The most thorough and up to date is Christa Lichtenstern's *Henry Moore: Work – Theory – Impact*, which examines the influence of morphological ideas on Moore's practice throughout his career. On Read's interwar and postwar organic aesthetic, see David Thistlewood's two chapters, 'Herbert Read's Organic Aesthetic: I, 1918–1950' and 'Herbert Read's Organic Aesthetic: II, 1950–1968', in D. Goodway (ed.), *Herbert Read Reassessed* (Liverpool: Liverpool University Press, 1998), pp. 215–32, 233–47.

9 On postwar reappraisals of Moore's reputation, see C. Stephens, 'Anything But Gentle – Henry Moore: Modern Sculptor', in C. Stephens (ed.), *Henry Moore* (London: Tate Publishing, 2010), p. 14.

10 S. Forgan, 'Festivals of Science and the Two Cultures: Science, Design and Display in the Festival of Britain, 1951', in *British Journal of the History of Science*, Vol. XXXI, No. 2, June 1998, p. 219.

11 *Ibid.*, p. 221.

12 PRO, Work 25/21, 'Festival of Britain, 1951', 1 April 1948, p. 2, quoted in B.E. Conekin, *'The Autobiography of a Nation': The 1951 Festival of Britain* (Manchester: Manchester University Press, 2003), p. 58.

13 J. Bronowski, 'The Commonsense of Science' (1951), reprinted in W. Eastwood (ed.), *A Book of Science Verse: The Poetic Relations of Science and Technology* (London: Macmillan, 1961), p. 271, quoted in, Conekin, *'The Autobiography of a Nation'*, p. 58.

14 See: J.M. Woodham, *Twentieth Century Ornament* (London: Studio Vista, 1990), p. 203; Conekin, *'The Autobiography of a Nation'*, p. 57.

15 See G. Wood, 'The Shapes of Life: Biomorphism and American Design', in G. Wood (ed.), *Surreal Things: Surrealism and Design* (London: V&A Publications, 2007), p. 81.

16 M.H. Thomas, 'Festival Pattern Group', in *Design*, Nos 29–30, May–June 1951, p. 13.

17 *Ibid.*, p. 17.

18 See Conekin, *'The Autobiography of a Nation'*, p. 62.

19 J. Bronowski, 'Introduction to the Guide Catalogue for the Science Exhibition at the Science Museum, South Kensington', in M. Banham & B. Hillier (eds), *A Tonic to the Nation: The Festival of Britain 1951* (London: Thames & Hudson, 1976), p. 144.

20 I. Moffat, '"A Horror of Abstract Thought": Postwar Britain and Hamilton's 1951 "Growth and Form" Exhibition', in *October*, Vol. XCIV, autumn 2000, pp. 91–2; Conekin, *'The Autobiography of a Nation'*, pp. 62–3.

21 Moffat, '"A Horror of Abstract Thought"', p. 100.

22 See, for example, the interwar and postwar Surrealist exhibitions discussed in A. Mahon, 'Staging Desire', in J. Mundy (ed.), *Surrealism: Desire Unbound* (London: Tate Publishing, 2001), pp. 277–91.

23 See Moffat, '"A Horror of Abstract Thought"', pp. 100–5.

24 See D. Mellor, 'The Pleasures and Sorrows of Modernity: Vision, Space and the Social Body in Richard Hamilton', in *Richard Hamilton* (London: Tate Publishing, 1992), p. 29.

25 Richard Hamilton quoted in Moffat, '"A Horror of Abstract Thought"', p. 101.

26 H. Read, 'Preface to the 1951 Edition', in L.L. Whyte (ed.), *Aspects of Form: A Symposium on Form in Nature and Art* (London: Lund Humphries, 1968), pp. xxi–xxii.

27 L.L. Whyte, 'Introduction', in L.L. Whyte (ed.), *Aspects of Form: A Symposium on Form in Nature and Art* (London: Lund Humphries, 1968), p. 2.

28 See M. Garlake, *New Art, New World: British Art in Postwar Society* (New Haven: Yale University Press, 1998), pp. 102–5.

29 A. Heath, 'Statement', in L. Alloway (ed.), *Nine Abstract Artists: Their Work and Theory* (London: Alec Tiranti, 1954), p. 25.

30 *Ibid.*, p. 26.

31 See D. Thistlewood, 'Organic Art and the Popularisation of a Scientific Philosophy', in *British Journal of Aesthetics*, Vol. XXII, No. 4, autumn 1982, pp. 311–21.

32 V. Pasmore, 'The Artist Speaks', *Art News and Review*, p. 3, reprinted in A. Grieve, *Constructed Art in England: A Neglected Avant-Garde* (New Haven: Yale University Press, 2005), appendix.

33 Heath, 'Statement', p. 26.

34 Pasmore, 'The Artist Speaks'.
35 *Ibid.*
36 J.P. Hodin, 'Expressionism', in *Horizon*, Vol. XIX, No. 109, January 1949, p. 38.
37 *Ibid.*, p. 38.
38 *Ibid.*, p. 42.
39 H. Read, 'The Fate of Modern Painting', in *Horizon*, Vol. XVI, No. 95, November 1947, p. 254.
40 H. Read, 'Goethe and Art', in *The Listener*, Vol. XLIII, No. 1093, 5 January 1950, p. 13.
41 *Ibid.*
42 L.L. Whyte, 'Goethe and the Formative Process', in *Horizon*, Vol. XIX, No. 112, April 1949, p. 240.
43 See C. Harrison & P. Wood, 'Introduction: Part V. The Individual and the Social', in C. Harrison & P. Wood (eds), *Art in Theory* (Oxford: Blackwell, 1992), pp. 549–52; Foster, Krauss, Bois & Buchloh, *Art Since 1900*, p. 337.
44 J.P. Hodin, 'Goethe's Succession', in *Horizon*, Vol. XIX, No. 112, April 1949, p. 247.
45 Read, 'The Fate of Modern Painting', p. 253.
46 C. Giedion-Welcker, 'Contemporary Sculptors: IV, Jean Arp', in *Horizon*, Vol. XIV, No. 82, October 1946, p. 236.
47 *Ibid.*, p. 237.

Bibliography

Adams, Mary, 'Perspective in Science', in *The Listener*, Vol. III, No. 52, 8 January 1930, p. 64

—, 'What Do We Owe to Our Forefathers?', in *The Listener*, Vol. IV, No. 97, 19 November 1930, pp. 838–9

—, 'Has Human Nature Ever Changed?', in *The Listener*, Vol. IV, No. 98, 26 November, 1930, pp. 883–4

—, 'Shall the Unfit Survive?', in *The Listener*, Vol. IV, No. 99, 3 December 1930, pp. 933–4

Ades, Dawn, 'Form', in D. Ades & S. Baker (eds), *Undercover Surrealism: Georges Bataille and Documents* (London: Hayward Gallery, 2006), pp. 152–62

—, 'Little Things: Close-Up in Photo and Film 1839–1963', in D. Ades & S. Baker (eds), *Close-Up: Proximity and Defamiliarisation in Art, Film and Photography* (Edinburgh: Fruitmarket Gallery, 2008), pp. 9–60

Albers, Joseph, 'Concerning Fundamental Design', in H. Bayer, W. Gropius & I. Gropius (eds), *Bauhaus: 1919–1928* (London: George Allen & Unwin, 1939), pp. 116–20

Alberti, Leon Battista, *On the Art of Building, in Ten Books*, J. Rykwert, N. Leach & R. Tavernor (trans.) (Cambridge: MIT Press, 1988)

Allen, Garland, *Life Science in the Twentieth Century* (Cambridge: Cambridge University Press, 1978)

Ames-Lewis, Francis, *The Intellectual Life of the Early Renaissance Artist* (New Haven: Yale University Press, 2000)

Anker, Peder, 'The Bauhaus of Nature', in MODERNISM/*modernity*, Vol. XII, No. 2, 2005, pp. 229–251

Anon., 'Editorial – Broadcasting and Popular Science', in *The Listener*, Vol. I, No. 18, 15 May 1929, p. 676

—, 'What Nature Knows of Art: *Art Forms in Nature*', in *The Listener*, Vol. II, No. 36, 18 September 1929, p. 370

—, 'Art Forms in Nature', in *The Times*, 5 October 1929, p. 13

—, 'Art and Nature: A Review of "Art Forms in Nature"', in *Times Literary Supplement*, No. 1148, 31 October 1929, p. 871

—, 'The Film World and Travel Pictures', in *The Times*, 18 December 1929, p. 12

—, 'The Film World and Travel Pictures', in *The Times*, 22 October 1930, p. 14

—, 'The Progress of Physical Science: Films of Living Cells, Study of Growth', in *The Times*, 6 July 1931, p. 19

—, 'Book Reviews – New Books', in *Close-Up*, Vol. IX, No. 3, September 1932, p. 208

—, 'The Positive Electron', in *The Times*, 11 March 1933, p. 12

—, 'Circle Review', in *The Listener*, Vol. XVIII, No. 447, 4 August 1937, pp. 259–60

—, 'Photographs of Atom Tracks', in *The Times*, 20 November 1937, p. 17

—, 'Features of London: Exhibition – Drawings by Sculptors', in the *Glasgow Herald*, 13 January 1938

—, 'Key to a Hidden World: New Light on the Infinitesimal, the Super-Microscope', in *The Times*, 26 November 1938, p. 9

—, 'Pond Life on the Screen', in *The Times*, 18 April 1939, p. 10

—, 'World in a Wineglass', *Monthly Film Bulletin*, Vol. VI, No. 69, 30 September 1939, p. 196

Antliff, Mark, *Inventing Bergson: Cultural Politics and the Parisian Avant-Garde* (Princeton: Princeton University Press, 1993)

Arber, Agnes, *The Natural Philosophy of Plant Form* (Cambridge: Cambridge University Press, 1950)

Arp, Hans, 'Notes from a Diary', in *Transition*, No. 21, 1932, pp. 190–4

Baker, John, 'Recovering the Missing Links', in *The Listener*, Vol. VI, No. 151, 2 December 1931, pp. 941–3

Baker, John R. & Haldane, J.S., *Biology in Everyday Life* (London: George Unwin, 1933)

Baker, S., 'Doctrine (The Appearance of Things)', in D. Ades & S. Baker (eds), *Undercover Surrealism: Georges Bataille and Documents* (London: Hayward Gallery, 2006), pp. 34–40

Banham, Mary & Hillier, Bevis (eds), *A Tonic to the Nation: The Festival of Britain 1951* (London: Thames & Hudson, 1976)

Barlow, Anne J., 'Barbara Hepworth and Science', in D. Thistlewood, *Barbara Hepworth Reconsidered* (Liverpool: Liverpool University Press, 1996), pp. 95–107

Barnett, Vivian Endicott, 'Kandinsky and Science: The Introduction of Biological Images in the Paris Period', in O. Botar & I. Wünsche (eds), *Biocentrism and Modernism* (Farnham: Ashgate, 2011), pp. 207–26

Barr, Alfred, *Cubism and Abstract Art* (1936) (London: Secker & Warburg, 1964)

Bataille, Georges, 'Le langage des fleurs', in *Documents*, No. 3, June 1929, pp. 160–4

—, 'Le gros orteil', in *Documents*, No. 6, November 1929, pp. 297–302

Beckett, Jane, 'Circle: The Theory and Patronage of Constructive Art in the Thirties', in J. Lewison (ed.), *Circle: Constructive Art in Britain, 1934–40* (Cambridge: Kettle's Yard Gallery, 1982), pp. 11–19

Beer, Gillian, *Darwin's Plots: Evolutionary Narrative in Darwin, George Eliot and Nineteenth-Century Fiction* (Cambridge: Cambridge University Press, 2000)

Bell, Clive, *Art* (Oxford: 1914)

Benjamin, Walter, 'New Things About Plants' (1928), in D. Mellor (ed.), *Germany: The New Photography, 1927–33* (London: Arts Council of Great Britain, 1978), pp. 20–1

Bentley, W.A. & Humphreys, W.J., *Snow Crystals* (New York: McGraw-Hill, 1931)

Benton, Tim, 'Modernism and Nature', in C. Wilk (ed.), *Modernism, 1914–1939: Designing a New World* (London: V&A Publications, 2006), pp. 311–26

Béothy, Etienne, 'L'abstraction est la qualité spécifique de l'homme', *abstraction-création: art non-figuratif*, No. 4, 1935, p. 4

Berg, Brigitte, 'Contradictory Forces: Jean Painlevé, 1902–1989', in A.M. Bellows & M. McDougall (eds), *Science Is Fiction: The Films of Jean Painlevé* (Cambridge: MIT Press, 2000), pp. 3–47

Bergson, Henri, *Creative Evolution* (1911) (New York: Dover, 1998)

Berkeley, George, *Philosophical Works – Including the Works on Vision* (London: Everyman's Library, 1983)

Bernal, J.D., 'Art and the Scientist', in J.L. Martin, B. Nicholson & N. Gabo (eds), *Circle: International Survey of Constructive Art* (London: Faber & Faber, 1937), pp. 119–23

—, 'Foreword to a Catalogue of Sculpture by Barbara Hepworth' (London: Alex, Reid & Lefevre, October 1937), pp. 2–3

Bertram, Anthony, 'The Art of Paul Nash', in *The Listener*, Vol. VIII, No. 200, 9 November 1932, pp. 662–3

—, 'Artists Indoors', in *The Listener*, Vol. X, No. 251, 1 November 1933, pp. 660–1

Beveridge, William, 'Nature and Nurture', in *The Listener*, Vol. VII, No. 167, 23 March 1932, pp. 417–18

Blakeston, Oswell, 'Personally About Percy Smith', in *Close-Up*, Vol. VIII, No. 2, June 1931, pp. 143–6

—, 'Book Reviews – New Books', in *Close-Up*, Vol. IX, No. 3, September 1932, pp. 207–8

Bleichmar, Daniela, 'Training the Naturalist's Eye in the Eighteenth Century: Perfect Global Visions and Local Blind Spots', in A. Graciano (ed.), *Visualising the Unseen, Imagining the Unknown, Perfecting the Natural: Art and Science in the 18th and 19th Centuries* (Newcastle: Cambridge Scholars, 2008), pp. 1–24

Blossfeldt, Karl, *Art Forms in Nature: Examples from the Plant World Photographed Direct from Nature*, with an introduction by Karl Nierendorf (London: A. Zwemmer, 1929)

Blunt, Anthony, 'Nature and Design', in *The Listener*, Vol. XI, No. 274, 11 April 1934, pp. 610–12

Bohr, Nils, 'Light and Life', in *Discovery*, No. 161, May 1933, pp. 151–5

Boon, Timothy, *Films of Fact: A History of Documentary Films and Television* (London: Wallflower Press, 2008)

Botar, Oliver, 'Ernö Kállai and the Hidden Face of Nature', in *Structurist*, Nos 23, 24, 1983/84, pp. 77–82

—, *Prolegomena to the Study of Biomorphic Modernism: Biocentrism, László Moholy-Nagy's 'New Vision', and Ernó Kállai's Bioromantik*, unpublished PhD thesis (Toronto: University of Toronto, 1998)

—, 'The Origins of László Moholy-Nagy's Biocentric Constructivism', in E. Kac (ed.), *Signs of Life: Bio Art and Beyond* (Cambridge: MIT Press, 2007), pp. 315–44

—, 'Defining Biocentrism', in O. Botar & I. Wünsche (eds), *Biocentrism and Modernism* (Farnham: Ashgate, 2011), pp. 15–45

— & Wünsche, Isabel, 'Introduction: Biocentrism as a Constituent Element of Modernism', in O. Botar & I. Wünsche (eds), *Biocentrism and Modernism* (Farnham: Ashgate, 2011), pp. 1–13

Bowler, Peter J., *Reconciling Science and Religion: The Debate in Early-Twentieth-Century Britain* (Chicago: University of Chicago Press, 2001)

—, *Science for All: The Popularization of Science in Early Twentieth-Century Britain* (Chicago: University of Chicago Press, 2009)

Bragg, William, 'Liquid Crystals', *Nature*, No. 3360, 24 March 1934, pp. 445–56

—, 'The Nature of Things', in *The Listener*, Vol. XVII, No. 436, 19 May 1937, pp. 985–6

Breton, André, *Surrealism and Painting* (1928) (Boston: MFA Publications, 2002)

—, 'La beauté sera convulsive', in *Minotaure*, No. 5, 1934, pp. 9–15

—, 'What Is Surrealism?', reprinted in F. Rosemont (ed.), *André Breton: What Is Surrealism? Selected Writings* (London: Pluto Press, 1978), pp. 112–41

—, *L'amour fou* (1937) (Paris: Éditions Gallimard, 2010)

Broks, Peter, *Media Science Before the Great War* (London: Macmillan, 1996)

Bronowski, Jacob, 'Introduction to the Guide Catalogue for the Science Exhibition at the Science Museum, South Kensington', in M. Banham & B. Hillier (eds), *A Tonic to the Nation: The Festival of Britain 1951* (London: Thames & Hudson, 1976), pp. 144–6

Brown, Andrew, *J.D. Bernal: The Sage of Science* (Oxford: Oxford University Press, 2005)

Burke, Edmund, *A Philosophical Enquiry into the Origin of Our Ideas of the Sublime and Beautiful* (1757) (Oxford: Oxford University Press, 1998)

Burnham, Jack, *Beyond Modern Sculpture: The Effects of Technology on the Sculpture of This Century* (New York: George Braziller, 1968)

Burwick, Frederick & Douglass, Paul, 'Introduction', in F. Burwick & P. Douglass (eds),

The Crisis in Modernism: Bergson and the Vitalist Controversy (Cambridge: Cambridge University Press, 1992), pp. 1–2

Canguilhem, G., 'On the History of the Life-Sciences Since Darwin', in *Ideology and Rationality in the History of the Life-Sciences*, A. Goldhammer (trans.) (Cambridge: MIT Press, 1988), pp. 103–24

Causey, Andrew, 'Formalism and the Figurative Tradition in British Painting', in S. Compton (ed.), *British Art in the 20th Century – The Modern Movement* (London: Royal Academy, 1987), pp. 15–30

—, 'Herbert Read and Contemporary Art', in D. Goodway (ed.), *Herbert Read Reassessed* (Liverpool: Liverpool University Press, 1998), pp. 123–44

—, 'Henry Moore and the Uncanny', in J. Beckett & F. Russell (eds), *Henry Moore: Critical Essays* (Aldershot: Ashgate, 2003), pp. 81–106

—, 'R.H. Wilenski and *The Meaning of Modern Sculpture*', in D.J. Getsy (ed.), *Sculpture and the Pursuit of a Modern Ideal in Britain, c.1880–1930* (Aldershot: Ashgate, 2004), pp. 266–85

Chadwick, Whitney, *Women, Art, and Society* (London: Thames & Hudson, 1990)

Chave, Anna C., *Constantin Brancusi: Shifting the Bases of Art* (New Haven: Yale University Press, 1993)

Cole, Margaret, 'A Petition to Scientists', in *The Listener*, Vol. III, No. 64, 2 April 1930, p. 602

Coleridge, Samuel Taylor, *Biographia Literaria*, J. Shawcross (ed.) (Oxford, Clarendon Press, 1907)

Collini, Stefan, *Absent Minds: Intellectuals in Britain* (Oxford: Oxford University Press, 2006)

Commission on Educational and Cultural Films, *The Film in National Life* (London: George Allen & Unwin, 1932)

Conekin, Becky, *'The Autobiography of a Nation': The 1951 Festival of Britain* (Manchester: Manchester University Press, 2003)

Cook, Theodore Andrea, *The Curves of Life* (1914) (New York: Dover, 1979)

Curtis, Penelope, *Barbara Hepworth* (London: Tate Publishing, 1998)

—, *Sculpture: 1900–1945* (Oxford: Oxford University Press, 1999)

Daston, Lorraine & Galison, Peter, *Objectivity* (New York: Zone Books, 2010)

Deepwell, Katy, 'Hepworth and Her Critics', in D. Thistlewood (ed.), *Barbara Hepworth Reconsidered* (Liverpool: Liverpool University Press, 1996), pp. 75–93

Derrida, Jacques, *The Truth in Painting*, Geoff Bennington & Ian McLeod (trans.) (Chicago: University of Chicago Press, 1987)

Dewsbury, Donald A., 'Psychobiology', in *American Psychologist*, Vol. XLVI, No. 3, March 1991, pp. 198–205

Driesch, Hans, *The Science and Philosophy of the Organism* (Aberdeen: Aberdeen University Press, 1908)

Eckermann, Johann Peter, *Conversations with Eckermann* (1836) (London: Dent & Sons, 1930)

Eibl-Eibesfeldt, Irenäus, 'Ernst Haeckel: The Artist in the Scientist', in *Art Forms in Nature: The Prints of Ernst Haeckel* (Munich: Prestel, 1998), pp. 19–30

Eliot, T.S, 'The Idea of a Literary Review', in *The New Criterion*, Vol. IV, No. 1, January 1926, pp. 1–6

Etchells, F., 'Nature and Art', in the *Architectural Review*, December 1929, p. 299

Evans, Myfanwy, 'Dead or Alive', in *Axis*, No. 1, January 1935, pp. 3–4

—, 'Hélion Today: A Personal Comment', in *Axis*, No. 4, November 1935, pp. 4–9

—, 'Order, Order!', *Axis*, No. 6, summer 1936, pp. 4–8

Fausset, Hugh I'A, 'The Insufficiency of Science', in *The Listener*, Vol. VII, No. 159, 27 January 1932, pp. 129–30

Fer, Briony, *On Abstract Art* (New Haven: Yale University Press, 1997)

Fernie, Eric, *Art History and Its Methods* (London: Phaidon, 1995)

Festing, Sally, *Barbara Hepworth: A Life of Forms* (London: Viking, 1995)

Focillon, Henri, *The Life of Forms in Art* (1934) (New York: Zone Books, 1989)

Forgan, Sophie, 'Festivals of Science and the Two Cultures: Science, Design and Display in the Festival of Britain, 1951', in *British Journal of the History of Science*, Vol. XXXI, No. 2, June 1998, pp. 217–40

Foster, Hal, Krauss, Rosalind, Bois, Yves-Alain & Buchloh, Benjamin, *Art Since 1900* (London: Thames & Hudson, 2004)

Francé, Raoul, *Plants as Inventors* (London: Simpkins, Marshall & Co., 1926)

Frankfort, Henri, 'New Works by Barbara Hepworth', in *Axis*, No. 3, July 1935, p. 14

Friedberg, Anne, 'Reading *Close-Up*, 1927–1933', in James Donald, Anne Friedberg & Laura Marcus, *Close-Up, 1927–1933: Cinema and Modernism* (London: Cassell, 1998), pp. 1–26

Fry, Roger, 'Art and Science' (1919), in R. Fry, *Vision and Design* (Harmondsworth: Penguin Books, 1961), pp. 69–74

Fuller, Peter, 'The Visual Arts', in B. Ford (ed.), *The Cambridge Guide to the Arts in Britain* (Cambridge: Cambridge University Press, 1988), Vol. IX, pp. 99–145

Gabo, Naum, 'Statement', in *abstraction-création: art non-figuratif*, No. 1, 1932, p. 14, reprinted in M. Hammer & C. Lodder (eds), *Gabo on Gabo* (Bury St Edmunds: Artists Bookworks, 2000), p. 83

—, 'Constructive Art', in *The Listener*, Vol. XVI, No. 408, 4 November 1936, pp. 846–8

—, 'Sculpture: Carving and Construction in Space', in J.L. Martin, B. Nicholson & N. Gabo (eds), *Circle: International Survey of Constructive Art* (London: Faber & Faber, 1937), pp. 103–111

—, 'The Constructive Idea in Art', in J.L. Martin, B. Nicholson & N. Gabo (eds), *Circle: International Survey of Constructive Art* (London: Faber & Faber, 1937), pp. 1–10

Gage, John, *Colour and Meaning: Art, Science and Symbolism* (London: Thames & Hudson, 1999)

Gale, Matthew, 'Brancusi: An Equal Among Rocks, Trees, People, Beasts and Plants', in C. Giménez & M. Gale (eds), *Constantin Brancusi: The Essence of Things* (London: Tate Publishing, 2004), pp. 21–35

Gamwell, Lynn, *Exploring the Invisible: Art, Science and the Spiritual* (Princeton: Princeton University Press, 2002)

Garlake, Margaret, *New Art, New World: British Art in Postwar Society* (New Haven: Yale University Press, 1998)

Garrould, Ann (ed.), *Henry Moore, Vol. II: Complete Drawings, 1930–39* (London: Henry Moore Foundation, 1998)

—, Friedman, Terry & Mitchinson, David (eds), *Henry Moore: Early Carvings, 1920–1940* (Leeds: Leeds City Art Galleries, 1982)

Gaunt, W., 'Introduction', in W. Watson-Baker, *World Beneath the Microscope* (London: The Studio, 1935),

Ghyka, Matila, *Le nombre d'or: rites et rythmes pythagoriciens dans le développement de la civilisation occidentale* (Paris: Gallimard, 1931)

Giedion-Welcker, Carola, 'Contemporary Sculptors: IV, Jean Arp', in *Horizon*, Vol. XIV, No. 82, October 1946, pp. 232–9

—, *Contemporary Sculpture: An Evolution in Form and Space* (1937) (London: Faber & Faber, 1954)

Gore, Frederick, 'Introduction', in S. Compton (ed.), *British Art in the 20th Century – The Modern Movement* (London: Royal Academy, 1987), pp. 9–14

Graciano, Andrew, 'Introduction', in A. Graciano (ed.), *Visualizing the Unseen, Imagining the Unkown, Perfecting the Natural: Art and Science in the 18th and 19th Centuries* (Newcastle: Cambridge Scholars, 2008), pp. xv–xxiv

Grierson, John, 'The New Generation in Sculpture', in *Apollo*, Vol. XII, November 1930, pp. 347–51

Grieve, Alastair, *Constructed Abstract Art in England: A Neglected Avant-Garde* (New Haven: Yale University Press, 2005)

Grigson, Geoffrey, 'Painting and Sculpture Today', in G. Grigson (ed.), *The Arts Today* (London: John Lane, 1935), pp. 71–109

—, 'Comment on England', in *Axis*, No. 1, January 1935, pp. 8–10

—, *Henry Moore* (Harmondsworth: Penguin Books, 1943)

Gropius, Walter, *The New Architecture and the Bauhaus* (London: Faber, 1935)

Gurney, Edmund, 'Monism', in *Mind*, Vol. VI, No. 22, April 1881, pp. 153–73

Haden-Guest, Stephen, 'A New Energy in the Arts', in *The Listener*, Vol. VII, No. 156, 6 January 1932, pp. 19–20

Haeckel, Ernst, 'Foreword of "Crystal Souls – Studies of Inorganic Life"' (1917), reprinted in *Forma* (special issue: 'Crystal Souls by Ernst Haeckel'), Vol. XIV, Nos 1, 2, 1999, pp. 31–4

Haldane, J.B.S., *The Causes of Evolution* (London: Longmans, Green & Co., 1932)

—, 'Health Before Wealth', in *The Listener*, Vol. VII, No. 161, 10 February 1932, pp. 189–92

— & Huxley, Julian, *Animal Biology* (Oxford: Clarendon Press, 1927)

Haldane, J.S., *The Philosophical Basis of Biology* (London: Hodder & Stoughton, 1931)

—, 'Life Beyond Individual Death', in *The Listener*, Vol. IX, No. 217, 8 March 1933, pp. 353–4, 375

—, *The Philosophy of a Biologist* (Oxford: Oxford University Press, 1936)

Halliday, Nigel, *More Than a Bookshop: Zwemmer's and Art in the Twentieth Century* (London: Philip Wilson, 1991)

Hambidge, Jay, *Dynamic Symmetry: The Greek Vase* (New Haven: Yale University Press, 1920)

Hammer, Martin & Lodder, Christina, 'Hepworth and Gabo: A Constructive Dialogue', in D. Thistlewood (ed.), *Barbara Hepworth Reconsidered* (Liverpool: Liverpool University Press, 1996), pp. 109–33

— & —, *Constructing Modernity: The Art and Career of Naum Gabo* (New Haven: Yale University Press, 2000)

— & — (eds), *Gabo on Gabo* (Forest Row: Artists Bookworks, 2000)

Haraway, Donna Jeanne, *Crystals, Fabrics, and Fields: Metaphors That Shape Embryos* (Berkeley: North Atlantic Books, 2004)

Häring, Hugo, 'Approaches to Form' (1925–26), reprinted in C. Benton, T. Benton & D. Sharp (eds), *Form and Function: A Sourcebook for the History of Architecture and Design, 1890–1939* (London: Open University, 1975), pp. 103–5

Harrington, Anne, *Reenchanted Science: Holism in German Culture from Wilhelm II to Hitler* (Princeton: Princeton University Press, 1996)

Harrison, Charles, *English Art and Modernism, 1900–1939* (New Haven: Yale University Press, 1994)

—, Francina, F. & Perry G. (eds), *Primitivism, Cubism, Abstraction: The Early Twentieth Century* (New Haven: Yale University Press, 1993)

— & Wood, P., 'Introduction: Part V. The Individual and the Social', in C. Harrison & P. Wood (eds), *Art in Theory* (Oxford: Blackwell, 1992), pp. 549–52

Harte, Verity, *Plato on Parts and Wholes: The Metaphysics of Structure* (Oxford: Oxford University Press, 2005)

Hauser, Kitty, *'Shadow Sites': Photography, Archaeology and the British Landscape 1927–1955* (Oxford: Oxford University Press, 2007)

Haynes, E.S.P., 'Law', in H. Kingsmill (ed.), *The English Genius* (London: Right Book Club, 1939), pp. 83–97

Heard, Gerald, 'Humanizing Science', in *The Listener*, Vol. III, No. 69, 7 May 1930, pp. 807–8

Heath, Adrian, 'Statement', in L. Alloway (ed.), *Nine Abstract Artists: Their Work and Theory* (London: Alec Tiranti, 1954), pp. 25–6

Hélion, Jean, 'From Reduction to Growth', in *Axis*, No. 2, April 1935, pp. 19–24

—, 'Avowals and Comments', in M. Evans (ed.), *The Painter's Object* (London: Gerald Howe, 1937), pp. 31–7

—, 'Poussin, Seurat and Double Rhythm', in M. Evans (ed.), *The Painter's Object* (London: Gerald Howe, 1937), pp. 94–107

Henderson, Linda Dalrymple, 'Editor's Introduction: I. Writing Modern Art and Science – An Overview; II. Cubism, Futurism and Ether Physics in the Early Twentieth Century', in *Science in Context*, Vol. XVII, No. 4, 2004, pp. 423–66

Hepworth, Barbara, Interview, in *The Studio*, Vol. CIV, July–December 1932, p. 332

—, 'Statement', in, *abstraction-création*, No. 2, 1933, p. 6

—, Untitled statement, in H. Read (ed.), *Unit 1: The Modern Movement in English Architecture, Painting and Sculpture* (London: Cassell, 1934), pp. 19–20

—, 'Sculpture', in J.L. Martin, B. Nicholson & N. Gabo (eds), *Circle: International Survey of Constructive Art* (London: Faber & Faber, 1937), pp. 113–16

Hildebrand, Adolf, *The Problem of Form in Painting and Sculpture* (London: G.E. Stechert & Co., 1932)

Hobbes, Thomas, *Leviathan* (1651), C.B. Macpherson (ed.) (Harmondsworth: Penguin Books, 1968)

Hodin, J.P., 'Expressionism', in *Horizon*, Vol. XIX, No. 109, January 1949, pp. 38–53

—, 'Goethe's Succession', in *Horizon*, Vol. XIX, No. 112, April 1949, pp. 241–81

Hogben, Lancelot, *Science for the Citizen* (London: George Allen & Unwin, 1943)

Holding, Eileen, 'London Shows', in *Axis*, No. 2, April 1935, pp. 29–30

Holt, Niles R., 'Ernst Haeckel's Monistic Religion', in *Journal of the History of Ideas*, Vol. XXXII, No. 2, April–June 1971, pp. 265–80

Honzík, Karel, 'A Note on Biotechnics', in J.L. Martin, B. Nicholson & N. Gabo (eds), *Circle: International Survey of Constructive Art* (London: Faber & Faber, 1937), pp. 256–62

Hooke, R., *Micrographia or Some Physiological Descriptions of Minute Bodies Made by Magnifying Glasses with Observations and Inquiries Thereupon* (1665) (New York: Dover, 1961)

Huxley, Aldous, 'Science – The Double-Edged Tool', in *The Listener*, Vol. VII, No. 158, 20 January 1932, pp. 77–9, 112

Huxley, Julian, *Essays of a Biologist* (1923) (Harmondsworth: Penguin Books, 1939)

—, 'The Age of Change', in *The Listener*, Vol. IX, No. 222, 12 April 1933, pp. 590–1

—, 'Tissue Culture and Human Habits', in *The Listener*, Vol. IX, No. 231, 14 June 1933, pp. 952–3

—, 'Organizers in Animal Development', in *The Listener*, Vol. X, No. 234, 5 July 1933, pp. 28–9

—, *Evolution: The Modern Synthesis* (London: George Allen & Unwin, 1942)

—, Wells, H.G. & Wells, G.P., *The Science of Life* (London: Cassell, 1938)

Jakovski, Anatole, 'Inscriptions Under Pictures', in *Axis*, No. 1, January 1935, pp. 14–20

—, 'Brancusi', in *Axis*, No. 3, July 1935, pp. 3–9

James, P. (ed.), *Henry Moore on Sculpture* (London: MacDonald, 1966)

Jarron, Matthew & Caudwell, Cathy, *D'Arcy Thompson and His Zoology Museum in Dundee* (Dundee: University of Dundee Museum Services and the College of Life Sciences, 2010)

Jeans, James, *The Mysterious Universe* (1930) (Cambridge: Cambridge University Press, 1931)

Jeffrey, Ian, *Photography: A Concise History* (London: Thames & Hudson, 1981)

Joad, C.E.M., *The Meaning of Life* (London: Watts & Co., 1928)

—, 'The Confusion of Modern Thought', in *The Listener*, Vol. X, No. 240, 16 August 1933, pp. 229–30

Johnson, James, *The Mechanism of Life* (London: Edward Arnold, 1921)

Jones, Caroline A. & Galison, Peter, 'Introduction: Picturing Science, Producing Art', in C.A. Jones & P. Galison (eds), *Picturing Science, Producing Art* (New York: Routledge, 1998), pp. 1–23

Jordanova, L.J., 'Natural Facts: A Historical Perspective on Science and Sexuality', in C. MacCormack & M. Strathern (eds), *Nature, Culture and Gender* (Cambridge: Cambridge University Press, 1980), pp. 42–69

Juler, Edward, 'The Key to a Hidden World: Photomicrography and Close-Up Nature Photography in Interwar Britain', in *History of Photography*, Vol. XXXVI, No. 1, February 2012, pp. 87–98

Jung, Carl Gustav, 'The Concept of the Collective Unconscious', in the *Journal of St Bartholomew's Hospital*, 1936/37, reprinted in H. Read, M. Fordham & G. Adler (eds), *The Archetypes and the Collective Unconscious: The Collected Works of C.G. Jung* (2nd edition) (London: Routledge, 1980), Vol. IX, Part 1, pp. 42–53

Kac, Eduardo (ed.), *Signs of Life: Bio Art and Beyond* (Cambridge: MIT Press, 2007)

Kandinsky, Wassily, 'Line and Fish', in *Axis*, No. 2, April 1935, p. 6

Kant, Immanuel, *The Critique of Judgment*, J.C. Meredith (trans.) (Oxford: Oxford University Press, 1969)

Keller, E.F., *Making Sense of Life: Explaining Biological Development with Models, Metaphors and Machines* (Cambridge: Harvard University Press, 2002)

Kelly, Julia, 'The Unfamiliar Figure: Henry Moore in French Periodicals of the 1930s', in J. Beckett & F. Russell (eds), *Henry Moore: Critical Essays* (Aldershot: Ashgate, 2003), pp. 43–62

Kerr, John Graham, 'Links in the Chain of Life: I. The Amoeba, an Old-Fashioned Creature', in *The Listener*, Vol. III, No. 55, 29 January 1930, pp. 188–9

—, 'Links in the Chain of Life: VI. The Animal World of Today', in *The Listener*, Vol. III, No. 60, 5 March 1930, pp. 420–1

Killian, J., *Crystals: Secrets of the Inorganic* (London: John Gifford, 1941)

Klee, Paul, 'Statement', in H. Bayer, W. Gropius & I. Gropius (eds), *Bauhaus, 1919–1928* (London: George Allen & Unwin, 1939), p. 172

Klingender, F.D., 'Content and Form in Art', in B. Rea (ed.), *5 on Revolutionary Art* (London: Wishart, 1935), pp. 25–44

Kosinski, D. (ed.), *Henry Moore: Sculpting the Twentieth Century* (New Haven: Yale University Press, 2001)

Krauss, Rosalind, *Passages in Modern Sculpture* (Cambridge: MIT Press, 1981)

Laitinen, Ahti & Maude, George, 'Biologism, Politics and International Politics', in *Acta Sociologica*, Vol. XXIX, No. 2, 1986, pp. 113–28

Lavallée, M., 'Art Nouveau', *Grove Art* online, Oxford Art online, Oxford University Press, www.oxfordartonline.com/subscriber/article/grove/art/T004438 (accessed 4 March 2013)

Lawrence, Christopher & Weisz, George, 'Medical Holism: The Context', in C. Lawrence & G. Weisz (eds), *'Greater than the Parts': Holism in Biomedicine, 1920–1950* (Oxford: Oxford University Press, 1998), pp. 1–22

Lawrence, Christoper & Mayer, Anna-K., 'Regenerating England: An Introduction', in C. Lawrence & A-K. Mayer (eds), *Regenerating England: Science, Medicine and Culture in Interwar Britain* (Amsterdam: Rodopi, 2000), pp. 1–23

Leavis, F.R. & Thompson, Denys, *Culture and Environment: The Training of Critical Awareness* (London: Chatto & Windus, 1933)

Leduc, Stéphane, *The Mechanism of Life* (London: Rebman, 1911)

Leiris, Michel, 'Exposition Hans Arp', in *Documents*, No. 6, 1929, pp. 340–2

Lessing, Gotthold, *Laocoön*, E. Frothingham (trans.) (New York: Noonday, 1957)

Lichtenstern, Christa, 'Henry Moore and Surrealism', in *The Burlington Magazine*, Vol. CXXIII, No. 944, November 1981, pp. 644–58

—, *Henry Moore: Work – Theory – Impact* (London: Royal Academy of Arts, 2008)

Lloyd, A.L., 'Modern Art and Modern Society', in B. Rea (ed.), *5 on Revolutionary Art* (London: Wishart, 1935), pp. 53–71

Lodge, Oliver, 'The Spirit of Science', in Adams, M. (ed.), *Science in This Changing World* (London: George Allen & Unwin, 1933), pp. 268–80

Loeb, Jacques, *The Dynamics of Living Matter* (London: Columbia University Press, 1906)

Low, Rachael, *The History of British Film, 1918–1929* (London: George Allen & Unwin, 1971)

—, *Documentary and Educational Films of the 1930s* (London: George Allen & Unwin, 1979)

Mabille, Pierre, 'Préface à l'éloge des préjugés populaires', in *Minotaure*, No. 6, 1935, pp. 1–3

MacColl, D.S., 'Object and Subject in Painting – I', in *The Listener*, Vol. VIII, No. 183, 13 July 1932, pp. 58–9

—, 'Visual and Vocal Art', in *The Listener*, Vol. XI, No. 278, 9 May 1934, pp. 799–800

Mahon, Alyce, 'Staging Desire', in J. Mundy (ed.), *Surrealism: Desire Unbound* (London: Tate Publishing, 2001), pp. 277–97

Maienschein, Jane, 'Cell Theory and Development', in R.C. Olby, G.N. Cantor, J.R.R. Christie & M.J.S. Hodge (eds), *Companion to the History of Science* (London: Routledge, 1990), pp. 357–73

Maldonado, Guitemie, *Le cercle et l'amibe: Le biomorphisme dans l'art des années 1930* (Paris: CTHS/INHA, 2006)

Malloy, V.V., 'Rethinking Alexander Calder's Universes and Mobiles: The Influences of Einsteinian Physics and Modern Astronomy', in *Immediations*, Vol. III, No. 1, 2012, pp. 9–25

Martin, J.L., 'The State of Transition', in J.L. Martin, B. Nicholson & N. Gabo (eds), *Circle: International Survey of Constructive Art* (London: Faber & Faber, 1937), pp. 215–19

Marwood, Kimberly, 'Shadows of Femininity: Women, Surrealism and the Gothic', in *re.bus*, No. 4, autumn/winter 2009, pp. 1–17

Mayer, A.-K., '"A Combative Sense of Duty": Englishness and the Scientists', in C. Lawrence & A.-K. Mayer (eds), *Regenerating England: Science, Medicine and Culture in Interwar Britain* (Amsterdam: Editions Rodopi, 2000), pp. 67–106

Mellor, David, 'London–Berlin–London: A Cultural History – The Reception and Influence of the New German Photography in Britain 1927–33', in D. Mellor (ed.), *Germany: The New Photography, 1927–33* (London: Arts Council, 1978), pp. 113–30

—, 'The Pleasures and Sorrows of Modernity: Vision, Space and the Social Body in Richard Hamilton', in *Richard Hamilton* (London: Tate Publishing, 1992), pp. 27–39

Michalski, Sergiusz, *New Objectivity: Neue Sachlichkeit – Painting in Germany in the 1920s* (Cologne: Taschen, 2003)

Mitchinson, D., '1930–1940', in *Henry Moore: Early Carvings, 1920–1940* (Leeds: Leeds City Art Galleries, 1982), pp. 30–7

Moffat, Isabelle, '"A Horror of Abstract Thought": Postwar Britain and Hamilton's 1951 "Growth and Form" Exhibition', in *October*, Vol. XCIV, autumn 2000, pp. 89–112

Moholy-Nagy, László, *The New Vision: Fundamentals of Bauhaus Design, Paintings, Sculpture, and Architecture* (London: Faber & Faber, 1934)

Moore, Henry, 'A View of Sculpture', in *Architectural Association Journal*, Vol. XLV, May 1930, pp. 408–13

—, 'On Carving', in *New English Weekly*, 5 May 1932, pp. 65–66, reprinted in A. Wilkinson (ed.), *Henry Moore: Writings and Conversations* (Aldershot: Lund Humphries, 2002), pp. 188–91

—, Untitled statement, in H. Read (ed.), *Unit 1: The Modern Movement in English Architecture, Painting and Sculpture* (London: Cassell, 1934), pp. 29–30

—, 'The Sculptor Speaks', in *The Listener*, Vol. XVIII, No. 449, 18 August 1937, pp. 334–8

— (in interview with John and Vera Russell), 'Moore Explains His Universal Shapes', in *New York Times Magazine*, 11 November 1962, reprinted in P. James (ed.), *Henry Moore on Sculpture* (London: MacDonald, 1966), pp. 198–203

—, *Henry Moore: Wood Sculptures* (London: Sidgwick & Jackson, 1983)

Morgan, C. Lloyd, *Instinct and Experience* (London: Methuen, 1912)

Mundy, Jennifer, 'Form and Creation: The Impact of the Biological Sciences on Modern Art', in *Creation: Modern Art and Nature* (Edinburgh: Scottish National Gallery of Modern Art, 1984), pp. 16–23

—, *Biomorphism*, unpublished PhD thesis (London: University of London, 1986)

—, 'Comment on England', in C. Stephens (ed.), *Henry Moore* (London: Tate Publishing, 2010), pp. 22–37

—, 'The Naming of Biomorphism', in O. Botar & I. Wünsche (eds), *Biocentrism and Modernism* (Farnham: Ashgate, 2011), pp. 61–73

Nash, Paul, 'Photography and Modern Art', in *The Listener*, Vol. VIII, No. 185, 27 July 1932, p. 130

—, 'A Painter and a Sculptor', in *Weekend Review*, 19 November 1932, p. 613

—, 'The Nest of Wild Stones', in M. Evans (ed.), *The Painter's Object* (London: Gerald Howe, 1937), pp. 38–42

—, 'The Life of the Inanimate Object', in *Country Life*, 1 May 1937, pp. 496–7, reprinted in A. Causey (ed.), *Paul Nash: Writings on Art* (Oxford: Oxford University Press, 2000), pp. 137–9

—, 'F.E. McWilliam', in the *London Bulletin*, March 1939, pp. 11–12

—, *Monster Field* (1939) (Oxford: Counterpoint Publications, 1946), pp. 244–7, reprinted in A. Causey (ed.), *Paul Nash: Writings on Art* (Oxford: Oxford University Press, 2000), pp. 150–2

Needham, Joseph, *Order and Life* (New Haven: Yale University Press, 1936)

Newbolt, Henry, in *The Listener*, Vol. V, No. 114, Supplement No. 12 (English Poetry Today), 18 March 1931, pp. i–viii

Nicholson, Ben, 'Interview', in *The Studio*, Vol. CIV, July–December 1932, p. 333

—, 'Points from Letters: The Art of Picasso', in *The Listener*, Vol. X, No. 238, 2 August 1933, p. 174

—, Untitled statement, in H. Read (ed.), *Unit 1: The Modern Movement in English Architecture, Painting and Sculpture* (London: Cassell, 1934), p. 89

Nierendorf, Karl, 'Preface to Karl Blossfeldt, *Urformen Der Kunst*' (1928), in D. Mellor, *Germany: The New Photography, 1927–33* (London: Arts Council, 1978), pp. 17–19

Offord, J. Milton, 'President's Address' (14 February 1933), in *Journal of the Quekett Microscopical Club*, Third Series, Vol. I, 1934–37, pp. 1–9

Olby, Robert, 'Structural and Dynamical Explanation in the World of Neglected Dimensions', in T.J. Horder (ed.), *A History of Embryology* (Cambridge: Cambridge University Press, 1986), pp. 275–93

Orsini, G.N., 'The Organic Concepts in Aesthetics', in *Comparative Literature*, Vol. XXI, No. 1, winter 1969, pp. 1–30

—, 'Organicism', in *The Dictionary of the History of Ideas* (Charlottesville: University of Virginia, 2003), Vol. III, pp. 422–7 (online edition http://etext.virginia.edu/cgi-local/DHI/dhi.cgi?id=dv3-52, accessed 19 September 2006)

Orwell, George, *Coming Up for Air* (1939) (London: Secker & Warburg, 1967)

Otterly, Charles D., *The Cinema in Education: A Handbook for Teachers* (London: George Routledge & Sons, 1935)

Overy, Richard, *The Morbid Age: Britain Between the Wars* (London: Allen Lane, 2009)

Ozenfant, Amedée, *The Foundations of Modern Art* (1931) (New York: Dover, 1952)

Papapetros, Spyros, 'On the Biology of the Inorganic: Crystallography and Discourses of Latent Life in Art and Architectural Historiography of the Early Twentieth Century', in O. Botar & I. Wünsche (eds), *Biocentrism and Modernism* (Farnham: Ashgate, 2011), pp. 77–106

Parkinson, Gavin, *Surrealism, Art and Modern Science* (New Haven: Yale University Press, 2008)

Pasmore, Victor, 'The Artist Speaks', in *Art News and Review*, p. 3, reprinted in A. Grieve, *Constructed Abstract Art in England: A Neglected Avant-Garde* (New Haven: Yale University Press, 2005), appendix

Peters Corbett, David, Holt, Ysanne & Russell, Fiona, 'Introduction', in D. Peters Corbett, Y. Holt & F. Russell (eds), *Geographies of Englishness: Landscape and the National Past, 1880–1940* (New Haven: Yale University Press, 2002), pp. ix–xix

Photography Yearbook (London: Fountain Publications, 1935)

Piaget, Jean, *Structuralism* (London: Routledge & Kegan Paul, 1971)

Pickstone, John, *Ways of Knowing: A New History of Science, Technology and Medicine* (Manchester: Manchester University Press, 2000)

Piper, John (J.P.), 'Review of *World Beneath the Microscope*', in *Axis*, No. 4, 1935, p. 28

Plato, *Phaedrus*, C.J. Rowe (trans.) (Warminster: Aris & Phillips, 1986)

Porteus, Hugh Gordon, 'The Painter Speaks', in *The Listener*, Vol. XI, No. 273, 4 April 1934, pp. 584–6

—, 'New Planets', in *Axis*, No. 3, July 1935, pp. 22–3

Potts, Alex, 'Carving and the Engendering of Sculpture: Adrian Stokes on Barbara Hepworth', in D. Thistlewood (ed.), *Barbara Hepworth Reconsidered* (Liverpool: Liverpool University Press, 1996), pp. 43–52

—, *The Sculptural Imagination: Figurative, Modernist, Minimalist* (New Haven: Yale University Press, 2000)

Powers, Alan, 'The Reluctant Romantics: *Axis* Magazine 1935–37', in D. Peters Corbett, Y. Holt & F. Russell (eds), *The Geographies of Englishness: Landscape and the National Past, 1880–1940* (New Haven: Yale University Press, 2002), pp. 249–74

Radford, Robert, *Art for a Purpose: The Artists' International Association, 1933–1953* (Winchester: Winchester School of Art Press, 1987)

Ramsden, E.H., 'Barbara Hepworth – Sculptor', in *Horizon*, Vol. VII, No. 42, June 1943, pp. 418–22

Rea, Betty, 'Foreword', in B. Rea (ed.), *5 on Revolutionary Art* (London: Wishart, 1935), p. 7

Read, Herbert, 'Psychoanalysis and the Critic', in *The Criterion*, Vol. III, No. 10, January 1925, pp. 214–30

—, 'Books of the Quarter: Review of *Science and the Modern World*', in *The New Criterion*, Vol. IV, No. 3, June 1926, pp. 581–6

—, 'Art and Decoration', in *The Listener*, Vol. III, No. 69, 7 May 1930, p. 805

—, *The Meaning of Art* (London, Faber & Faber, 1931)

—, 'The Golden Section', in *The Listener*, Vol. V, No. 107, 28 January 1931, pp. 142–3

—, 'Henry Moore', in *The Listener*, Vol. V, No. 119, 22 April 1931, pp. 688–9

—, *Art Now: An Introduction to the Theory of Modern Painting and Sculpture* (1933) (London: Faber & Faber, 1948)

—, *Art and Industry* (London: Faber & Faber, 1934)

—, *Henry Moore* (London: A. Zwemmer, 1934)

—, 'Introduction', in H. Read (ed.), *Unit 1: The Modern Movement in English Architecture, Painting and Sculpture* (London: Cassell, 1934)

—, 'Mr MacColl on Abstract Art', in *The Listener*, 16 May 1934, pp. 844–5

—, 'What Is Revolutionary Art?', in B. Rea (ed.), *5 on Revolutionary Art* (London: Wishart, 1935), pp. 11–22

—, 'Our Terminology', in *Axis*, No. 1, January 1935, pp. 6–8

—, 'Ben Nicholson and the Future of Painting', in *The Listener*, Vol. XIV, No. 352, 9 October 1935, pp. 604–5

—, 'Jean Hélion', in *Axis*, No. 4, November 1935, pp. 3–4

—, *Art and Society* (1937) (London: Faber & Faber, 1956)

—, 'Abstract Art: A Note for the Uninitiated', in *Axis*, No. 5, spring 1936, p. 3

—, 'The Faculty of Abstraction', in J.L. Martin, B. Nicholson & N. Gabo (eds), *Circle: International Survey of Constructive Art* (London: Faber & Faber, 1937), pp. 61–6

—, 'An Art of Pure Form', in *London Bulletin*, No. 14, 1 May 1939, pp. 6–9

—, 'The Universal Harmony', in *The Listener*, Vol. XXVIII, No. 708, 6 August 1942, p. 187

—, *Annals of Innocence and Experience* (1940) (London: Faber & Faber, 1946)

—, 'The Fate of Modern Painting', in *Horizon*, Vol. XVI, No. 95, November 1947, pp. 242–54

—, 'Realism and Abstraction in Modern Art' (1948), in H. Read, *The Philosophy of Modern Art* (London: Faber & Faber, 1952), pp. 88–104

—, 'Goethe and Art', in *The Listener*, Vol. XLIII, No. 1093, 5 January 1950, pp. 13–14

—, *Icon and Idea: The Function of Art in the Development of Human Consciousness* (London: Faber & Faber, 1955)

—, 'Preface to the 1951 Edition', in L.L. Whyte (ed.), *Aspects of Form: A Symposium on Form in Nature and Art* (London: Lund Humphries, 1968), pp. xxi–xxii

Remy, Michel, *Surrealism in Britain* (Aldershot: Ashgate, 1999)

Reynolds, Andrew, 'The Theory of the Cell State and the Question of Cell Autonomy in Nineteenth and Early Twentieth-Century Biology', in *Science in Context*, Vol. XX, No. 1, 2007, pp. 71–95

Reynolds, Bernard, 'Letters', in *The Listener*, Vol. XVIII, No. 453, 15 September 1937, p. 575

Ridley, Mark, 'Embryology and Classical Zoology in Great Britain', in T.J. Horder (ed.), *A History of Embryology* (Cambridge: Cambridge University Press, 1986), pp. 35–67

Ritterbush, Philip C., *The Art of Organic Forms* (Washington, DC: Smithsonian Institution, 1968)

—, 'Aesthetics and Objectivity in the Study of Form in the Life Sciences', in G.S. Rousseau (ed.), *Organic Form: The Life of an Idea* (London: Routledge & Kegan Paul, 1972), pp. 26–59

Robertson, Eric, *Arp: Painter, Poet, Sculptor* (New Haven: Yale University Press, 2006)

Rousseau, George, 'The Perpetual Crises of Modernism and the Traditions of Enlightenment Vitalism: With a Note on Mikhail Bakhtin', in F. Burwick & P. Douglass (eds),

The Crisis in Modernism: Bergson and the Vitalist Controversy (Cambridge: Cambridge University Press, 1992), pp. 15–75

Rose, William (ed.), *Letters of Wyndham Lewis* (London: New Directions, 1963)

Rosenberg, Charles E., 'Holism in Twentieth-Century Medicine', in C. Lawrence & G. Weisz (eds), *'Greater than the Parts': Holism in Biomedicine 1920–1950* (Oxford: Oxford University Press, 1998), pp. 335–55

Ruskin, John, *The Seven Lamps of Architecture* (1849), reprinted in E.T. Cook & A. Wedderburn (eds), *The Works of John Ruskin* (Vol. VIII) (London: George Allen & Unwin, 1903)

Russell, A.S., 'Mechanism and Biology', in *The Listener*, Vol. III, No. 60, 5 March 1930, p. 411

—, 'Science and Its Popularization', in *The Listener*, Vol. III, No. 67, 23 April 1930, p. 731

—, 'Should Scientists and Philosophers Confer?', in *The Listener*, Vol. III, No. 75, 18 June 1930, p. 1076

—, 'Science Notes: The Proton, the Electron and the Penetrating Radiation', in *The Listener*, Vol. V, No. 106, 21 January 1931, p. 95

—, 'Science Notes: The New Zealand Earthquake, the Golden Section', in *The Listener*, Vol. V, No. 110, 18 February 1931, pp. 273–4

—, 'The Origin of Life', in *The Listener*, Vol. X, No. 245, 20 September 1933, p. 411

Russell, Bertrand, *The Scientific Outlook* (London: George Allen & Unwin, 1931)

—, 'How Science Has Changed Society', in *The Listener*, Vol. VII, No. 157, 13 January 1932, pp. 39–42

—, 'The Scientific Society', in M. Adams (ed.), *Science in This Changing World* (London: George Allen & Unwin, 1933), pp. 201–8

—, *Power: A New Social Analysis* (London: Allen & Unwin, 1938)

Russell, E.S., *Form and Function* (London: John Murray, 1916)

Sanger, Margaret, *The Pivot of Civilization* (London: Jonathon Cape, 1923)

Saxon Mills, G.H., 'Modern Photography: Its Development, Scope and Possibilities', in *Modern Photography* (London: The Studio, 1931), pp. 4–14

Schaaf, L.J., 'Invention and Discovery: First Images', in A. Thomas (ed.), *Beauty of Another Order: Photography in Science* (New Haven: Yale University Press, 1997), pp. 26–59

Simaika, John P. & Samways, Michael J., 'Biophilia as a Universal Ethic for Conserving Biodiversity', in *Conservation Biology*, Vol. XXIV, No. 3, June 2010, pp. 903–6

Simpson, J.Y., 'Science and the Idea of God', in *The Listener*, Vol. IX, No. 217, 8 March 1933, pp. 377–8

Skinner, David, 'Racialized Features: Biologism and the Changing Politics of Identity', in *Social Studies of Science*, Vol. XXXVI, No. 3, June 2006, pp. 459–88

Smith, Cyril Leeston, 'Present Day Technical Apparatus and Its Applications', in *Modern Photography* (London: The Studio, 1931), pp. 111–18

Smith, Percy & Field, Mary, *Secrets of Nature* (London: Faber, 1934)

Snow, C.P., *The Two Cultures* (1959) (Cambridge: Cambridge University Press, 1998)

Stebbing, Susan L., *Philosophy and the Physicists* (London: Methuen, 1937)

Stephens, C., 'Anything but Gentle – Henry Moore: Modern Sculptor', in C. Stephens (ed.), *Henry Moore* (London: Tate Publishing, 2010)

Stokes, Adrian, 'Miss Hepworth's Carving', in *The Spectator*, November 1933, reprinted in *The Critical Writings of Adrian Stokes: Vol. I, 1930–1937* (London: Thames & Hudson, 1978), pp. 309–10

Stump, Ulrike Meyer, 'Karl Blossfeldt's Working Collages – A Photographic Sketchbook', in A. Wilde & J. Wilde (eds), *Karl Blossfeldt: Working Collages* (Cambridge: MIT Press, 2001), pp. 7–22

Sullivan, J.W.N., 'Science and Literature', in *Athenaeum*, 13 June 1919, p. 464

—, 'Science: The Entente Cordial', in *Athenaeum*, 9 April 1920, p. 482

Thistlewood, David, 'Organic Art and the Popularization of a Scientific Philosophy', in *Journal of Aesthetics*, Vol. XXII, No. 4, autumn 1982, pp. 311–21

—, 'Herbert Read's Organic Aesthetic: I, 1918–1950', in D. Goodway (ed.), *Herbert Read Reassessed* (Liverpool: Liverpool University Press, 1998), pp. 215–32

—, 'Herbert Read's Organic Aesthetic: II, 1950–1968', in D. Goodway (ed.), *Herbert Read Reassessed* (Liverpool: Liverpool University Press, 1998), pp. 233–47

Thomas, Ann, 'The Search for Pattern', in A. Thomas (ed.), *Beauty of Another Order: Photography in Science* (New Haven: Yale University Press, 1997), pp. 76–119

Thomas, Mark Hartland, 'Festival Pattern Group', in *Design*, Nos 29–30, May–June 1951, pp. 12–25

Thompson, D'Arcy Wentworth, *On Growth and Form* (1917) (New York: Dover, 1992)

Thwaites, J. & Thwaites, M., 'Surrealism and Abstraction – The Search for Subjective Form', in *Axis*, No. 6, summer 1936, pp. 21–5

Tzara, T., 'Un Art nouveau: deux solutions sur le principle de l'immédiat, postulées par H. Arp et formulées par Tristan Tzara' (c.1917), reprinted in Tristan Tzara, *Oeuvres completes, Vol. I: 1912–1924* (Paris: Flammarion, 1975), pp. 556–8

van Eck, Caroline, *Organicism in Nineteenth-Century Architecture: An Inquiry into Its Theoretical and Philosophical Background* (Amsterdam: Architectura & Natura Press, 1994)

Wagner, Anne Middleton, 'Miss Hepworth's Stone *Is* a Mother', in D. Thistlewood, *Barbara Hepworth Reconsidered* (Liverpool: Liverpool University Press, 1996), pp. 53–74

—, *Mother Stone: The Vitality of Modern British Sculpture* (New Haven: Yale University Press, 2005)

Walker, Ian, *City Gorged with Dreams: Surrealism and Documentary Photography in Interwar Paris* (Manchester: Manchester University Press, 2002)

Warehime, Marja, *Brassaï: Images of Culture and the Surrealist Observer* (Baton Rouge: Louisiana State University Press, 1996)

Warren, Howard C., *A History of Association Psychology* (New York: Charles Scribner's Sons, 1921)

Wasmuth, Ewald, untitled response to the question of how the individual and metaphysics will develop in a collectivist epoch, in *transition*, No. 21, March 1932, pp. 142–4

Watson-Baker, W., *World Beneath the Microscope* (London: The Studio, 1935)

Weiss, Trude, 'Film Review: *Das Keimende Leben*', in *Close-Up*, Vol. IX, No. 3, September 1932, pp. 207–8

—, 'The International Exhibition of Photography in Brussels', in *Close-Up*, Vol. IX, No. 3, September 1932, pp. 188–9

Wellman, K., 'Materialism and Vitalism', in J.L. Heilbron (ed.), *The Oxford Companion to the History of Modern Science* (Oxford: Oxford University Press, 2003) (Oxford Reference Online, www.oxfordreference.com/views/ENTRY.html?subview=Mainandentry=t124.e0443, accessed 2 July 2007)

Werskey, Gary, *The Visible College* (London: Free Association Books, 1988)

Wescher, Herta, 'New Work in Paris', in *Axis*, No. 6, summer 1936, pp. 28–9

Whitehead, Alfred North, *An Enquiry Concerning the Principles of Natural Knowledge* (Cambridge: Cambridge University Press, 1919)

—, *Science and the Modern World* (Cambridge: Cambridge University Press, 1926)

—, *Nature and Life* (Cambridge: Cambridge University Press, 1934)

Whitworth, Michael, 'The Clothbound Universe: Popular Physics Books, 1919–1939', in *Publishing History*, Vol. XL, 1996, pp. 53–82

—, *Einstein's Wake: Relativity, Metaphor and Modernist Literature* (Oxford: Oxford University Press, 2001)

Whyte, Lancelot Law, 'Goethe and the Formative Process', in *Horizon*, Vol. XIX, No. 112, April 1949, p. 240

—, 'Introduction', in L.L. Whyte (ed.), *Aspects of Form: A Symposium on Form in Nature and Art* (London: Lund Humphries, 1968), pp. 1–7

Wilde, A. & Wilde, J., *Karl Blossfeldt: Working Collages* (Cambridge: MIT Press, 2001)

Wilenski, R.H., *The Modern Movement in Art* (London: Faber & Gwyer, 1927)

—, 'Ruminations on Sculpture and the Work of Henry Moore', in *Apollo*, Vol. XII, December 1930, pp. 409–13

—, *The Meaning of Modern Sculpture* (London: Faber & Faber, 1932)

—, 'The Place of Sculpture Today and Tomorrow', in *The Studio*, Vol. CX, July–December 1935, pp. 216–23

—, *Modern French Painters* (London: Faber & Faber, 1940)

Wilk, Christopher, 'Introduction: What Was Modernism?', in C. Wilk (ed.), *Modernism: Designing a New World* (London: V&A Publications, 2006), pp. 11–21

Williams, Raymond, *Culture and Society, 1780–1950* (London: Penguin Books, 1961)

—, *Keywords: A Vocabulary of Culture and Society* (Guildford: Fontana, 1983)

Williams-Ellis, Clough, *England and the Octopus* (London: Geoffrey Bles, 1928)

Windsor, Alan, 'Frederick Edward McWilliam', in Alan Windsor (ed.), *British Sculptors of the Twentieth Century* (Aldershot: Ashgate, 2003), p. 140

Wood, Ghislaine, 'The Shapes of Life: Biomorphism and American Design', in G. Wood (ed.), *Surreal Things: Surrealism and Design* (London: V&A Publications, 2007), pp. 81–99

Woodcock, George, *Herbert Read: The Stream and the Source* (London: Faber & Faber, 1972)

Woodger, J.H., *Biological Principles: A Critical Study* (London: Kegan Paul, 1929)

Woodham, J.M., *Twentieth-Century Ornament* (London: Studio Vista, 1990)

Woods, S. John, 'Time to Forget Ourselves', in *Axis*, No. 6, summer 1936, pp. 19–21

Worringer, Wilhelm, *Abstraction and Empathy* (1908), M. Bullok (trans.) (London, 1953), reprinted in C. Harrison & P. Wood (eds), *Art in Theory: 1900–1990* (Oxford: Blackwell, 1992), pp. 68–72

Young, Edward, *Conjectures on Original Composition*, E.J. Morley (ed.) (Manchester and London: Manchester University Press and Longmans, Green & Co., 1918)

Index

EU authorised representative for GPSR:
Easy Access System Europe, Mustamäe tee 50,
10621 Tallinn, Estonia
gpsr.requests@easproject.com